Intelligent Systems Reference Library

Volume 144

The aim of this series is to publish a Reference Library, including novel advances and developments in all aspects of Intelligent Systems in an easily accessible and well structured form. The series includes reference works, handbooks, compendia, textbooks, well-structured monographs, dictionaries, and encyclopedias. It contains well integrated knowledge and current information in the field of Intelligent Systems. The series covers the theory, applications, and design methods of Intelligent Systems. Virtually all disciplines such as engineering, computer science, avionics, business, e-commerce, environment, healthcare, physics and life science are included. The list of topics spans all the areas of modern intelligent systems such as: Ambient intelligence, Computational intelligence, Social intelligence, Computational neuroscience, Artificial life, Virtual society, Cognitive systems, DNA and immunity-based systems, e-Learning and teaching, Human-centred computing and Machine ethics, Intelligent control, Intelligent data analysis, Knowledge-based paradigms, Knowledge management, Intelligent agents, Intelligent decision making, Intelligent network security, Interactive entertainment, Learning paradigms, Recommender systems, Robotics and Mechatronics including human-machine teaming, Self-organizing and adaptive systems, Soft computing including Neural systems, Fuzzy systems, Evolutionary computing and the Fusion of these paradigms, Perception and Vision, Web intelligence and Multimedia.

More information about this series at http://www.springer.com/series/8578

Boris Kovalerchuk

Visual Knowledge Discovery and Machine Learning

Boris Kovalerchuk
Central Washington University
Ellensburg, WA
USA

ISSN 1868-4394 ISSN 1868-4408 (electronic)
Intelligent Systems Reference Library
ISBN 978-3-319-89230-6 ISBN 978-3-319-73040-0 (eBook)
https://doi.org/10.1007/978-3-319-73040-0

Softcover re-print of the Hardcover 1st edition 2018

Printed on acid-free paper

This Springer imprint is published by Springer Nature
The registered company is Springer International Publishing AG
The registered company address is: Gewerbestrasse 11, 6330 Cham, Switzerland

To my family

Preface

Emergence of Data Science placed knowledge discovery, machine learning, and data mining in multidimensional data, into the forefront of a wide range of current research, and application activities in computer science, and many domains far beyond it.

Discovering patterns, in multidimensional data, using a combination of visual and analytical machine learning means are an attractive visual analytics opportunity. It allows the injection of the unique human perceptual and cognitive abilities, directly into the process of discovering multidimensional patterns. While this opportunity exists, the long-standing problem is that we cannot see the n-D data with a naked eye. Our cognitive and perceptual abilities are perfected only in the 3-D physical world. We need enhanced visualization tools ("n-D glasses") to represent the n-D data in 2-D completely, without loss of information, which is important for knowledge discovery. While multiple visualization methods for the n-D data have been developed and successfully used for many tasks, many of them are non-reversible and lossy. Such methods *do not represent the n-D data fully* and do not allow the restoration of the n-D data completely from their 2-D representation. Respectively, our abilities to discover the n-D data patterns, from such incomplete 2-D representations, are limited and potentially erroneous. The number of available approaches, to overcome these limitations, is quite limited itself. The Parallel Coordinates and the Radial/Star Coordinates, today, are the most powerful reversible and lossless n-D data visualization methods, while suffer from occlusion.

There is a need to *extend* the class of *reversible and lossless n-D data visual representations*, for the knowledge discovery in the n-D data. A new class of such representations, called the **General Line Coordinate (GLC)** and several of their specifications, are the focus of this book. This book describes the GLCs, and their advantages, which include analyzing the data of the Challenger disaster, World hunger, semantic shift in humorous texts, image processing, medical computer-aided diagnostics, stock market, and the currency exchange rate predictions. Reversible methods for visualizing the n-D data have the advantages as *cognitive enhancers,* of the human cognitive abilities, to discover the n-D data patterns. This book reviews the state of the art in this area, outlines the challenges, and describes the solutions in the framework of the General Line Coordinates.

This book expands the methods of the visual analytics for the knowledge discovery, by presenting the *visual and hybrid methods*, which combine the analytical *machine learning and the visual means*. New approaches are explored, from both the theoretical and the experimental viewpoints, using the modeled and real data. The inspiration, for a new large class of coordinates, is twofold. The first one is the marvelous success of the Parallel Coordinates, pioneered by Alfred Inselberg. The second inspiration is the absence of a "silver bullet" visualization, which is perfect for the pattern discovery, in the all possible n-D datasets. Multiple GLCs can serve as a collective "silver bullet." This multiplicity of GLCs increases the chances that the humans will reveal the hidden n-D patterns in these visualizations.

The topic of this book is related to the prospects of both the *super-intelligent machines* and the *super-intelligent humans,* which can far surpass the current human intelligence, significantly lifting the human cognitive limitations. This book is about a technical way for reaching some of the aspects of super-intelligence, which are beyond the current human cognitive abilities. It is to overcome the inabilities to analyze a large amount of abstract, numeric, and *high-dimensional* data; and to find the complex *patterns*, in these data, with a *naked eye, supported by the analytical means of machine learning*. The new algorithms are presented for the reversible GLC visual representations of high-dimensional data and knowledge discovery. The advantages of GLCs are shown, both mathematically and using the different datasets. These advantages form a basis, for the future studies, in this super-intelligence area.

This book is organized as follows. Chapter 1 presents the goal, motivation, and the approach. Chapter 2 introduces the concept of the General Line Coordinates, which is illustrated with multiple examples. Chapter 3 provides the rigorous mathematical definitions of the GLC concepts along with the mathematical statements of their properties. A reader, interested only in the applied aspects of GLC, can skip this chapter. A reader, interested in implementing GLC algorithms, may find Chap. 3 useful for this. Chapter 4 describes the methods of the simplification of visual patterns in GLCs for the better human perception.

Chapter 5 presents several GLC case studies, on the real data, which show the GLC capabilities. Chapter 6 presents the results of the experiments on discovering the visual features in the GLCs by multiple participants, with the analysis of the human shape perception capabilities with over hundred dimensions, in these experiments. Chapter 7 presents the linear GLCs combined with machine learning, including hybrid, automatic, interactive, and collaborative versions of linear GLC, with the data classification applications from medicine to finance and image processing. Chapter 8 demonstrates the hybrid, visual, and analytical knowledge discovery and the machine learning approach for the investment strategy with GLCs. Chapter 9 presents a hybrid, visual, and analytical machine learning approach in text mining, for discovering the incongruity in humor modeling. Chapter 10 describes the capabilities of the GLC visual means to enhance evaluation of accuracy and errors of machine learning algorithms. Chapter 11 shows an approach, to how the GLC visualization benefits the exploration of the multidimensional Pareto front, in multi-objective optimization tasks. Chapter 12 outlines the vision of a virtual data scientist and the super-intelligence with visual means. Chapter 13 concludes this book with a comparison and the fusion of methods and the discussion of the future research. The final note is on the topics, which are outside of this book. These topics are "goal-free" visualizations that are not related to the specific knowledge discovery tasks of supervised and unsupervised learning, and the Pareto optimization in the n-D data. The author's Web site of this book is located at http://www.cwu.edu/~borisk/visualKD, where additional information and updates can be found.

Ellensburg, USA Boris Kovalerchuk

elegance and power. As we know now, Parallel Coordinates were originated in nineteenth century. However, for almost 100 years, they have been forgotten. Mathematics, in Cartesian Coordinates, continues to dominate in science for the last 400 years, providing tremendous benefits, while other known coordinate systems play a much more limited role. The emergence of Data Science requires going beyond the Cartesian Coordinates. Alfred Inselberg likely was the first person to recognize this need, long before even the term Data Science was coined. This book is a further step in Data Science beyond the Cartesian Coordinates, in this long-term journey.

Contents

List of Abbreviations

APC	Anchored Paired Coordinates
ATC	Anchored Tripled Coordinates
CF	Chernoff Face
CPC	Collocated Paired Coordinates
CTC	Collocated Tripled Coordinates
CV	Cross Validation
DM	Data Mining
GLC	General Line Coordinates
GLC-AL	GLC-L algorithm for automatic discovery
GLC-B	Basic GLC graph-constructing algorithm
GLC-CC1	Graph-constructing algorithm that generalizes CPC
GLC-CC2	Graph-constructing algorithm that generalizes CPC and SC
GLC-DRL	GLC-L algorithm for dimension reduction
GLC-IL	Interactive GLC-L algorithm
GLC-L	GLC for linear functions
GLC-PC	Graph-constructing algorithm that generalizes PC
GLC-SC1	Forward graph-constructing algorithm that generalizes SC
GLC-SC2	Backward graph-constructing algorithm that generalizes SC
ILC	In-Line Coordinates
IPC	In-Plane Coordinates
LDA	Linear Discriminant Analysis
MDF	Multiple Disk Form
MDS	Multidimensional Scaling
ML	Machine Learning
MOO	Multiobjective Optimization
PC	Parallel Coordinates
PCA	Principal Component Analysis
PCC	Partially Collocated Coordinates

PF	Pareto Front
P-to-G representation	Mapping an n-D point to a graph
P-to-P representation	Mapping an n-D point to a 2-D point
PWC	Paired Crown Coordinate
SC	Star Coordinate
SF	Stick Figure
SME	Subject Matter Expert
SOM	Self-Organized Map
SPC	Shifted Paired Coordinate
STP	Shifted Tripled Coordinate
SVM	Support Vector Machine
URC	Unconnected Radial Coordinates

Abstract

This book combines the advantages of the high-dimensional data visualization and machine learning for discovering complex n-D data patterns. It vastly expands the class of reversible lossless 2-D and 3-D visualization methods which preserve the n-D information for the knowledge discovery. This class of visual representations, called the General Lines Coordinates (GLCs), is accompanied by a set of algorithms for n-D data classification, clustering, dimension reduction, and Pareto optimization. The mathematical and theoretical analyses and methodology of GLC are included. The usefulness of this new approach is demonstrated in multiple case studies. These case studies include the Challenger disaster, the World hunger data, health monitoring, image processing, the text classification, market prediction for a currency exchange rate, and computer-aided medical diagnostics. Students, researchers, and practitioners in the emerging Data Science are the intended readership of this book.

Chapter 1
Motivation, Problems and Approach

The noblest pleasure is the joy of understanding.
Leonardo da Vinci

1.1 Motivation

High-dimensional data play an important and growing role in knowledge discovery, modeling, decision making, information management, and other areas. Visual representation of high-dimensional data opens the opportunity for understanding, comparing and analyzing visually hundreds of features of complicated multidimensional relations of n-D points in the multidimensional data space. This chapter presents motivation, problems, methodology and the approach used in this book for Visual Knowledge Discovery and Machine Learning. The chapter discussed the difference between reversible lossless and irreversible lossy visual representations of n-D data along with their impact on efficiency of solving Data Mining/Machine Learning tasks. The approach concentrates on reversible representations along with the hybrid methodology to mitigate deficiencies of both representations. This book summarizes a series of new studies on Visual Knowledge Discovery and Machine Learning with General Line Coordinates, that include the following conference and journal papers (Kovalerchuk 2014, 2017; Kovalerchuk and Grishin 2014, 2016, 2017; Grishin and Kovalerchuk 2014; Kovalerchuk and Smigaj 2015; Wilinski and Kovalerchuk 2017; Smigaj and Kovalerchuk 2017; Kovalerchuk and Dovhalets 2017). While visual shape perception supplies 95–98% of information for pattern recognition, the visualization techniques do not use it very efficiently (Bertini et al. 2011; Ward et al. 2010). There are multiple long-standing challenges to deal with high-dimensional data that are discussed below.

Many procedures for n-D data analysis, knowledge discovery and visualization have demonstrated efficiency for different datasets (Bertini et al. 2011; Ward et al. 2010; Rübel et al. 2010; Inselberg 2009). However, the *loss of information and occlusion*, in visualizations of n-D data, continues to be a challenge for knowledge discovery (Bertini et al. 2011; Ward et al. 2010). The *dimension scalability* challenge for visualization of n-D data is already present at a low dimension of $n = 4$.

B. Kovalerchuk, *Visual Knowledge Discovery and Machine Learning*,
Intelligent Systems Reference Library 144,
https://doi.org/10.1007/978-3-319-73040-0_1

Since only 2-D and 3-D data can be directly visualized in the physical 3-D world, visualization of n-D data becomes more difficult with higher dimensions. Further progress in data science require greater involvement of end users in constructing machine learning models, along with more scalable, intuitive and efficient visual discovery methods and tools that we discuss in Chap. 12.

In Data Mining (DM), Machine Learning (ML), and related fields one of these challenges is ineffective heuristic initial *selection of a class of models*. Often we do not have both (1) *prior knowledge* to select a class of these models directly, and (2) *visualization tools* to facilitate model selection losslessly and without occlusion.

In DM/ML often we are in essence *guessing* the class of models in advance, e.g., linear regression, decision trees, SVM, linear discrimination, linear programming, SOM and so on. In contrast the success is evident in model selection in low-dimensional 2-D or 3-D data that we can observe with a naked eye as we illustrate later. While identifying a class of ML models for a given data is rather an art than science, there is a progress in automating this process. For instance, a method to learn a kernel function for SVM automatically is proposed in (Nguyen et al. 2017).

In visualization of multi-dimensional data, the major challenges are (1) occlusion, (2) loss of significant n-D information in 2-D visualization of n-D data, and (3) difficulties of finding visual representation with clear and meaningful 2-D patterns.

While n-D data visualization is a well-studied area, none of the current solutions fully address these long-standing challenges (Agrawal et al. 2015; Bertini, et al. 2011; Ward et al. 2010; Inselberg 2009; Simov et al. 2008; Tergan and Keller 2005; Keim et al. 2002; Wong and Bergeron 1997; Heer and Perer 2014; Wang et al. 2015). In this book, we consider the problem of the loss of information in visualization as a problem of developing reversible lossless visual representation of multidimensional (n-D) data in 2-D and 3-D. This challenging task is addressed by generalizing Parallel and Radial coordinates with a new concept of **General Line Coordinates (GLC)**.

1.2 Visualization: From n-D Points to 2-D Points

The simplest method to represent n-D data in 2-D is splitting n-D space $X_1 \times X_2 \times \ldots \times X_n$ into all 2-D projections $X_i \times X_j$, $i, j = 1, \ldots, n$ and showing them to the user. It produces a large number of fragmented visual representations of n-D data and *destroys the integrity* of n-D data. In each projection $X_i \times X_j$, this method maps each n-D point to a single 2-D point. We will call such mapping as *n-D point to 2-D-point* mapping and denote is as *P-to-P representation* for short.

Multidimensional scaling (MDS) and other similar nonreversible lossy methods are such point-to-point representations. These methods aim preserving the *proximity* of n-D points in 2-D using specific metrics (Jäckle et al. 2016; Kruskal and Wish 1978; Mead 1992). It means that n-D information beyond proximity can be *lost* in 2-D in general, because its preservation is not controlled. Next, the proximity captured by these methods may or may not be relevant to the user's task, such as classification of n-D points, when the proximity measure is imposed on the task externally not derived from it. As a result, such methods can drastically *distort* initial data structures (Duch et al. 2000) that were relevant to the user's task. For instance, a formal proximity measure such as the Euclidean metric can contradict meaningful similarity of n-D points known in the given domain. Domain experts can know that n-D points a and b are closer to each other than n-D points c and d, $|a, b| < |c, d|$, but the formal externally imposed metric F may set up an opposite relation, $F(a, b) > F(c, d)$. In contrast, lossless data displays presented in this book provide opportunity to improve interpretability of visualization result and its understanding by subject matter experts (SME).

The common expectation of metric approaches is that they will produce relatively simple *clouds of 2-D points* on the plane with distinct lengths, widths, orientations, crossings, and densities. Otherwise, if patterns differ from such clouds, these methods do not help much to use other unique human visual perception and shape recognition capabilities in visualization (Grishin 1982; Grishin et al. 2003). Together all these deficiencies lead to a *shallow understanding* of complex n-D data.

To cope with abilities of the vision system to observe directly only 2-D/3-D spaces, many other common approaches such as Principal Components Analysis (PCA) also project every n-D data point into a *single* 2-D or 3-D point. In PCA and similar dimension reduction methods, it is done by plotting the two main components of these n-D points (e.g., Jeong et al. 2009). These two components show only a *fraction* of all information contained in these n-D points. There is no way to restore completely n-D points from these two components in general beyond some very special datasets. In other words, these methods do not provide an *isomorphic* (*bijective*, *lossless, reversible*) *mapping* between an n-D dataset and a 2-D dataset. These methods provide only a *one-way irreversible mapping* from an n-D dataset to a 2-D data set.

Such lossy visualization algorithms may not find *complex relations* even after multiple time-consuming adjustments of parameters of the visualization algorithms, because they cut out needed information from entering the visualization channel. As a result, decisions based on such truncated visual information can be *incorrect*. Thus, we have two major types of 2-D visualizations of n-D data available to be combined in the hybrid approach:

(1) each n-D *point* is mapped to a 2-D *point* (P-to-P mapping), and
(2) each n-D *point* is mapped to a 2-D structure such as a *graph* (we denote this mapping as P-to-G), which is the focus of this book.

Both types of mapping have their own advantages and disadvantages.

Principal Component Analysis (PCA) (Jolliffe 1986; Yin 2002), Multidimensional Scaling (MDS) (Kruskal and Wish 1978), Self-Organized maps (SOM) (Kohonen 1984), RadVis (Sharko et al. 2008) are examples of (1), and Parallel Coordinates (PC) (Inselberg 2009), and General Line Coordinates (GLC) presented in this book are examples of (2). The P-to-P representations (1) are not reversible (lossy), i.e., in general there is no way to restore the n-D point from its 2-D representation. In contrast, PC and GLC graphs are reversible as we discuss in depth later.

The next issue is **preserving n-D *distance* in 2-D**. While such P-to-P representations as MDS and SOM are specifically designed to meet this goal, in fact, they only minimize the mean difference in distance between the points in n-D and their representations in 2-D. PCA minimizes the mean-square difference between the original points and the projected ones (Yin 2002). For individual points, the difference can be quite large. For a 4-D hypercube SOM and MDS have Kruskal's stress values $S_{SOM} = 0.327$ and $S_{MDS} = 0.312$, respectively, i.e., on average the distances in 2-D differ from distances in n-D over 30% (Duch et al. 2000).

Such high distortion of n-D distances (loss of the actual distance information) can lead to misclassification, when such corrupted 2-D distances are used for the classification in 2-D. This problem is well known and several attempts have been made to address by controlling and decreasing it, e.g., for SOM in (Yin 2002). It can lead to disasters and loss of life in tasks with high cost of error that are common in medical, engineering and defense applications.

In current machine learning practice, 2-D representation is commonly used for illustration and explanation of the ideas of the algorithms such as SVM or LDA, but much less for actual discovery of n-D rules due to the difficulties to adequately represent the n-D data in 2-D, which we discussed above. In the *hybrid approach* that combined analytical and visual machine learning presented in this book the visualization guides both:

- Getting the information about the *structure* of data, and pattern discovery,
- Finding most informative splits of data into the training–validation pairs for evaluation of machine learning models. This includes *worst, best* and *median* split of data.

1.3 Visualization: From n-D Points to 2-D Structures

While mapping n-D points to 2-D points provides an intuitive and simple visual metaphor for n-D data in 2-D, it is also a major source of the loss of information in 2-D visualization. For visualization methods discussed in the previous section, this mapping is a self-inflicted limitation. In fact, it is *not mandatory* for visualization of n-D data to represent each n-D point as a single 2-D point.

Each n-D point can be represented as a *2-D structure* or a *glyph*. Some of them can be reversible and lossless. Several such representations are already well-known for a long time, such as *radial coordinates* (star glyphs), *parallel coordinates* (PC), *bar- and pie-graphs*, and *heat maps*. However, these methods have different limitations on the size and dimension of data that are illustrated below.

Figure 1.1 shows two 7-D points *A* and *B* in Bar (column)-graph chart and in Parallel Coordinates. In a bar-graph each value of coordinates of an n-D point is represented by the height of a rectangle instead of a point on the axis in the Parallel Coordinates.

The PC lines in Fig. 1.1b can be obtained by connecting tops of the bars (columns) 7-D points A and B. The backward process allows getting Fig. 1.1a from Fig. 1.1b.

The major difference between these visualizations is in *scalability*. The length of the Bar-graph will be 100 times wider than in Fig. 1.1a if we put 100 7-D points to the Bar graph with the same width of the bars. It will not fit the page. If we try to keep the same size of the graph as in Fig. 1.1, then the width of bars will be 100 times smaller, making bars invisible.

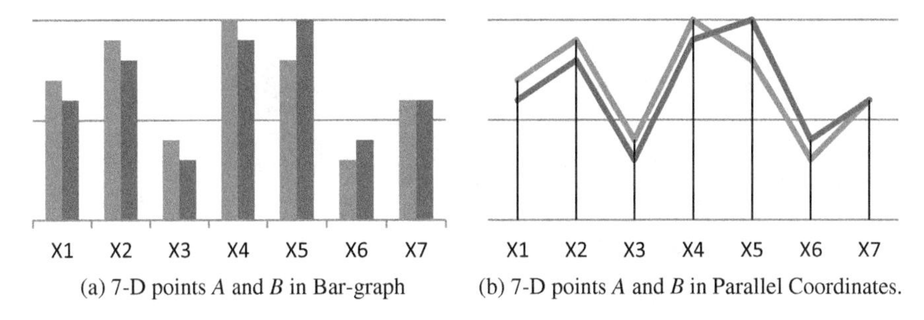

(a) 7-D points A and B in Bar-graph (b) 7-D points A and B in Parallel Coordinates.

Fig. 1.1 7D points $A = (7, 9, 4, 10, 8, 3, 6)$ in red and $B = (6, 8, 3, 9, 10, 4, 6)$ in blue in a Bar-graph chart (**a**) and in Parallel coordinates (**b**)

In contrast, PC and Radial coordinates (see Fig. 1.2a) can accommodate 100 lines without increasing the size of the chart, but with significant occlusion. An alternative Bar-graph with bars for point B drawn on the same location as A (on the top of A without shifting to the right) will keep the size of the chart, but with severe occlusion.

The last three bars of point A will be completely covered by bars from point B. The same will happen if lines in PC will be represented as filled areas. See Fig. 1.2b. Thus, when we visualize only a *single* n-D point a bar-graph is equivalent to the lines in PC. Both methods are lossless in this situation. For more n-D points, these methods are not equivalent in general beyond some specific data.

Figure 1.2a shows points A and B in Radial (star) Coordinates and Fig. 1.3 shows 6-D point $C = (2, 4, 6, 2, 5, 4)$ in the Area (pie) chart and Radial (star) Coordinates. The pie-chart uses the height of sectors (or length of the sectors) instead of the length of radii in the radial coordinates.

Tops of the pieces of the pie in Fig. 1.3a can be connected to get visualization of point C in Radial Coordinates. The backward process allows getting Fig. 1.3a from Fig. 1.3b. Thus, such pie-graph is equivalent to its representation in the Radial Coordinates.

As was pointed out above, more n-D points in the same plot occlude each other very significantly, making quickly these visual representations inefficient. To avoid

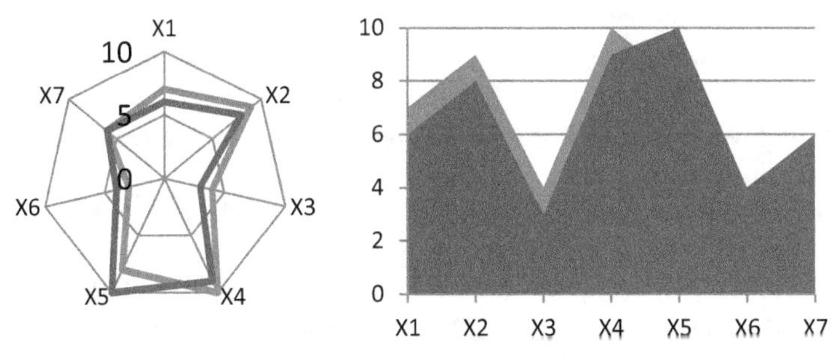

(a) 7-D points A and B in Radial Coordinates. (b) 7-D points A and B in Area chart based on PC.

Fig. 1.2 7D points $A = (7, 9, 4, 10, 8, 3, 6)$ in red and $B = (6, 8, 3, 9, 10, 4, 6)$ in Area-Graph based on PC (**b**) and in Radial Coordinates (**a**)

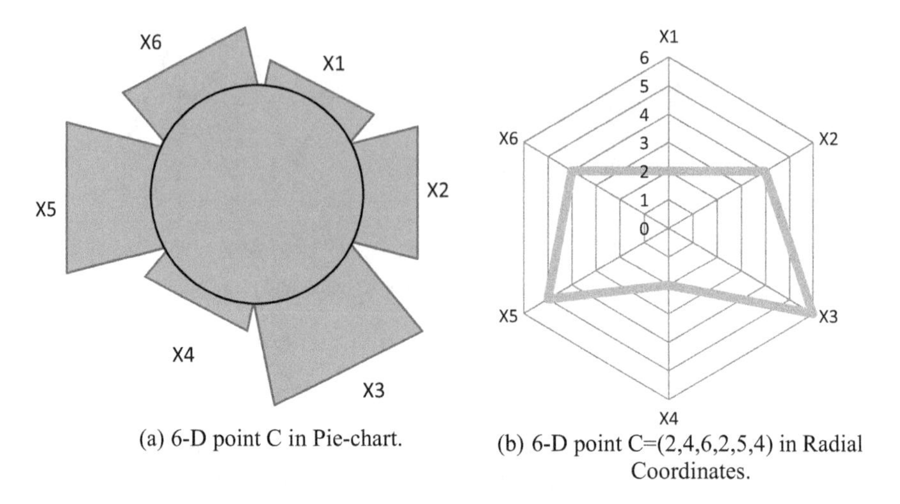

(a) 6-D point C in Pie-chart.

(b) 6-D point C=(2,4,6,2,5,4) in Radial Coordinates.

Fig. 1.3 6-D point C = (2, 4, 6, 2, 5, 4) in Pie-chart (**a**) and Radial Coordinates (**b**)

occlusion, n-D points can be shown *side-by-side* in multiple plots not in a single plot. In this case, we are limited by the number of the plots that can be shown side-by-side on the screen and by perceptual abilities of humans to analyze multiple plots at the same time.

Parallel and radial coordinates have fundamental advantage over bar- and pie-charts allowing the visualization of larger n-D datasets with less occlusion. However parallel and radial coordinates suffer from occlusion just for larger datasets.

To visualize each n-D data point $\mathbf{x} = (x_1, x_2, \ldots x_n)$ the *heat map* uses a line of *n*-bars (cells) of the same size with values of color intensity of the bar (cell) matched to the value of x_i. While the heat map does not suffer from the occlusion, it is limited in the number of n-D *points* and *dimension n* that can be presented to the user on a single screen. It is also unable to show *all* n-D points that are close to the given n-D point next to that n-D point. Only two n-D points can be shown on the adjacent rows.

The discussed visualization approaches can be interpreted as specific *glyph-based approaches* where each glyph is a sequence of bars (cells), segments, or connected points specifically located on the plane in the parallel or radial way. These visual representations provide *homomorphism* or *isomorphism* of each n-D data point into visual features of some figures, e.g., a "star".

Homomorphic mapping is a source of one of the difficulty of these visualizations, because it maps two or more equal n-D points to a single visual representation (e.g., to a single polyline in the parallel coordinates). As a result, the information about *frequencies* of n-D points in the dataset is lost in 2-D visualization. Commonly it is addressed by drawing wider lines to represent more often n-D points, but with higher occlusion. In the heat map all equal points can be preserved at the cost of less number of different n-D points shown.

The capabilities of lossless visual analytics based on shape perception have been shown in (Grishin 1982; Grishin et al. 2003), and are widely used now in technical and medical diagnostics, and other areas with data dimensions up to a few thousands with the use of a sliding window to show more attributes than can be represented in a static screen.

In this book, Chap. 6 demonstrates shape perception capabilities in experimental setting. While moving to datasets with millions of records and many thousands of dimensions is a challenge for both lossless and lossy algorithms, lossless representations are very desirable due to preservation of information. The combination of both types of algorithms is most promising.

1.4 Analysis of Alternatives

An important advantage of lossless visualizations is that an analyst can compare much more data attributes than in lossy visualizations. For instance, multidimensional scaling (MDS) allows comparing only a few attributes such as a relative distance, because other data attributes are not presented in MDS.

Despite the fundamental difference between lossy and lossless visual representations of n-D data and needs in more lossless representations, the research publications on developing new lossless methods are scarce.

The positive moment is that the importance of this issue is recognized, which is reflected in appearance of both terms "lossy" and "lossless" in the literature and conference panel discussions (Wong and Bergeron 1997; Jacobson et al. 2007; Ljung et al. 2004; Morrissey and Grinstein 2009; Grinstein et al. 2008; Belianinov et al. 2015).

In (Morrissey and Grinstein 2009) the term *lossless* is specifically used for Parallel Coordinates. In (Belianinov et al. 2015) the term *lossless visualization* is also applied to parallel coordinates and its enhancement, to contrast it with PCA and similar techniques ("Parallel coordinates avoid the loss of information afforded by dimensionality reduction technique"). Multiple aspects of dimension reduction for visualization are discusses in (Gorban et al. 2008).

There is a link between lossy image/volume compression and lossy visualization. In several domains such as medical imaging and remote sensing, large subsets of the image/volume do not carry much information. This motivates lossy compression of some parts of them (to a lower resolution) and lossless representation of other parts (Ljung et al. 2004). Rendering such images/volumes is a form of visualization that is partially lossy.

In (Jacobson et al. 2007) a term *lossy visualization* is used to identify the visualization where each n-D data point is mapped to a single color. In fact, this is a mapping of each n-D point to a 3-D point, because this "fused" color is represented by three basis color functions. It is designed for lossy fusing and visualizing large image sets with many highly correlated components (e.g., hyperspectral images), or relatively few non-zero components (e.g., the passive radar video).

The loss can be controlled by selecting an appropriate fused color (3-D point) depending on the task. In the passive radar data, the noisy background is visualized as a lossy textured gray area. In both these examples, the visualization method does not cause the loss of information. The uncontrolled lossy image/volume compression that precedes such visualization/rendering could be the cause. This is the major difference from lossy visualizations considered above.

A common main idea behind Parallel, Radial and Paired Coordinates defined in Chap. 2 is the exchange *of a simple n-D point that has no internal structure for a 2-D line (graph) that has the internal structure*. In short, this is the exchange of the *dimensionality* for a *structure*. Every object with an internal structure includes two or more points. 2-D points do not overlap if they are not equal. Any other unequal 2-D objects that contain more than one point can overlap. Thus, clutter is a direct result of this exchange.

The only way to avoid clutter fundamentally is locating structured 2-D objects side-by-side as it is done with Chernoff faces (Chernoff 1973). The price for this is more difficulty in correlating features of the faces relative to objects that are stacked (Schroeder 2005).

A *multivariate dataset* consists of n-tuples (n-D points), where each element of an n-D point is a nominal or ordinal value corresponding to an independent or dependent variable. The techniques to display multivariate data are classified in (Fua et al. 1999) as it is summarized below:

(1) *Axis reconfiguration* techniques, such as parallel coordinates (Inselberg 2009; Wegman 1990) and radial/star coordinates (Fienberg 1979),
(2) *Glyphs* (Andrews 1972; Chernoff 1973; Ribarsky et al. 1994; Ward 2008),
(3) *Dimensional embedding techniques*, such as dimensional stacking (LeBlanc et al. 1990) and worlds within worlds (Feiner and Beshers 1990),
(4) *Dimensional subsetting*, such as scatterplots (Cleveland and McGill 1988),
(5) *Dimensional reduction* techniques, such as multidimensional scaling (Kruskal and Wish 1978; Mead 1992; Weinberg 1991), principal component analysis (Jolliffe 1986) and self-organizing maps (Kohonen 1984).

Axis reconfiguration and *Glyphs* map axis into another coordinate system. Chernoff faces map axis onto facial features (icons). *Glyphs/Icons* are a form of multivariate visualization in orthogonal 2-D coordinates that augment each spatial point with a vector of values, in the form of a visual icon that encodes the values coordinates (Nielson et al. 1990). The glyph approach is more limited in dimensionality than parallel coordinates (Fua et al. 1999).

There is also a type of glyph visualization where each number in the n-D point is *visualized individually*. For instance, an n-D point (0, 0.25, 0.5, 0.75, 1) is represented by a string of Harvey balls or by color intensities. This visualization is not scaled well for large number of points and large dimensions, but it is interesting conceptually because it is does not use any line to connect values in the visualization. These lines are a major source of the clutter in visualizations based on Parallel and Radial coordinates. It is easy to see that Harvey balls are equivalent to heat maps.

Parallel and Radial coordinates are *planar representations* of an n-D space that map points to polylines. The transformation to the planar representation means that *axis reconfiguration* and *glyphs* trade a structurally simple n-D object to a more complex object, but in a lower dimension (complex 2-D face, or polyline versus a simple n-D string of numbers). *Pixel oriented techniques* map n-D points to a pixel-based area of certain properties such as color or shape (Ankerst et al. 1996).

Dimensional subsetting literally means that a set of dimensions (attributes) is sliced into subsets, e.g., pairs of attributes (X_i, X_j) and each pair is visualized by a scatterplot with total n^2 of scatterplots that form a matrix of scatterplots. *Dimensional embedding* also is based on subsets of dimensions, but with specific roles. The dimensions are divided into those that are in the slice and those that create the *wrapping* space where these slices are then embedded at their respective position (Spence 2001).

Technically (1)–(3) are *lossless transformations*, but (4) can be a lossy or a lossless transformation depending on completeness of the set of subsets, and the *dimensional reduction* (5) is a lossy transformation in general. Among lossless representations, only (1) and (2) *preserve n-D integrity of data*. In contrast, (3) and (4) split each n-D record adding a new perceptual task of assembling low-dimensional visualized pieces of each record to the whole record. Therefore, we are interested in enhancing (1) and (2).

The examples of (1) and (2) listed above fundamentally try to represent visually actual values of *all attributes* of an n-D point. While this ensures lossless representation, it fundamentally *limits the size* of the dataset that can be visualized (Fua et al. 1999). The good news is that visualizing *all attributes is not necessary* for lossless representation. The *position* of the visual element on 2-D plane can be *sufficient* to restore completely the n-D vector as it was shown for Boolean vectors in (Kovalerchuk and Schwing 2005; Kovalerchuk et al. 2012).

The major advantage of PC and related methods is that they are lossless and reversible. We can restore an n-D point from its 2-D PC polyline. This ensures that we do not throw the baby out with the bathwater. i.e., we will be able to discover n-D patterns in 2-D visualization that are present in the n-D space. This advantage comes with the price.

The number of pixels needed to draw a polyline is much more than in "n-D point to 2-D point" visualizations such as PCA. For instance, for 10-D data point in PC, the use of only 10 pixels per line that connects adjacent nodes will require $10 \times 10 = 100$ pixels, while PCA may require only one pixel. As a result, reversible methods suffer from occlusion much more than PCA. For some datasets, the existing n-D pattern will be completely hidden under the occlusion (e.g., (Kovalerchuk et al. 2012) for breast cancer data).

Therefore, we need new or enhanced methods that will be reversible (lossless), but with smaller footprint in 2-D (less pixels used). The General Line Coordinates (GLC) such as Collocated Pared Coordinates (CPC) defined in Chap. 2 have the footprint that is two times smaller than in PC (two times less nodes and edges of the graph).

Parallel and Radial Coordinates provide lossless representation of each n-D point visualized individually. However, their ability to represent losslessly a set of n-D points in a single coordinate plot is limited by occlusion and overlapping values. The same is true for other General Line Coordinates presented in this book. While full losslessness is an ideal goal, the actual level of losslessness allows discovering complex patterns as this book demonstrates.

1.5 Approach

The approach taken in this book to enhance visual discovery in n-D data consists of three major components:

(A1) Generating *new reversible lossless visualization methods* of n-D data.
(A2) Combining *lossless and lossy visualizations* for the knowledge discovery, when each of them separately is not sufficient.
(A3) Combining *analytical and visual data mining/machine learning knowledge discovery* means.

The *generation* of new reversible lossless visual representations includes:

(G1) Mapping n-D data points into separate 2-D *figures* (*graphs*) providing better pattern recognition in correspondence with Gestalt laws and recent psychological experiments with more effective usage of human vision capabilities of shape perception.
(G2) Ensuring *interpretation* of features of visual representations in the original n-D data properties.
(G3) *Generating* n-D data of given *mathematical structures* such as hyper-planes, hyper-spheres, hyper–tubes, and
(G4) *Discovering mathematical structures* such as hyper-planes, hyper-spheres, hyper–tubes and others in real n-D data in individual and collaborative settings by using a combination of visual and analytical means.

The motivation for G3 is that visualization results for n-D data with *known in advance structure* (*modeled data*) are applicable for a whole class of data with this structure. In contrast, a popular approach of inventing visualizations for *specific empirical data* with *unknown math properties* may not be generalizable.

In other words, inventions of specific visualization for specific data do not help much for visualization of other data. In contrast, if we can establish that new data have the same structure that was explored on the modeled data we can use the derived properties for these new data to construct the efficient visualization of these new data. The implementation of this idea is presented in Chap. 6 with hyper-cylinders (hyper-tubes).

Example Consider modeled n-D data with the following structural property. All n-D points of class 1 are in the one hypercube and all n-D points of class 2 are in another hypercube, and the distance between these hyper-cubes is greater or equal to *k* lengths of these hyper-cubes.

Assume that it was established by a *mathematical proof* that, for any n-D data with this structure, a lossless visualization method *V*, produces visualizations of n-D points of classes 1 and 2, which do not overlap in 2-D. Next, assume also that this property was tested on new n-D data and was confirmed. Then the visualization method *V* can be applied with the confidence that it will produce a desirable visualization without occlusion.

The *combination* of lossless and lossy visual representations includes

(CV1) Providing means for evaluating the weaknesses of each representation and
(CV2) Mitigating weaknesses by sequential use of these representations for knowledge discovery.

The results of this combination, fusion of methods are *hybrid methods*. The motivation for the fusion is in the opportunity to combine the abilities of lossy methods to handle larger data sets and of larger dimensions with abilities of the lossless methods to preserve better n-D information in 2-D.

The goal of hybrid methods is handling the same large data dimensions as lossy methods, but with radically improved quality of results by analyzing more information. It is possible by applying first lossy methods to reduce dimensionality with *acceptable and controllable loss of information*, from, say, 400 dimensions to 30 dimensions, and then applying lossless methods to represent 30 dimensions in 2-D *losslessly*. This approach is illustrated in Chap. 7 in Sect. 7.3.3, where 484 dimensions of the image were reduced to 38 dimensions by a lossy method and then then 38-D data are visualized losslessly in 2-D and classified with high accuracy.

The future wide scope of applications of hybrid methods is illustrated by the large number of activities in lossless Parallel Coordinates and lossy PCA captured by Google search: 268,000 records for "Parallel Coordinates" and 3,460,000 records for "Principal Component Analysis" as of 10/20/2017.

The progress in PC took multiple directions (e.g., Heinrich and Weiskopf 2013; Viau et al. 2010; Yuan et al. 2009) that include unstructured and large datasets with millions of points, hierarchical, smooth, and high order PC along with reordering, spacing and filtering PC, and others. The GLC and hybrid methods can progress in the same way to address Big Data knowledge discovery challenges. Some of these ways are considered in this book.

The third component of our approach (A3) is combining analytical and visual data mining/machine learning knowledge discovery means. This combination is in line with the methodology of visual analytics (Keim et al. 2008). Chapter 8 illustrates it, where analytical means search for profitable patterns in the lossless visual representation of n-D data for USD-Euro trading. Chapter 9 illustrates it too, where the incongruity model is combined with visual of texts representations to distinguish jokes from non-jokes.

References

Agrawal, R., Kadadi, A., Dai, X., Andres, F.: Challenges and opportunities with big data visualization. In Proceedings of the 7th International Conference on Management of computational and collective intElligence in Digital EcoSystems 2015 Oct 25 (pp. 169–173). ACM

Andrews, D.: Plots of high dimensional data. Biometrics **28**, 125–136 (1972)

Ankerst, M., Keim, D.A., Kriegel, H.P.: Circle segments: a technique for visually exploring large multidimensional data sets. In: Visualization (1996)

Belianinov, A., Vasudevan, R., Strelcov, E., Steed, C., Yang, S.M., Tselev, A., Jesse, S., Biegalski, M., Shipman, G., Symons, C., Borisevich, A.: Big data and deep data in scanning and electron microscopies: deriving functionality from multidimensional data sets. Adv. Struct. Chem. Imag. **1**(1) (2015)

Bertini, E., Tatu, A., Keim, D.: Quality metrics in high-dimensional data visualization: an overview and systematization, IEEE Tr. on Vis. Comput. Graph. **17**(12), 2203–2212 2011

Chernoff, H.: The use of faces to represent points in k-dimensional space graphically. J. Am. Stat. Assoc. **68**, 361–368 (1973)

Cleveland, W., McGill, M.: Dynamic graphics for statistics. Wadsworth, Inc. (1988)

Duch, W., Adamczak, R., Grąbczewski, K., Grudziński, K., Jankowski, N., Naud, A.: Extraction of knowledge from Data using Computational Intelligence Methods. Copernicus University, Toruń, Poland (2000). https://www.fizyka.umk.pl/~duch/ref/kdd-tut/Antoine/mds.htm

Feiner, S., Beshers, C.: Worlds within worlds: metaphors for exploring n-dimensional virtual worlds. In Proceedings of the 3rd annual ACM SIGGRAPH symposium on User interface software and technology, pp. 76–83 (1990)

Fienberg, S.E.: Graphical methods in statistics. Am. Stat. **33**, 165–178 (1979)

Fua, Y.,Ward, M.O., Rundensteiner, A.: Hierarchical parallel coordinates for exploration of large datasets. Proc. of IEEE Vis. 43–50 (1999)

Gorban, A.N., Kégl, B., Wunsch, D.C., Zinovyev, A.Y. (eds.): Principal Manifolds for Data Visualization and Dimension Reduction. Springer, Berlin (2008)

Grinstein G., Muntzner T., Keim, D.: Grand challenges in information visualization. In: Panel at the IEEE 2008 Visualization Conference (2008). http://vis.computer.org/VisWeek2008/session/panels.html

Grishin V., Kovalerchuk, B.: Stars advantages vs, parallel coordinates: shape perception as visualization reserve. In: SPIE Visualization and Data Analysis 2014, Proc. SPIE 9017, 90170Q, p. 8 (2014)

Grishin V., Kovalerchuk, B.: Multidimensional collaborative lossless visualization: experimental study, CDVE 2014, Seattle, Sept 2014. Luo (ed.) CDVE 2014. LNCS, vol. 8683, pp. 27–35. Springer, Switzerland (2014)

Grishin, V., Sula, A., Ulieru, M.: Pictorial analysis: a multi-resolution data visualization approach for monitoring and diagnosis of complex systems. Int. J. Inf. Sci. **152**, 1–24 (2003)

Grishin, V.G.: Pictorial analysis of Experimental Data. pp. 1–237, Nauka Publishing, Moscow (1982)

Heer, J., Perer, A.: Orion: a system for modeling, transformation and visualization of multidimensional heterogeneous networks. Inf. Vis. **13**(2), 111–133 (2014)

Heinrich, J., Weiskopf, D.: State of the Art of Parallel Coordinates, EUROGRAPHICS 2013/ M. Sbert, L Szirmay-Kalos, 95–116 (2013)

Inselberg, A.: Parallel coordinates: Visual Multidimensional Geometry and its Applications. Springer, Berlin, (2009)

Jäckle, D., Fischer, F., Schreck, T., Keim, D.A.: Temporal MDS plots for analysis of multivariate data. IEEE Trans. Vis. Comput. Graph. **22**(1), 141–150 (2016)

Jacobson, N., Gupta, M., Cole, J.: Linear fusion of image sets for display. IEEE Trans. Geosci. Remote Sens. **45**(10), 3277–3288 (2007)

Jeong, D.H., Ziemkiewicz, C., Ribarsky, W., Chang, R., Center, C.V.: Understanding Principal Component Analysis Using a Visual Analytics Tool. Charlotte visualization center, UNC Charlotte (2009)

Jolliffe, J.: Principal of Component Analysis. Springer, Berlin (1986)

Keim, D.A., Hao, M.C., Dayal, U., Hsu, M.: Pixel bar charts: a visualization technique for very large multi-attribute data sets. Information Visualization. **1**(1), 20–34 (2002 Mar)

Keim, D., Mansmann, F., Schneidewind, J., Thomas, J., Ziegler, H.: Visual analytics: scope and challenges. In: Visual Data Mining, pp. 76–90 (2008)

Kohonen, T.: Self-organization and Associative Memory. Springer, Berlin (1984)

Kovalerchuk, B.: Visualization of multidimensional data with collocated paired coordinates and general line coordinates. In: SPIE Visualization and Data Analysis 2014, Proc. SPIE 9017, Paper 90170I, 2014, https://doi.org/10.1117/12.2042427, 15 p

Kovalerchuk, B.: Super-intelligence Challenges and Lossless Visual Representation of High-Dimensional Data. International Joint Conference on Neural Networks (IJCNN), pp. 1803–1810. IEEE (2016)

Kovalerchuk, B.: Visual cognitive algorithms for high-dimensional data and super-intelligence challenges. Cognitive Systems Research **45**, 95–108 (2017)

Kovalerchuk, B, Schwing, J, eds.: Visual and Spatial Analysis: Advances in Data Mining, Reasoning, and Problem Solving. Springer Science & Business Media (2005)

Kovalerchuk B., Grishin V., Collaborative lossless visualization of n-D data by collocated paired coordinates, CDVE 2014, Seattle, Sept 2014, Y. Luo (ed.) CDVE 2014, LNCS vol. 8683, pp. 19–26, Springer, Switzerland (2014)

Kovalerchuk, B., Smigaj A.: Computing with words beyond quantitative words: in-congruity modelling. In Proc. of NAFIPS, 08–17-19, 2015, Redmond, WA, pp. 226–233. IEEE (2015)

Kovalerchuk, B., Grishin V.: Visual Data Mining in Closed Contour Coordinates, IS&T International Symposium on Electronic Imaging 2016, San Francisco, CA, Visualization and Data Analysis 2016, VDA-503.1- VDA-503.10

Kovalerchuk, B., Dovhalets, D.: Constructing Interactive Visual Classification, Clustering and Dimension Reduction Models for n-D Data. Informatics **4**(3), 23 (2017)

Kovalerchuk, B., Grishin, V.: Adjustable general line coordinates for visual knowledge discovery in n-D data. Inf. Vis. (2017). 10.1177/1473871617715860

Kovalerchuk, B., Kovalerchuk, M.: Toward virtual data scientist. In Proceedings of the 2017 International Joint Conference On Neural Networks, Anchorage, AK, USA, 14–19 May 2017 (pp. 3073–3080)

Kovalerchuk, B., Perlovsky, L., Wheeler, G.: Modeling of phenomena and dynamic logic of phenomena. J. Appl. Non-class. Log. **22**(1), 51–82 (2012)

Kruskal, J., Wish, M.: Multidimensional scaling. SAGE, Thousand Oaks, CA (1978)

LeBlanc, J., Ward, M., Wittels, N.: Exploring n-dimensional databases. Proc. Vis. '90. pp. 230–237 (1990)

Ljung, P., Lundstrom, C., Ynnerman, A., Museth, K.: Transfer function based adaptive decompression for volume rendering of large medical data sets. In: IEEE Symposium on Volume Visualization and Graphics, pp. 25–32 (2004)

Mead, A.: Review of the development of multidimensional scaling methods. J. Roy. Stat. Soc. D: Sta. **41**, 27–39 (1992)

Morrissey, S. Grinstein, G.: Visualizing firewall configurations using created voids. In: Proceedings of the IEEE 2009 Visualization Conference, VizSec Symposium, Atlantic City, New Jersey

Nguyen, T.D., Le, T., Bui, H., Phung, D.: Large-scale online kernel learning with random feature reparameterization. In: Proceedings of the 26th International Joint Conference on Artificial Intelligence (IJCAI-17) 2017 (pp. 2543–2549)

Nielson, G., Shriver, B., Rosenblum, L.: Visualization in scientific computing. IEEE Comp. Soc. (1990)

Ribarsky, W., Ayers, E., Eble, J., Mukherjea, S.: Glyphmaker: Creating customized visualization of complex Data. IEEE Comput. **27**(7), 57–64 (1994)

Rübel, O., Ahern, S., Bethel, E.W., Biggin, M.D., Childs, H., Cormier-Michel, E., DePace, A., Eisen, M.B., Fowlkes, C.C., Geddes, C.G., et al.: Coupling visualization and data analysis for knowledge discovery from multi-dimensional scientific data. Procedia Comput. Sci. **1**, 1757–1764 (2010)

Schroeder, M.: Intelligent information integration: from infrastructure through consistency management to information visualization. In: Dykes, J., MacEachren, A.M., Kraak, M.J. (eds.) Exploring Geovisualization, pp. 477–494, Elsevier, Amsterdam, (2005)

Sharko, J., Grinstein, G., Marx, K.: Vectorized radviz and its application to multiple cluster datasets. IEEE Trans. Vis. Comput. Graph. **14**(6), 1427–1444 (2008)

Simov, S., Bohlen, M., Mazeika, A. (Eds), Visual data mining, Springer, Berlin (2008). https://doi.org/10.1007/978-3-540-71080-61

Smigaj, A., Kovalerchuk, B.: Visualizing incongruity and resolution: Visual data mining strategies for modeling sequential humor containing shifts of interpretation. In: Proceedings of the 19th International Conference on Human-Computer Interaction (HCI International). Springer, Vancouver, Canada (9–14 July 2017)

Spence, R., Information Visualization, Harlow, London: Addison Wesley/ACM Press Books, 206 pp., (2001)

Tergan, S., Keller, T. (eds.) Knowledge and information visualization, Springer, Berlin (2005)

Viau, C., McGuffin, M.J., Chiricota, Y., Jurisica, I.: The FlowVizMenu and parallel scatterplot matrix: hybrid multidimensional visualizations for network exploration. IEEE Trans. Visual Comput. Graph. **16**(6), 1100–1108 (2010)

Wang, L., Wang, G., Alexander, C.A.: Big data and visualization: methods, challenges and technology progress. Digit. Technol. **1**(1), 33–38 (2015)

Ward, M.: Multivariate data glyphs: principles and practice, handbook of data visualization, pp. 179–198. Springer, Berlin (2008)

Ward, M. Grinstein, G., Keim, D.: Interactive data visualization: foundations, techniques, and applications. A K Peters, Natick (2010)

Wegman, E.: Hyperdimensional data analysis using parallel coordinates. J. Am. Stat. Assoc. **411** (85), 664 (1990)

Weinberg, S.: An introduction to multidimensional Scaling. Meas. Eval. Couns. Dev. **24**, 12–36 (1991)

Wilinski, A., Kovalerchuk, B.: Visual knowledge discovery and machine learning for investment strategy. Cognitive Systems Research **44**, 100–114 (2017)

Wong, P., Bergeron, R.D.: 30 Years of multidimensional multivariate visualization. In: Nielson, G.M., Hagan, H., Muller, H. (Eds.), Scientific Visualization—Overviews, Methodologies and Techniques, vol. 3–33, IEEE Computer Society Press (1997)

Yin, H.: ViSOM-a novel method for multivariate data projection and structure visualization. IEEE Trans. Neural Netw. **13**(1), 237–243 (2002)

Yuan, X., Guo, P., Xiao, H., Zhou, H., Qu, H.: Scattering points in parallel coordinates. IEEE Trans. Visual Comput. Graph. **15**(6), 1001–1008 (2009)

Chapter 2
General Line Coordinates (GLC)

Descartes lay in bed and invented the method of co-ordinate geometry.

Alfred North Whitehead

This chapter describes various types of General Line Coordinates for visualizing multidimensional data in 2-D and 3-D in a reversible way. These types of GLCs include n-Gon, Circular, In-Line, Dynamic, and Bush Coordinates, which directly generalize Parallel and Radial Coordinates. Another class of GLCs described in this chapter is a class of reversible Paired Coordinates that includes Paired Orthogonal, Non-orthogonal, Collocated, Partially Collocated, Shifted, Radial, Elliptic, and Crown Coordinates. All these coordinates generalize Cartesian Coordinates. In the consecutive chapters, we explore GLCs coordinates with references to this chapter for definitions. The discussion on the differences between reversible and non-reversible visualization methods for n-D data concludes this chapter.

2.1 Reversible General Line Coordinates

2.1.1 Generalization of Parallel and Radial Coordinates

The radial arrangement of n coordinates with a common origin is used in several 2-D visualizations of n-D data. The first has multiple names [e.g., *star glyphs* (Fanea et al. 2005), and *star plot* (Klippel et al. 2009)], the name *Radar plot* is used in Microsoft Excel. We call this *lossless representation* of n-D data as the **Traditional Radial (Star) Coordinates (TRC)**. In the TRC, the axes for variables radiate in equal angles from a common origin. A line segment can be drawn along each axis starting from the origin and the length of the line (or its end) represents the value of the variable (Fig. 2.1).

Often the tips of the star's beams are connected in order to create a closed contour, star (Ahonen-Rainio and Kraak 2005). In the case of the closed contour, we will call the Traditional Radial Coordinates as **Traditional Star Coordinates**

B. Kovalerchuk, *Visual Knowledge Discovery and Machine Learning*,
Intelligent Systems Reference Library 144,
https://doi.org/10.1007/978-3-319-73040-0_2

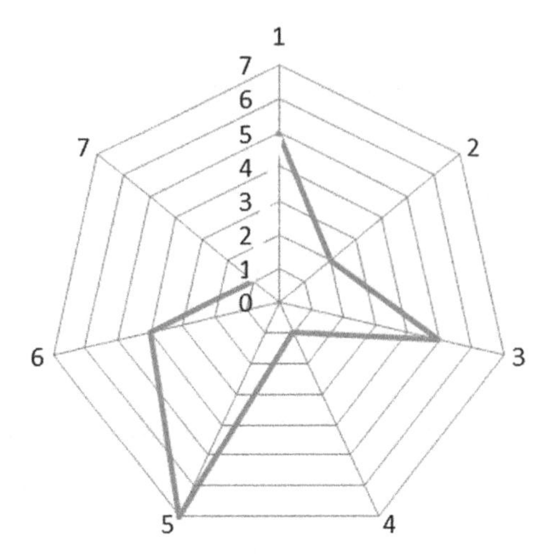

Fig. 2.1 7-D point D = (5, 2, 5, 1, 7, 4, 1) in radial coordinates

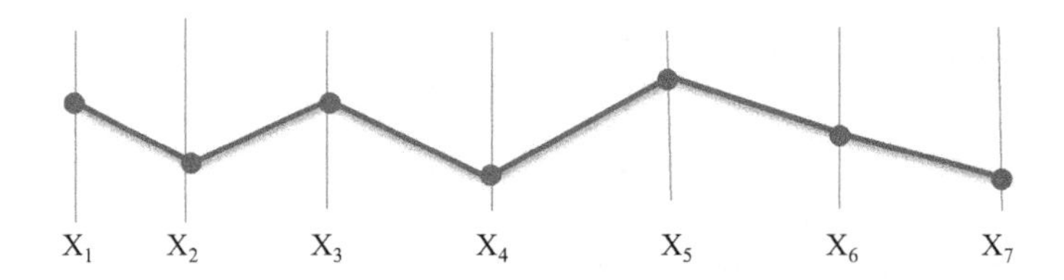

Fig. 2.2 7-D point D = (5, 2, 5, 1, 7, 4, 1) in parallel coordinates

(**TSC**), or **Star Coordinates** for short if there is no confusion with others. The closed contour is not required to have a full representation of the n-D point. A link between x_n and x_1 can be skipped.

Without closing the line, TRC and **Parallel Coordinates (PC)** (Fig. 2.2) are *mathematically equivalent* (homomorphic). For every point p on radial coordinate X, a point q exists in the parallel coordinate X that has the same value as p. The difference is in the geometric layout (radial or parallel) of n-D coordinates on the 2D plane. The next difference is that sometimes, in the Radial Coordinates, each n-D point is shown as a *separate small plot*, which serves as an *icon* of that n-D point.

In the parallel coordinates, all n-D points are drawn on the same plot. To make the use of the radial coordinates less occluded at the area close to the common origin of the axis, a non-linear scale can be used to spread data that are close to the origin as is shown later in Chap. 4. Radial and Parallel Coordinates above are examples of generalized coordinates, called **General Line Coordinates (GLC)**.

These GLC coordinates can be of *different length*, *curvilinear, connected* or *disconnected*, and *oriented to any direction* (see Fig. 2.3a, b). The methods for constructing curves with Bezier curves are explained later for In-Line Coordinates.

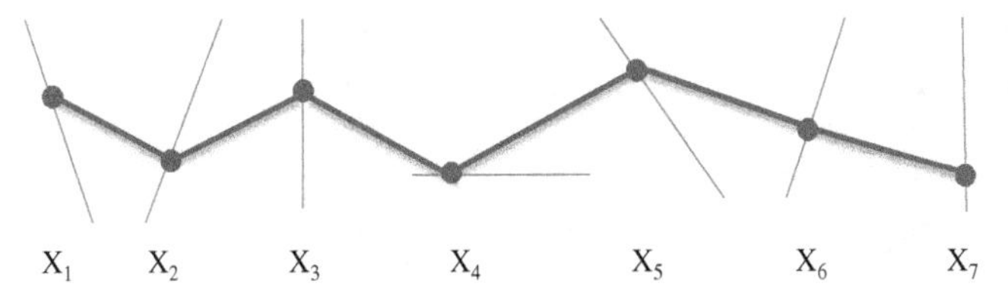

(a) 7-D point D in General Line Coordinates with straight lines.

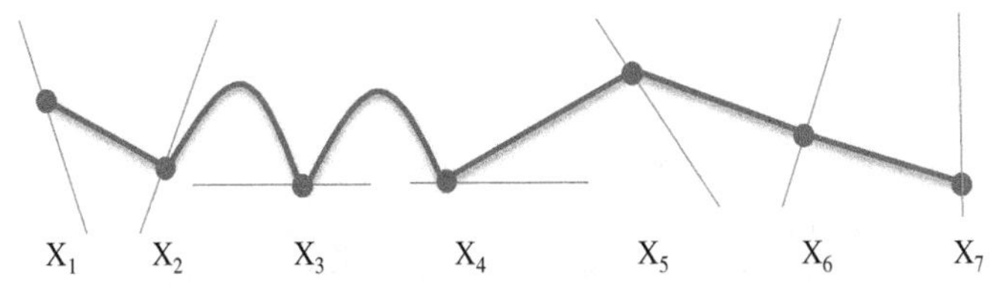

(b) 7-D point D in General Line Coordinates with curvilinear lines.

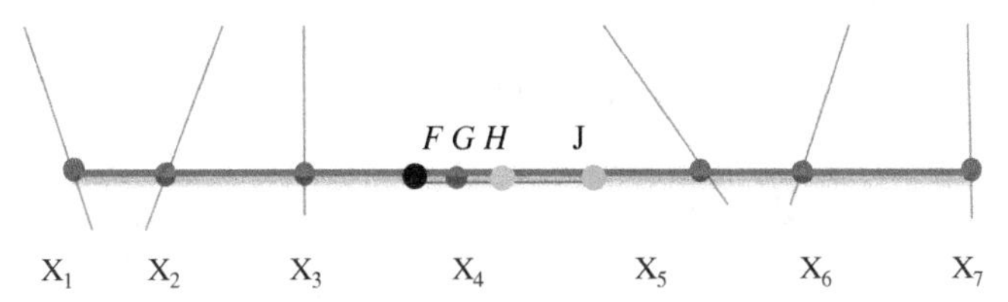

(c) 7-D points F-J in General Line Coordinates that form a simple single straight line.

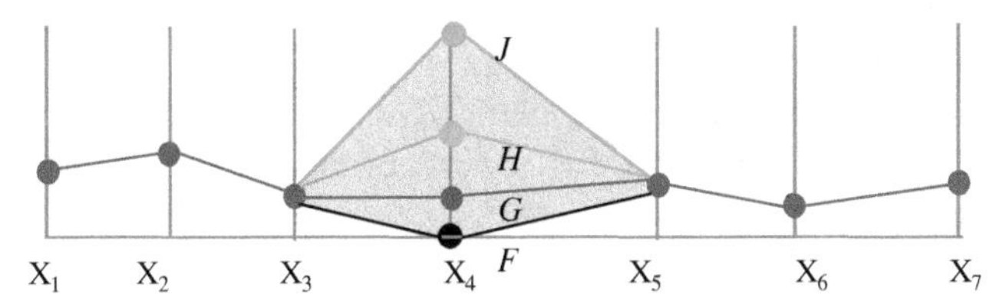

(d) 7-D points F-J in Parallel Coordinates that do not form a simple single straight line.

Fig. 2.3 7-D points in general line coordinates with different directions of coordinates $X_1, X_2, \ldots, X_7$ in comparison with parallel coordinates

The 7-D points shown in Fig. 2.3 are

$$F = (3, 3.5, 2, 0, 2.5, 1.5, 2.5), \quad G = (3, 3.5, 2, 2, 2.5, 1.5, 2.5),$$
$$H = (3, 3.5, 2, 4, 2.5, 1.5, 2.5), \quad J = (3, 3.5, 2, 8, 2.5, 1.5, 2.5),$$

where G is shown with red dots. Here F, G and J differ from G only in the values of x_4. Now let $\{(g_1, g_2, g_3, x_4, g_5, g_6, g_7)\}$ be a set of 7-D points with the same coordinates as in G, but x_4 can take any value in [0, 8].

This set is fully represented in Fig. 2.3c by the simple red line with dots completely covering X_4 coordinate. In contrast, this dataset is more complex in Parallel Coordinates as Fig. 2.3d shows.

This example illustrates the important issue that each GLC has its own set of n-D data that are simpler than in other GLC visualizations. This explains the need for developing:

(1) *Multiple* GLCs to get options for simpler visualization of a wide variety of n-D datasets,
(2) *Mathematical description* of classes of n-D data, where particular GLC is simpler than other GLCs, and
(3) *Algorithms* to visualize those n-D sets in simpler forms.

Several chapters of this book address these needs for a number of GLCs and can serve as a guide for development (1)–(3) for other GLCs in the future.

2.1.2 n-Gon and Circular Coordinates

The lines of some coordinates in the generalized coordinates can also form other shapes and continue straight after each other without any turn between them.

Figure 2.4 shows a form of the GLC, where coordinates are connected to form the **n-Gon Coordinates**. The n-Gon is divided into segments and each segment encodes a coordinate, e.g., in a normalized scale within [0, 1]. If $x_i = 0.5$ in an n-D point, then it is marked as a point on X_i segment. Next, these points are connected to form the directed graph starting from x_1.

Figure 2.5 shows examples of circular coordinates in comparison with Parallel Coordinates. **Circular Coordinates** is a form of the GLC where coordinates are connected to form a circle. Similarly, to n-Gon the circle is divided into segments,

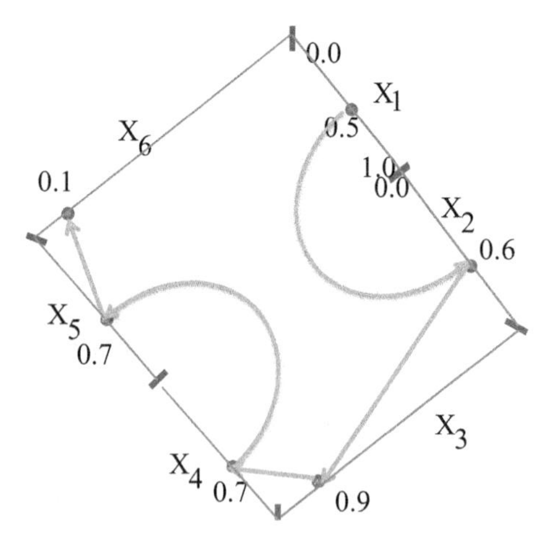

Fig. 2.4 n-Gon (rectangular) coordinates with 6-D point (0.5, 0.6, 0.9, 0.7, 0.7, 0.1)

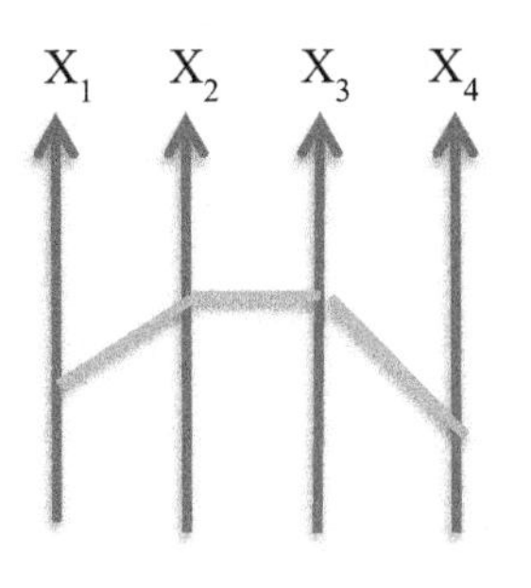

(a) Parallel Coordinates display.

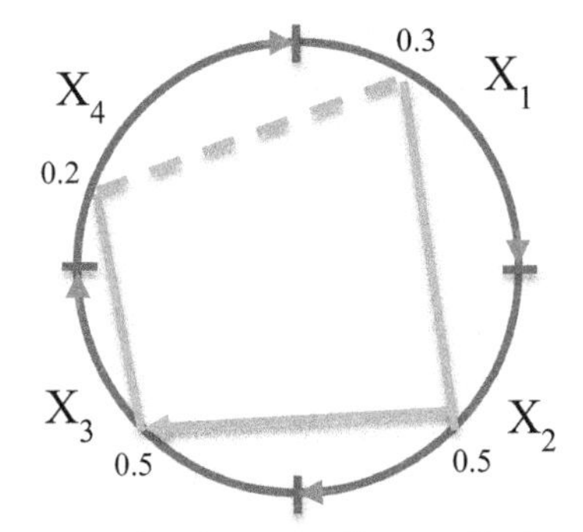

(b) Circular Coordinates display.

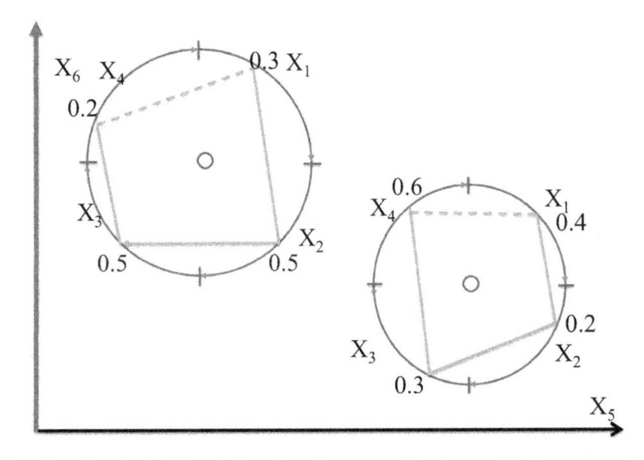

(c) Spatially distributed objects in circular coordinates with two coordinates X_5 and X_6 used as a location in 2-D and X_7 is encoded by the sizes of circles.

Fig. 2.5 Examples of circular coordinates in comparison with parallel coordinates

each segment encodes a coordinate, and points on the coordinates are connected to form the directed graph starting from x_1.

Circular coordinates also can be used with splitting coordinates, where two coordinates out of *n* coordinates identify the *location* of the center of the circle and remaining *n*-2 coordinates are encoded on the circle (Fig. 2.5).

This is a way to represent geospatial data. Multiple circles can be scaled to avoid their overlap. The size of the circle can encode additional coordinates (attributes). In the same way, n-Gon can be used in locational setting for representing geospatial information.

Figure 2.6 shows other examples of n-Gon coordinates, where the n-Gon is not arbitrary selected, but the use of a pentagon that reflects 5 trading days of the stock market.

Figure 2.7 shows stock data in Radial Coordinates. While visuals in Figs. 2.6 and 2.7 are different, both show that in this example the stock price did not change significantly during the week.

This circular setting of coordinates provides a convenient way to observe the change from the first trading data (Monday) to the last trading data (Friday) that are located next to each other. Parallel coordinates lack this ability due to linear location of coordinates.

Figure 2.8 presents 3-D point $A = (0.3, 0.7, 0.4)$ in 3-Gon (triangular) and in radial coordinates. It shows that they have the same expressiveness and can be used equally in the same applications.

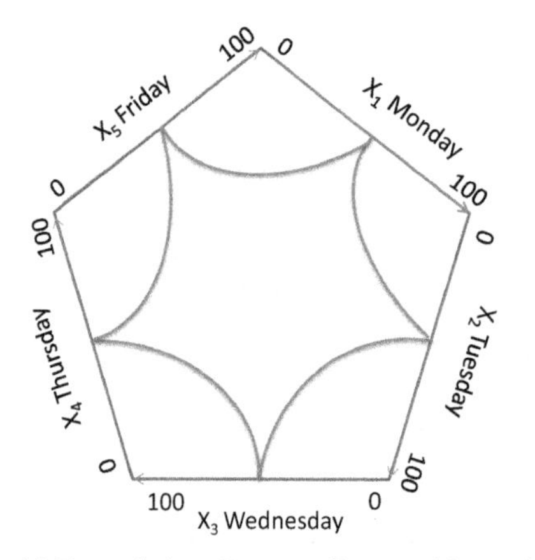

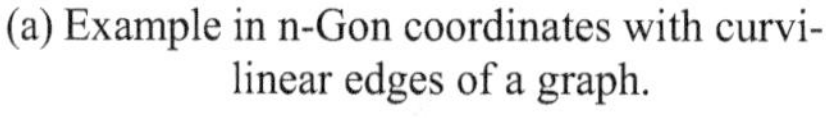
(a) Example in n-Gon coordinates with curvilinear edges of a graph.

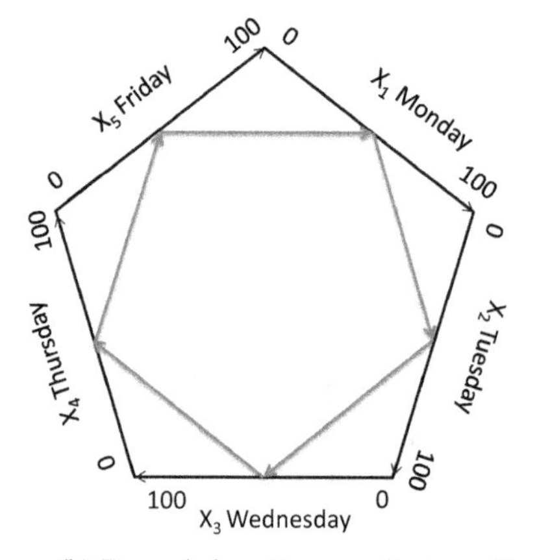

(b) Example in n-Gon coordinates with straight edges of a graph.

Fig. 2.6 Example of weekly stock data in n-Gon (pentagon) coordinates

Fig. 2.7 Weekly stock data in radial coordinates

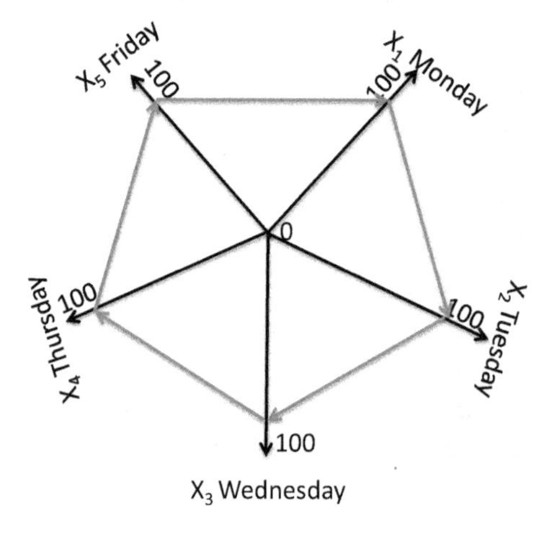

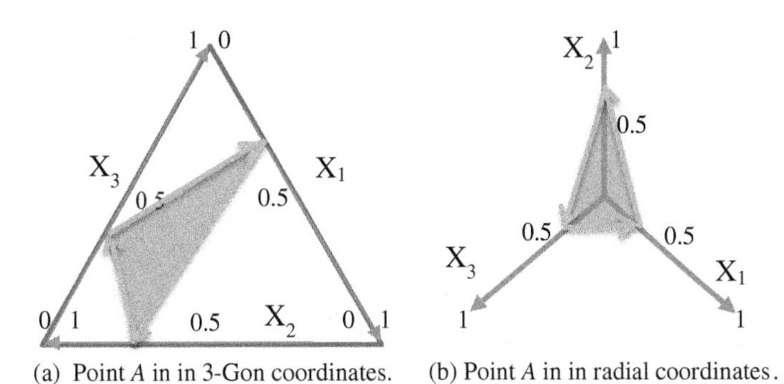

(a) Point A in in 3-Gon coordinates. (b) Point A in in radial coordinates.

Fig. 2.8 3-D point A = (0.3, 0.7, 0.4) in 3-Gon (triangular) coordinates and in radial coordinates

2.1.3 Types of GLC in 2-D and 3-D

Tables 2.1 and 2.2 summarize different types of General Line Coordinates in 2-D and 3-D, respectively. Some of them already have been explained and illustrated other will be in further sections and chapters.

Table 2.1 2-D line coordinates

Type	Characteristics
2-D General Line Coordinates (GLC)	Drawing n coordinate axes in 2-Din variety of ways: curved, parallel, unparalleled, collocated, disconnected, etc.
Collocated Paired Coordinates (CPC)	Splitting an n-D point x into pairs of its coordinates $(x_1, x_2),\ldots, (x_{n-1}, x_n)$; drawing each pair as a 2-D point in the collocated axes; and linking these points to form a directed graph. For odd n coordinate X_n is repeated to make n even
Basic Shifted Paired Coordinates (SPC)	Drawing each next pair in the shifted coordinate system by adding (1, 1) to the second pair, (2, 2) to the third pair, $(i-1, i-1)$ to the ith pair, and so on. More generally, shifts can be a function of some parameters
2-D Anchored Paired Coordinates (APC)	Drawing each next pair in the shifted coordinate system, i.e., coordinates shifted to the location of a given pair (anchor), e.g., the first pair of a given n-D point. Pairs are shown relative to the anchor easing the comparison with it
2-D Partially Collocated Coordinates (PCC)	Drawing some coordinate axes in 2D collocated and some coordinates not collocated
In-Line Coordinates (ILC)	Drawing all coordinate axes in 2D located one after another on a single straight line
Circular and n-Gon coordinates	Drawing all coordinate axes in 2D located on a circle or an n-Gon one after another
Elliptic Coordinates	Drawing all coordinate axes in 2D located on ellipses
GLC for Linear Functions (GLC-L)	Drawing all coordinates in 2D dynamically depending on coefficients of the linear function and value of n attributes
Paired Crown Coordinates (PWC)	Drawing odd coordinates collocated on the closed convex hull in 2-D and even coordinates orthogonal to them as a function of the odd coordinate

Table 2.2 3-D line coordinates

Type	Characteristics
3-D General Line Coordinates (GLC)	Drawing n coordinate axes in 3-D in variety of ways: curved, parallel, unparalleled, collocated, disconnected, etc.
Collocated Tripled Coordinates (CTC)	Splitting n coordinates into triples and representing each triple as 3-D point in the same three axes; and linking these points to form a directed graph. If n mod 3 is not 0 then repeat the last coordinate X_n one or two times to make it 0
Basic Shifted Tripled Coordinates (STC)	Drawing each next triple in the shifted coordinate system by adding (1, 1, 1) to the second tripple, (2, 2, 2) to the third tripple ($i-1$, $i-1$, $i-1$) to the ith triple, and so on. More generally, shifts can be a function of some parameters
Anchored Tripled Coordinates (ATC) in 3-D	Drawing each next triple in the shifted coordinate system, i.e., coordinates shifted to the location of the given triple of (anchor), e.g., the first triple of a given n-D point. Triple are shown relative to the anchor easing the comparison with it
3-D Partially Collocated Coordinates (PCC)	Drawing some coordinate axes in 3-D collocated and some coordinates not collocated
3-D In-Line Coordinates (ILC)	Drawing all coordinate axes in 3D located one after another on a single straight line
In-Plane Coordinates (IPC)	Drawing all coordinate axes in 3D located on a single plane (2-D GLC embedded to 3-D)
Spherical and Polyhedron Coordinates	Drawing all coordinate axes in 3D located on a sphere or a polyhedron
Ellipsoidal Coordinates	Drawing all coordinate axes in 3D located on ellipsoids
GLC for Linear Functions (GLC-L)	Drawing all coordinates in 3D dynamically depending on coefficients of the linear function and value of n attributes
Paired Crown Coordinates (PWC)	Drawing odd coordinates collocated on the closed convex hull in 3-D and even coordinates orthogonal to them as a function of the odd coordinate value

The last type of GLCs in Tables 2.1 and 2.2 called GLC for linear functions (GLC-L) is designed to deal with *linear* functions of n variables. In detail, this type of GLC is presented in Chap. 7.

In contrast with other GLCs listed in these tables, in GLC-L the *location of coordinates* is not static, but it is *dynamically build* from original positions of coordinates as a function of n-D points visualized. Another difference is that the candidate classification relation between coordinates is *built in* to the visualization. For other GLCs listed in these tables, it is not the case. In those visual representations, we are looking for any relations and features that might be informative for the given task.

In general, not all GLCs must have immediate applications. The situation here is similar to a situation with the definition of general linear equations of n variables x_1, $x_2,\ldots,x_n$ with arbitrary coefficients. Some equations are in use every day, some will be used tomorrow, but some may never be of any use. The value of both general definitions is in the fact that equations and visualizations do not cover all possible tasks. There is no "silver bullet" among them. Therefore, we need a collection of other equations and other visualization methods where we can look for the right one for new tasks.

2.1.4 In-Line Coordinates

The GLC include **In-Line Coordinates** (ILC) shown in Fig. 2.9 that are similar to Parallel Coordinates, except that the axes $X_1, X_2, \ldots, X_n$ are horizontal, not vertical. All coordinates are collocated on the same line, but may or may not overlap. Each pair is represented by a directed curve. Any curve from x_i to x_{i+1} will satisfy the requirement of lossless representation, but the curves of different heights and shapes can show additional information such as the distance between adjacent values, $|x_i - x_{i+1}|$. In Fig. 2.9 the height of the curve represents the distance between the two adjacent values, e.g., for point (5, 4, 0, 6, 4, 10), the heights represent 1, 4, 6, 2, 6. Respectively, the quadratic curve from x_1 to x_2 is constructed by fitting a quadratic equation to three points $(x_1, 0)$, $(O_2 + x_2, 0)$, and $((x_1 + O_2 + x_2)/2, |x_1 - x_2|)$, where O_2 is the coordinate of the origin of X_2 on the joint line that starts from the origin of X_1.

A quadratic Bezier curve based on these three points approximates this quadratic equation. A set of such Bezier curves is shown in Fig. 2.9. Consider curves for the points x_i, x_{i+1}. We treat these x_i and x_{i+1} on X_i and X_{i+1} as 2-D points $A = (a_1, a_2)$ and $B = (b_1, b_2)$, respectively and find the midpoint, $C = (A + B)/2$. Then the normal line N to the line (A, B) is constructed from point C and the point D on N is found such that its distance from C is $|x_i - x_{i+1}|$. The points A, B and D are used to construct the quadratic Bezier curve from x_i to x_{i+1}. Figure 2.9 shows that In-Line Coordinates require the same number of nodes as Parallel Coordinates, which makes the scopes of applicability of these methods similar.

An alternative to smooth curves are simpler "triangular" edges that connect point A, B and D, i.e., the top of the curve to points on coordinates, for short we will call these lines *triangles*. The curves are directed edges, but the direction is clear, we omit arrowheads. To decrease occlusion we draw n-D points of one class above the coordinate line and another class below it as is shown in Fig. 2.10.

Below we elaborate algorithmic options for constructing in-line coordinates and representing n-D data in ILC. Later in Chap. 5 some of these options are illustrated in the case study with real data.

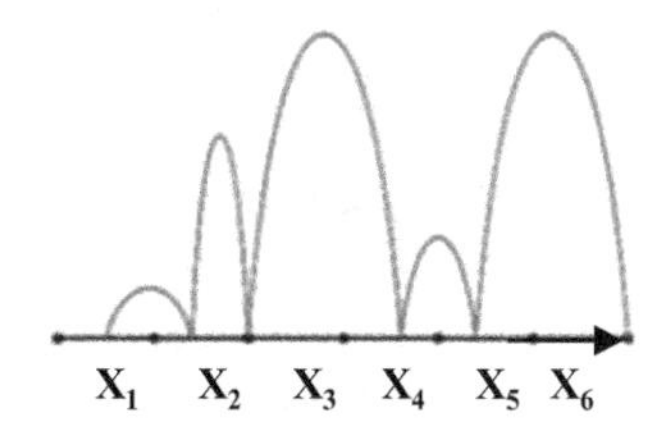

Fig. 2.9 6-D (5, 4, 0, 6, 4, 10) point in in-line coordinates

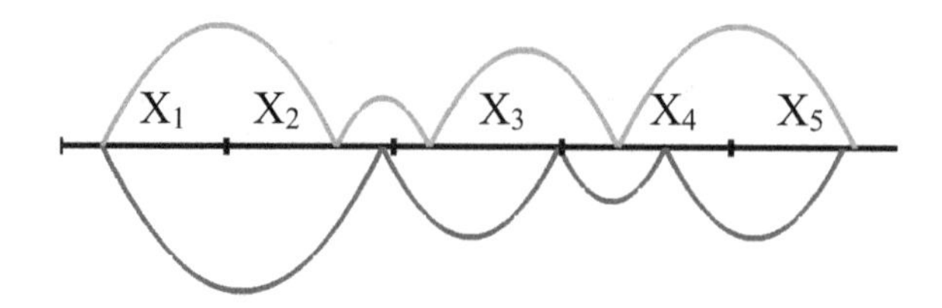

Fig. 2.10 Two 5-D points of two classes in sequential in-line coordinates

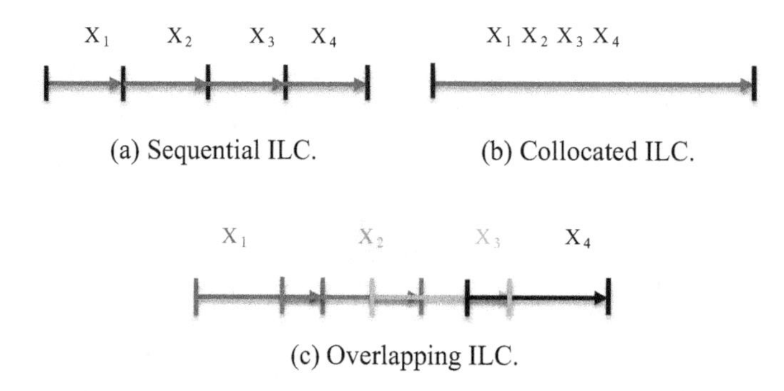

Fig. 2.11 Options to locate coordinates in on-line coordinates

In designing of ILC, we have several options to *locate coordinates* X_1–X_n:

(L1) All coordinated X_1–X_n are drawn sequentially one after another overlapping only in a single point where they are connected. See Fig. 2.11a.

(L2) All coordinated X_1–X_n are collocated, i.e., are drawn at the same location with full overlap. See Fig. 2.11b.

(L3) All coordinated X_1–X_n are drawn sequentially with partial overlap. See Fig. 2.11c. This overlap can depend on a parameter.

(L4) All coordinated X_1–X_n are dynamically sequentially located as it is shown in the next section on Dynamic Coordinates.

Note that in L2 and L3 the curves can go backward and arrowheads must be drawn to avoid confusion. A given n-D point $\mathbf{a} = (a_1,a_2,\ldots,a_n)$ can serve as a parameter for L3, where each a_i will indicate the value on coordinate X_i where the coordinate X_{i+1} will start.

Alternatively, given n-D point **a**, we set up the origin point O_2 for coordinate X_2 at the value a_1-a_2 on coordinate X_1. In this case, the value a_2 must be plotted at the position $O_2 + a_2$, which is a_1 given $O_2 = a_1-a_2$. Thus, a_1 and a_2 will be located at the same 2-D point. Next, we similarly assign all other origin points $O_i = a_1-a_i$. As a result, we get all a_i points in the same location as a_1. In other words, we will collapse n-D point **a** to a single 2-D point. This mapping is *reversible and* lossless because having all origins O_i we can restore n-D point **a**. For instance, for $\mathbf{a} = (1,1,\ldots,1)$ all $O_i = a_1-a_i = 0$. Thus, all coordinates X_i are collocated with the same origin as in Fig. 2.11b and point **a** can be restored.

Next, we have several options for *ordering/reordering coordinates* X_1–X_n and respective Bezier curves and triangles:

(O1) Keeping the original order of coordinates as in Fig. 2.10.
(O2) Ordering by increasing values of x_i for a given n-D point **x**.
(O3) Ordering by increasing distance $|x_i - x_{i+1}|$ between consecutive values x_i and x_{i+1} for a given n-D point **x**.
(O4) Ordering by increasing variance of values of x_i for a given set of n-D points, not a single n-D point as in (O2) and (O3).

In addition, we have several options to *position* triangles and Bezier or other curves that connect points on X_1–X_n.

(P1) Curves and triangles of all classes are above the ILC base line.
(P2) Curves and triangles of class 1 are above and class 2 is below the ILC base line growing downwards instead. See Fig. 2.10.
(P3) Curves and triangles of class 1 are above and all other classes are below the ILC base line growing downwards instead.

We have also several options to *construct* triangles and curves that connect points on X_1–X_n by assigning the *height* of triangles and curves as follows:

(C1) The distance between connected values x_i and x_{i+1} that repeats their distance on the base line of ILC).
(C2) The value x_i of the first connected coordinate in each pair (x_i, x_{i+1}). The last one x_n will not be represented in this way.
(C3) The value x_{i+1} of the second connected coordinate in each pair (x_i, x_{i+1}). The first one x_1 will not be represented in this way.
(C4) Another value, e.g., a constant or x_{i+2} for the pair (x_i, x_{i+1}). In the case of x_{i+2}, each third coordinate X_{i+2} can be omitted on the base line of ILC, because that value will be encoded as a height of the curve or the tringle for (x_i, x_{i+1}). See Fig. 2.12. In this case, the base line will be shortened containing only 2/3 of all coordinates. If x_i, x_{i+1}, and x_{i+2} are correlated the shape of the curve or the triangle will make it visible in ILC.
(C5) The value x_{i+2} is the height of the curve with an additional property that the *width* of the curve is x_{i+3} for each pair (x_i, x_{i+1}). See an example in Fig. 2.13 for 7-D point $\mathbf{x} = (x_1, x_2, x_3, x_4, x_5, x_6, x_7) = (1, 2, 3, 5, 3, 4, 2)$ with values of x_3 and x_4 encoded as the height and width of the curve that connects (x_1, x_2) and values of x_6 and x_7 the height and width of the curve that connects (x_2, x_5). As a result

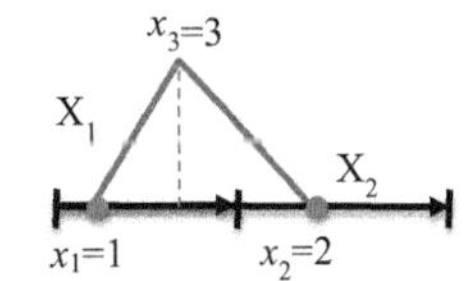

Fig. 2.12 3-D point $\mathbf{x} = (x_1, x_2, x_3) = (1, 2, 3)$ in in-line coordinates with the value of x_3 encoded in the height of triangle that connects (x_1, x_2)

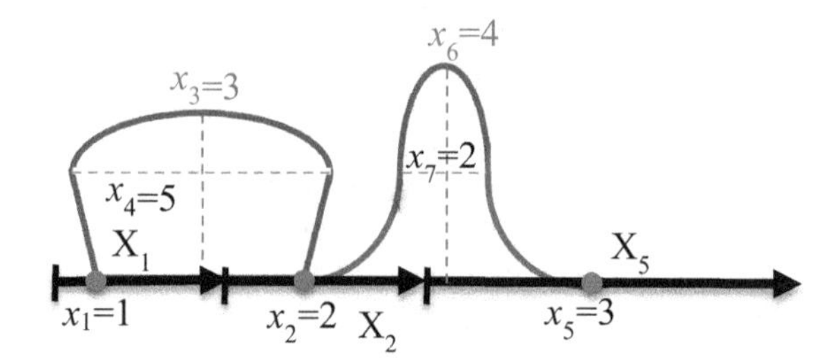

Fig. 2.13 7-D point $\mathbf{x} = (x_1, x_2, x_3, x_4, x_5, x_6, x_7) = (1, 2, 3, 5, 3, 4, 2)$ in in-line coordinates with values of x_3, x_4, x_6 and x_7 encoded in the height and width of the curves that connect (x_1, x_2) and (x_2, x_5)

out of seven coordinates only three coordinates X_1, X_2 and X_5 are directly encoded in the base line of ILC. Thus, in general with (C5) we keep 3/5 of the total number of coordinates on the base line making it shorter.
Examples of several ILC options defined above are presented in Chap. 5 for real world data.

2.1.5 *Dynamic Coordinates*

In Parallel, Radial and other coordinate systems described above the location of all coordinates X_1–X_n is fixed and values of each coordinate x_i of the n-D point $\mathbf{x}$ is located on X_i. We will call this mapping of n-D points to coordinates **static mapping**.

In the **dynamic mapping** of the given n-D point **x**, the location of the next value x_{i+1} depends on the location and value of x_i. Thus, the location of x_{i+1} will change dynamically with change of x_i. In Tables 2.1 and 2.2, we identified GLC for linear functions (GLC-L) and Paired Crown Coordinates (PWC) that do not use static mapping but dynamic one. PWC will be described later in this chapter and GLC-L in Chap. 7.

A 2-D graph $\mathbf{x}^*$ of an n-D point $\mathbf{x} = (x_1, x_2, \ldots, x_n)$ is created in *dynamic mapping* by connecting the consecutive edges as follows. The first edge has length x_1 and is located on the first coordinate X_1, starting from origin and ending on point x_1 on X_1. The second edge starts at the end of the first edge and is going *parallel* to the second coordinate X_2. It has length x_2. Similarly, the edge j is going parallel to coordinate X_j starting at the end of edge j-1. In general, this graph is not a closed contour. A *closed contour* is made by adding the edge that connects the last node with the origin node. In the cases below 16 coordinates X_1–X_{16} are located in four different ways: as radii from the common origin (Kandogan 2000) in Fig. 2.14a, as a star zig-zag in Fig. 2.14b, as a linear zigzag in Fig. 2.15a, b and as a single line in Fig. 2.15c forming Dynamic In-line Coordinates. All of them encode the same 16-D point $\mathbf{a} = (1, 1, 2, 2, 1, 1, 2, 2, 1, 1, 2, 2, 1, 1, 2, 2)$.

The first edge of the graph $\mathbf{a}^*$ is located on X_1 coordinate forming the edge of length $a_1 = 1$. The second edge starts at the point where x_1 is located and goes

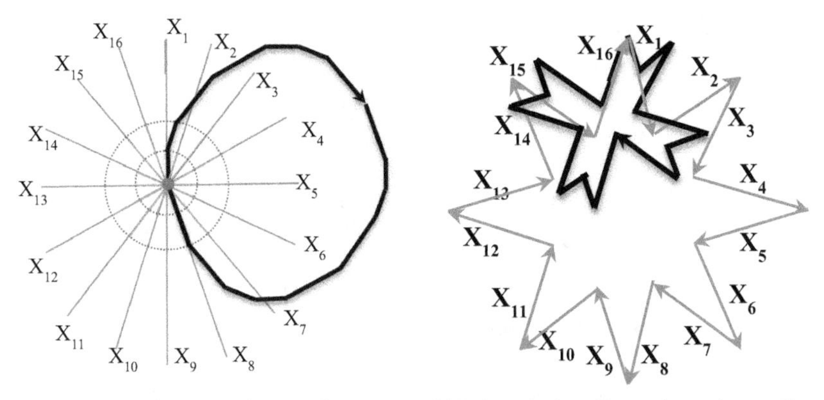

(a) Point **a** in Radial Dynamic Coordinates. (b) Point **a** in Star Zigzag dynamic coordinates.

Fig. 2.14 16-D point **a** = (1, 1, 2, 2, 1, 1, 2, 2, 1, 1, 2, 2, 1, 1, 2, 2) in two Dynamic Coordinates

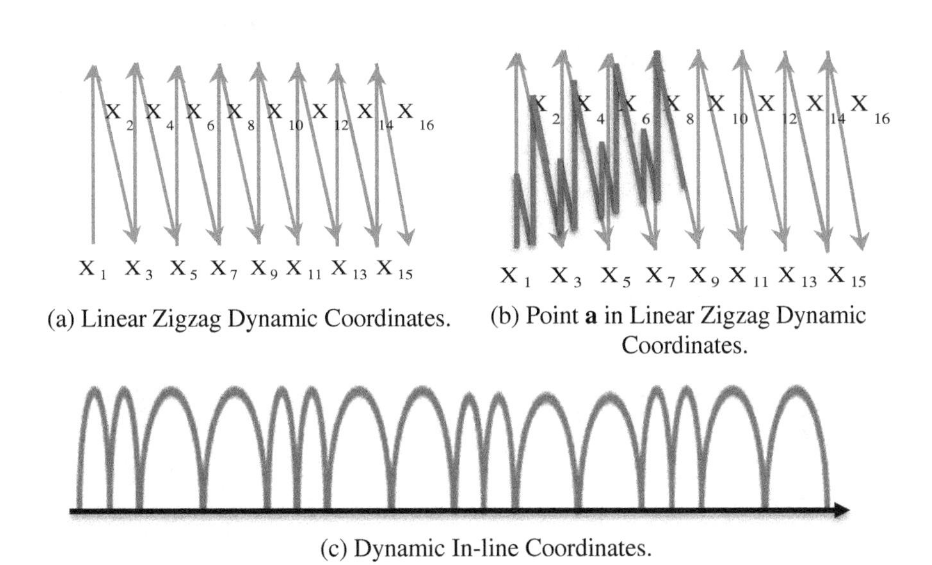

(a) Linear Zigzag Dynamic Coordinates. (b) Point **a** in Linear Zigzag Dynamic Coordinates.

(c) Dynamic In-line Coordinates.

Fig. 2.15 Point **a** = (1, 1, 2, 2, 1, 1, 2, 2, 1, 1, 2, 2, 1, 1, 2, 2) in 16-D Linear Zigzag Dynamic Coordinates

parallel to X_2 coordinate. For Fig. 2.14a it is with the angle 360°/16 = 22.5°. The length of this edge is $a_2 = 1$. Similarly, all other edges are generated.

Chapter 3 presents several other ways to implement dynamic mapping. All dynamic coordinates are lossless 2-D representations of n-D data. Their multiplicity expands opportunities to discover visual patterns on the same n-D data.

2.1.6 Bush and Parallel Coordinates with Shifts

The Bush Coordinates (BC) are modified Parallel Coordinates (PC) where only the middle coordinate in vertical. All other coordinates are tilted increasingly to form a "bush".

Figure 2.16a, b shows both Parallel and Bush coordinates for the same data. It shows how simple preattentive straight blue line in PC is converted to a more complex line in BC, while Fig. 2.16c, d show the opposite effect a preattentive blue line in BC is converted to a more complex line in PC.

This example along several other examples in this book shows the abilities to select a specific GLC that simplifies the visualization of a given n-D point. Chapter 4 presents the simplification in more detail and with more examples.

We also can shift Coordinates in Bush and Parallel Coordinates to the central coordinate. This shift can be in both horizontal and vertical directions. A specific shift of a given n-D point will collapse it to a single 2-D point as it is shown in Fig. 2.17 for the green line from Fig. 2.16.

While the given n-D point will be encoded as a single 2-D point, other n-D points will still have n nodes and n-1 edges in the graph as is shown in Fig. 2.17. However the graphs of the n-D points that are close to the given n-D point $\mathbf{x} = (x_1, x_2, \ldots, x_n)$ will be smaller (see Fig. 2.17 for the red line and curves).

When this shifting is applied to Parallel Coordinates in Fig. 2.16a we use curves to connect the points to be able to see n-D points as shown in Fig. 2.17b for blue and read 6-D points. This is similar to drawing lines with Bezier curves and triangles in ILC above (Fig. 2.10).

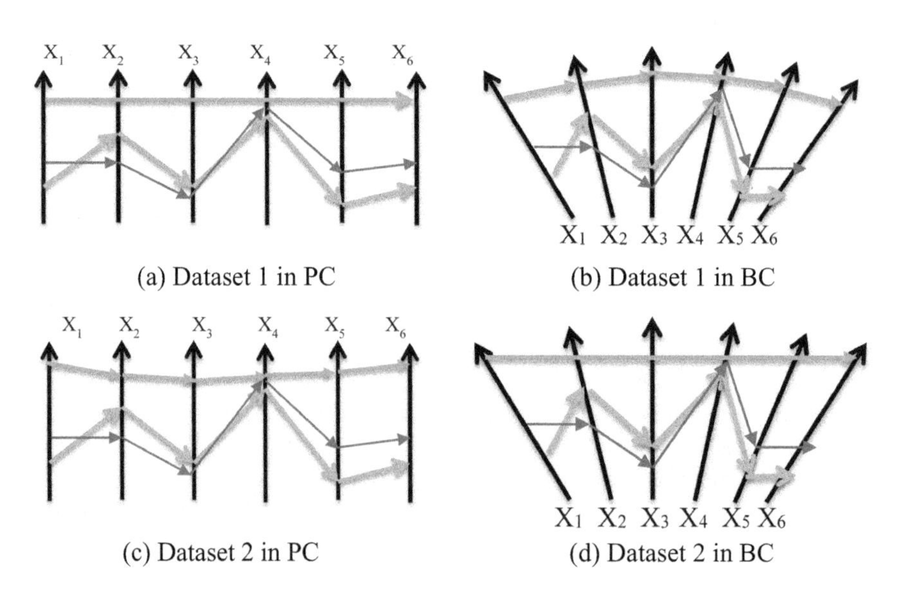

Fig. 2.16 Three 6-D points in parallel and bush coordinates

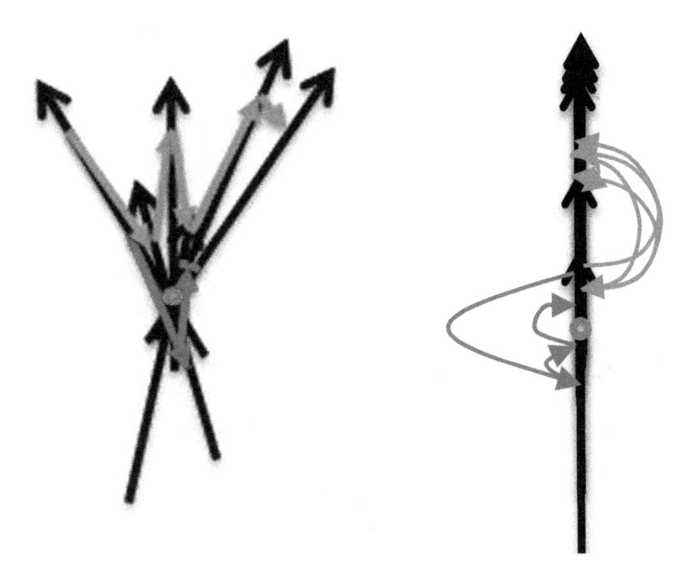

(a) Shifted Collapsing Bush Coordinates (b) Shifted Collapsing Parallel Coordinates

Fig. 2.17 Three 6-D points after shifting coordinates X_1, X_2, X_4, X_5, X_6 to X_3 with collapsing the green line from Fig. 2.16 to a single 2-D point (small green circle in the center) in BC and PC

2.2 Reversible Paired Coordinates

2.2.1 Paired Orthogonal Coordinates

The algorithm for representing n-D points in 2-D using lossless **Collocated Paired Coordinates** (**CPC**) is presented below. We use an example in 6-D with a state vector $\mathbf{x} = (x, y, x', y', x'', y'')$, where x and y are location of the object, x' and y' are velocities (first derivatives), and x'' and y'' are accelerations (second derivatives) of this object.

The main steps of the algorithm are:

1. Normalization of all dimensions to the same interval, e.g., [0, 1],
2. Grouping attributes into consecutive pairs (x, y) (x', y') (x'', y''),
3. Plotting each pair in the same orthogonal normalized Cartesian coordinates X and Y, and
4. Plotting a *directed graph* (*digraph*): $(x, y) \rightarrow (x', y') \rightarrow (x'', y'')$ with directed edges (arrows) from (x, y) to (x', y') and from (x', y') to (x'', y'').

In Fig. 2.18a CPC algorithm is applied to a 6-D point (5, 4, 0, 6, 4, 10) showing two arrows: $(5, 4) \rightarrow (0, 6) \rightarrow (4, 10)$.

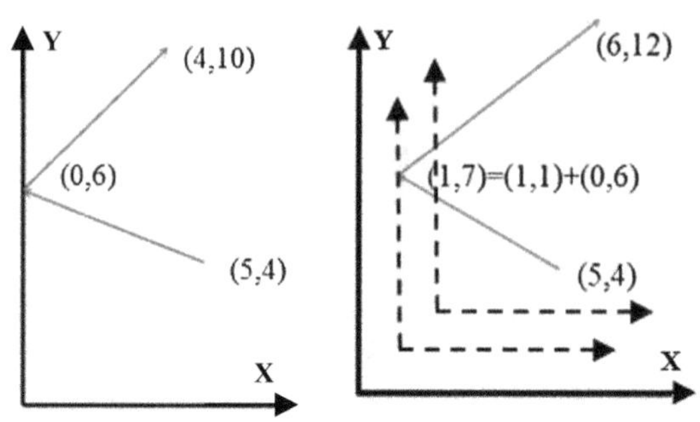

Fig. 2.18 6-D point (5, 4, 0, 6, 4, 10) in paired coordinates

The **Shifted Paired Coordinates** (SPC) show each next pair in the shifted coordinate system. The first pair (5, 4) is drawn in the (X, Y) system, pair (0, 6) is drawn in the (X + 1, Y + 1) coordinate system, and pair (4, 10) is drawn in the (X + 2, Y + 2) coordinate system. For point **x** = (5, 4, 0, 6, 4, 10), the graph consists of the arrows: from (5, 4) to (1, 1) + (0, 6) = (1, 7) then from (1, 7) to (2, 2) + (4, 10) = (6, 12) in the original (X, Y) coordinates. See Fig. 2.18b. In the original (X, Y) coordinate system the nodes of the graph are points (5, 4), (1, 7) and (6, 12).

In the SPC version presented above, the shift from each coordinate system to the next one is the constant 1. SPC allows generalizations by using:

(1) *any constants* as shifts from each coordinate system to the next one that differ from 1, e.g., 2, 3 and so on,
(2) *dynamic shifts* that depend on an n-D point **x** itself, e.g., the second pair (x′, y′) = (0, 6) is drawn in the (X + x, Y + y) coordinate system, (0, 6) + (5, 4) = (5, 10) with $x = 5$, $y = 4$. Similarly, the third pair (x″, y″) = (4, 10) is drawn in the (X + x + x', Y + y + y') coordinate system (4, 10) + (5 + 0, 4 + 6) = (9, 20) with $x' = 0$, $y' = 6$.

Note that this mapping is lossless. It allows restoring the original 6-D point (5, 4, 0, 6, 4, 10) from 6-D point (5, 4, 5, 10, 9, 20) by subtracting respective numbers.

The **Anchored Paired Coordinates** (APC) starts at the given pair (a_1, a_2) that serves an "anchor". Below we present two version of APC: APC1 and APC2.

In APC-1 pair (a_1, a_2) is drawn in a given (X, Y) coordinate system. Then a new coordinate system is created (X + a_1, Y + a_2) that has its origin in (a_1, a_2). Next this new system is used as a collocated coordinate system for all n-D points. In other words, APC1 is *CPC shifted* to the point (a_1, a_2). If anchor pair (a_1, a_2) is selected to be the first pair coordinates of a given n-D point **x**, then for this n-D point APC is CPC, because there is no shift for the first pair (x_1, x_2) of **x**.

In APC2 pair (a_1, a_2) is also drawn in a given (X, Y) coordinate system, but all pairs (x_1, x_2), (x_3, x_4),…,($x_{n\text{-}1}$, x_n) are represented by directed edges from the anchor (a_1, a_2).

$$(a_1, a_2) \to (x_1, x_2),\ (a_1, a_2) \to (x_3, x_4), \ldots,\ (a_1, a_2) \to (x_{n-1}, x_n)$$

In APC2 directed edges are labeled (numbered) to indicate the order of the edges to be able to restore the n-D point from the graph. These labels can be turned off to decrease occlusion to see a generalized visual pattern.

APC2 is illustrated below with two examples. Let the anchor be $(a_1, a_2) = (1, 2)$ in (X, Y) coordinates, then the 6-D point (5, 4, 0, 6, 4, 10), used in the example above, has three edges drawn in the given (X, Y) coordinates:

$$(1,2) \rightarrow (5+1,4+2),\ (1,2) \rightarrow (0+1,6+2),\ (1,2) \rightarrow (4+1,10+2)$$

These edges directly show velocity and acceleration vectors of the state vector that the 6-D point (5, 4, 0, 6, 4, 10) represents. This is an advantage of the APC2 for modeling state vectors. In contrast, in the coordinate systems such as standard Radial Coordinates the directions have no such physical meaning. Figure 2.19 illustrates APC2 for another 6-D point, $\mathbf{x} = (0.2, 0.4, 0.1, 0.6, 0.4, 0.8)$ when the anchor pair in the first pair of this 6-D point.

The general idea of the paired coordinates is converting a simple string of elements of n-D point $\mathbf{x} = (x_1, x_2, \ldots, x_n)$ in coordinates $X_1, X_2, \ldots, X_n$ to a more complex *structure* with consecutive 2-D elements (*pairs*) for even n:

$$\{(x_1, x_2)(x_3, x_4), \ldots, (x_i, x_{i+1}), \ldots, (x_{n-3}, x_{n-2}), (x_{n-1}, x_n)\}$$

For the odd n this structure is slightly different with x_{n-1} used in both last two pairs to give a pair to x_n:

$$\{(x_1, x_2)(x_3, x_4), \ldots, (x_i, x_{i+1}), \ldots, (x_{n-2}, x_{n-1}), (x_{n-1}, x_n)\}$$

Alternatively, x_n can form a pair with itself for an odd n:

$$\{(x_1, x_2), (x_3, x_4), \ldots, (x_i, x_{i+1}), \ldots, (x_{n-2}, x_{n-1}), (x_n, x_n)\}$$

An example shown in Fig. 2.20a illustrates the *advantages of Paired Coordinates* over Parallel Coordinates (Fig. 2.20b) using 6-D point $\mathbf{x} = (x, y, x', y',$

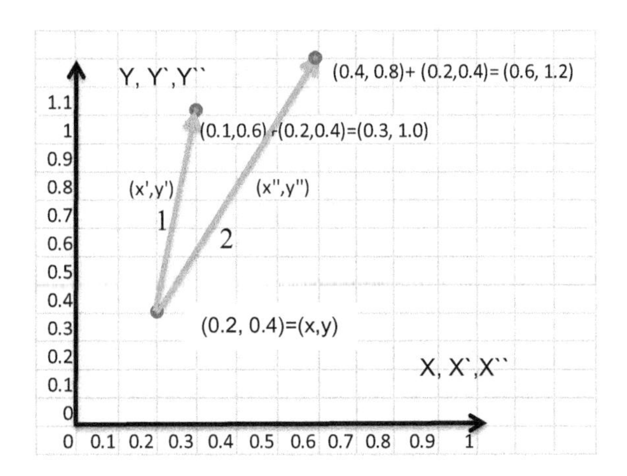

Fig. 2.19 6-D point $\mathbf{x} = (x, y, x', y', x'', y'') = (0.2, 0.4, 0.1, 0.6, 0.4, 0.8)$ in anchored paired coordinates with numbered arrows

x'', y'') = (0.2, 0.4, 0.1, 0.6, 0.4, 0.8). Parallel Coordinates require 5 lines to show **x** (Fig. 2.20b). In contrast, all collocated coordinates require only 2 lines, which leads to *less clutter* when multiple n-D points are visualized on the same (X, Y) coordinate plane. It is a general property of all paired coordinates to require two times fewer lines than the Parallel Coordinates require.

The 3-D version of the Collocated Paired Coordinates is a natural generalization of the visual representations shown above by adding the Z coordinate, and showing lines in 3-D.

Linearly correlated data in CPC. Figure 2.21a shows linearly correlated data from Table 2.3 in the Collocated Paired Coordinates. All shapes are identical and existence of the shifts is visible. This visualization required only 3 lines for each 8-D point instead of 7 lines in the parallel coordinates.

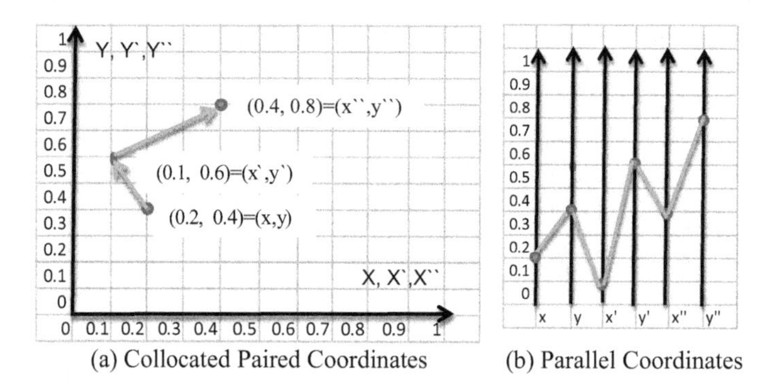

Fig. 2.20 State vector **x** = (x, y, x′, y′, x″, y″) = (0.2, 0.4, 0.1, 0.6, 0.4, 0.8) in Collocated Paired and Parallel Coordinates

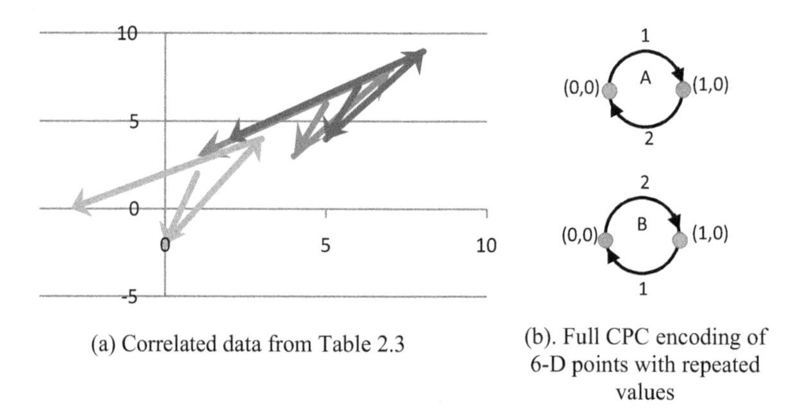

Fig. 2.21 CPC visualizations

Table 2.3 Correlated 8-D points that differ in a shift

	x_1	x_2	x_3	x_4	x_5	x_6	x_7	x_8
$\mathbf{x}_{red}$	5	6	4	3	7	8	1	3
$\mathbf{x}_{blue}$	10	11	9	8	12	13	6	8
$\mathbf{x}_{green}$	1	2	0	−2	3	4	−3	0

Typically, CPC visualization without labels is sufficient to observe the visual pattern. For long graphs produced for higher dimensional data we use **animation** of the graph with a point moving on the edges of the graph or jumping from node to node for faster traversing of the graph. One of the agents in the collaborative visualization can *drive the animation* for other agents. In this case the interactive CPC visualization uses coloring or blinking of these graphs. This **interactive visualization** allows:

(i) turning off all other graphs showing only such ambiguous graphs,
(ii) presenting these graphs in a separate window (CPC plane), and
(iii) showing actual numeric values.

Figure 2.21b illustrates a complete CPC graph visualization with labels 1 and 2 for two 6-D points $A = (0, 0, 1, 0, 0, 0)$ and $B = (1, 0, 0, 0, 1, 0)$ denoted as (A) and (B) respectively.

In both the green node indicates the start point of the graph, and labels 1 and 2 indicate the sequence of traversing the edges of graph. For point A the sequence is (0, 0) → (1, 0) → (0, 0) and for point B it is (1, 0) → (0, 0) → (1,0).

2.2.2 *Paired Coordinates with Non-linear Scaling*

Rescaling of coordinates changes visual pattern of n-D data that can be used to simplify visualization of n-D data and to make the pattern clearer. Figure 2.22a shows 4-D point (1, 1, 1, 0.5) in Shifted Paired Coordinates in regular linear

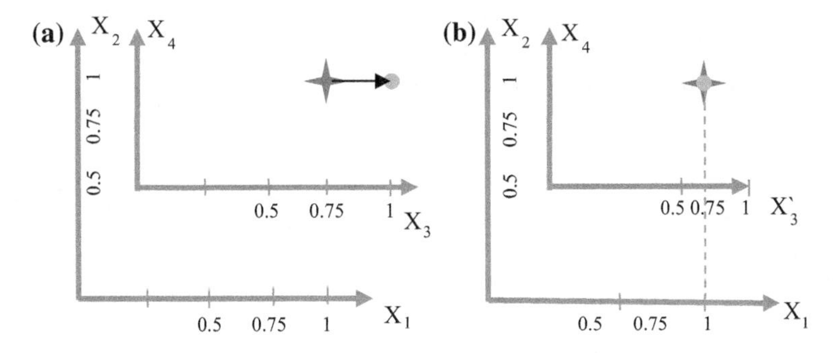

Fig. 2.22 4-D point (1, 1, 1, 0.5) in shifted paired coordinates in **a** regular (linear) scaling, **b** in disproportional (non-linear) scaling of coordinate X_3

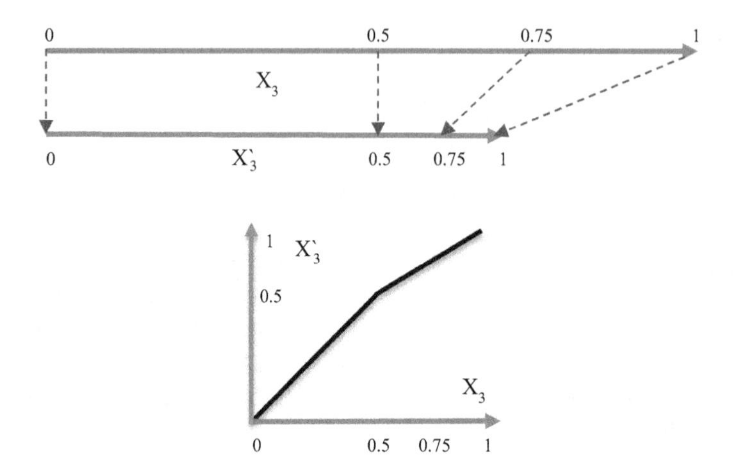

Fig. 2.23 Rescaling X_3 to X'_3

scaling, and Fig. 2.22b shows it with disproportional non-linear scaling of coordinate X_3.

This rescaling of X_3 converts the graph (arrow) to a single point that is a simpler visual representation. The rescaling formula of X_3 to X'_3 is below and it is illustrated in Fig. 2.23,

$$x'_3 = \begin{cases} x_3, & \text{if} \quad x_3 < 0.5 \\ 0.5x_3 + 0.25, & \text{if} \quad 0.5 \le x_3 < 1 \end{cases}$$

2.2.3 *Partially Collocated and Non-orthogonal Collocated Coordinates*

Collocated Paired Coordinates defined above in Sect. 2.2.1 require *full collocation* of *orthogonal pairs* of Cartesian coordinates. In this section, we present coordinate systems where these requirements for pairs are relaxed.

Figure 2.24 shows *Partially Collocated Orthogonal (Ortho) Coordinates.* Here X_1 and X_3 are partially collocated, i.e., X_3 starts at the blue dot in X_1 coordinate, but

Fig. 2.24 Partially Collocated Orthogonal (Ortho) Coordinates

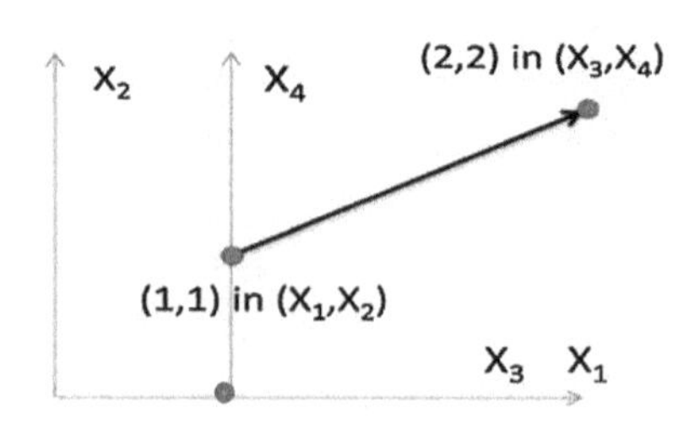

X_2 and X_4 do not collocate at all, X_4 is shifted relative to X_2. Thus, this is a mixture of CPC and SPC coordinates. The 4-D point (1, 1, 0, 1) in these coordinates collapses to a *single* 2-D point. This 2-D point is located at the place of 2-D point (1, 1) in (X_1, X_2) coordinates. This makes the visual representation of this point simple and *pre-attentive.*

Figure 2.25 shows *Partially Collocated Ortho and non-Ortho Coordinates.* Here (X_1, X_2) is an orthogonal pair and (X_3, X_4) is a non-orthogonal pair, where X_1 and X_3 are partially collocated in the same way as in Fig. 2.24, but X_4 is shifted and rotated relative to X_2. In these coordinates another 4-D point (1, 1, 0, 1.2) will collapse to a single 2-D point. This 2-D point is located at the place of 2-D point (1, 1) in (X_1, X_2) coordinates making it pre-attentive.

Figure 2.26 shows *Fully Collocated non-Ortho Coordinates,* where both (X_1, X_2) and (X_3, X_4) are non-orthogonal pairs. In these coordinates the third 4-D point (1, 1, 1, 1) will collapse to a single 2-D point. This 2-D point is located at the place of 2-D point (1, 1) in (X_1, X_2) coordinates making it pre-attentive.

Pairing of radially located coordinates is considered in the next section. In general, multiple different pairing can be designed to simplify visualization of given n-D points of interest.

2.2.4 Paired Radial (Star) Coordinates

Traditional Radial (Star) Coordinates shown on Fig. 2.1 use *n* nodes to represent each n-D point. Below we present the **Paired Radial (Star) Coordinates** that use a half of the nodes to get a reversible representation of an n-D point.

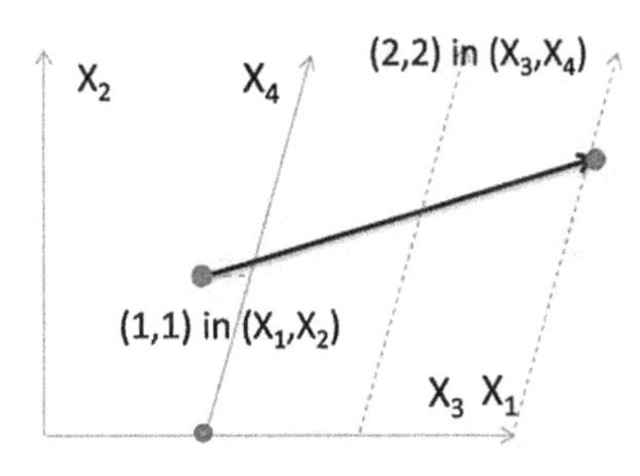

Fig. 2.25 Partially Collocated Ortho and non-Ortho Coordinates

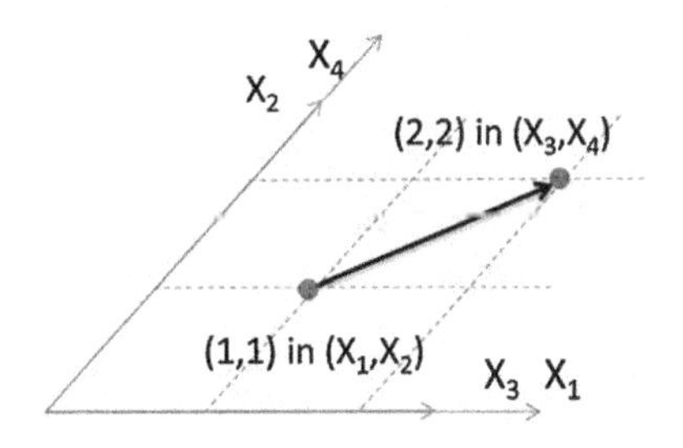

Fig. 2.26 Collocated Paired non-Ortho Coordinates

A 6-D point $\mathbf{x} = (x_1, x_2, x_3, x_4, x_5, x_6) = (1, 1, 2, 2, 1, 1)$ with non-orthogonal Cartesian mapping is shown in Fig. 2.27 in Paired Radial Coordinates. It is split to three 2-D pairs with the 2-D point $(x_1, x_2) = (1, 1)$ located in the first sector in coordinates (X_1, X_2), point $(x_3, x_4) = (2, 2)$ located in the second sector in coordinates (X_3, X_4), and point $(x_5, x_6) = (1, 1)$ located in the third sector in coordinates (X_5, X_6).

Then these points are connected sequentially to form a directed graph. As a result a 6-D point is represented losslessly not by six 2-D points as in Parallel Coordinated but by three 2-D points, i.e., two times less 2-D points.

Figure 2.28 shows other examples of 16-D and 192-D points represented in 2-D in Paired Radial (Star) Coordinates as *closed contours* ("stars"). These paired coordinates belong to the class of *Paired Coordinates*. For short, below we call these representations **CPC stars** or CPC-S.

CPC stars are presented here in two versions: CPC-SC and CPC-SP that use, respectively, *Cartesian* and *polar* encodings for pairs. Figure 2.28a shows that coordinates X_2 and X_3 are collocated. In the same way, X_4 and X_5 are collocated as well as all other X_j, X_{j+1} are collocated. At the end, coordinates X_1 and X_{16} are collocated due to the radial location of coordinates.

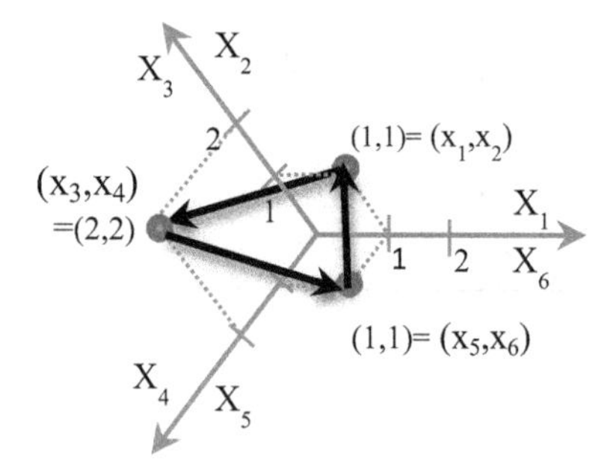

Fig. 2.27 Example of n-D point represented as a closed contour in 2-D where a 6-D point $\mathbf{x}$ = (1, 1, 2, 2, 1, 1) is forming a tringle from the edges of the graph in Paired Radial Coordinates with non-orthogonal Cartesian mapping

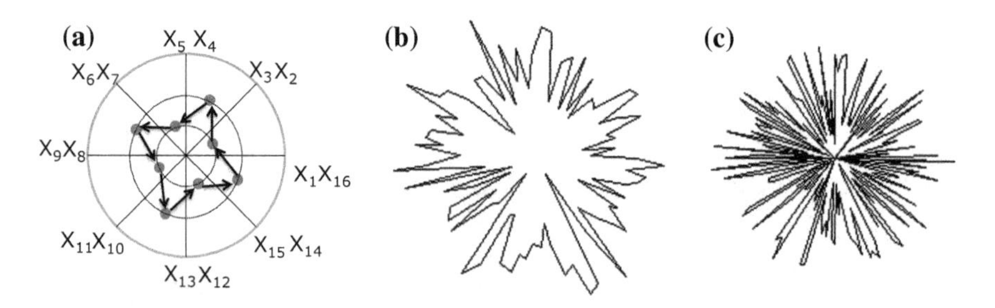

Fig. 2.28 Examples of n-D points as closed contours in 2-D: **a** 16-D point (1, 1, 2 ,2, 1, 1, 2, 2, 1, 1, 2, 2, 1, 1, 2, 2) in Partially Collocated Radial Coordinates with Cartesian encoding, **b** CPC star of a 192-D point in Polar encoding, **c** the same 192-D point as a traditional star in Polar encoding

The CPC stars are generated as follows: a full 2π circle is divided on $n/2$ equal sectors. Each pair of values of coordinates (x_j, x_{j+1}) of an n-D point **x** is displayed in its own sector as a 2-D point.

The graph of a 6-D point (1, 1, 1, 1, 1, 1) in *Partially Collocated Radial Coordinates* is shown in Fig. 2.29 on the left as a blue triangle. The same 6-D point in the *Cartesian Collocated Paired Coordinates* on the right produced a much simpler graph as a single point. Thus, this figure illustrates the perceptual and cognitive differences between the alternative 2-D representations of the same n-D data. Here a 2-D point is much simpler perceptually and cognitively than a triangle for the same 6-D point.

Figure 2.29 shows pair $(x_1, x_2) = (1, 1)$ as a point in the sector (X_1, X_2) using Cartesian mapping to these *non-orthogonal oblique coordinates*. Similarly the next pair $(x_3, x_4) = (2, 2)$ is shown in the sector (X_3, X_4). Note that coordinates X_2 and X_3 are collocated. In the same way, X_4 and X_5 are collocated as well as all other X_j, X_{j+1} are collocated including X_1 and $X_{16.}$ This collocation is a critical innovation of this method allowing having $n/2$ sectors and 2-D points instead of n sectors and 2-D points in the traditional star coordinates. This method dramatically decreases clutter as Fig. 2.28b shows in contrast with the traditional star in Fig. 2.28c.

In the *polar mapping,* pair (x_j, x_{j+1}) is mapped to the point $p = (r, \alpha)$ that is located at the distance $r = (x_j^2 + x_{j+1}^2)^{1/2}$ from the star center with the angle α of the ray to this point from the sector start. Here r is the Euclidean length of the projection of vector **x** on the plane of two coordinates (X_j, X_{j+1}). The angle α is proportional to the normalized value of x_j computed relative to the angle of the sector, $2\pi/(n/2)$.

In this way, we get $n/2$ points and connect them by straight lines (or arrows) to generate a star. This is a polar representation of *all* 2-D projections of **x** on plane. It is a **lossless display** forming a single connected figure (directed graph) *without crossing lines*. It also satisfies Gestalt Laws that support an effective use of human shape perception capabilities.

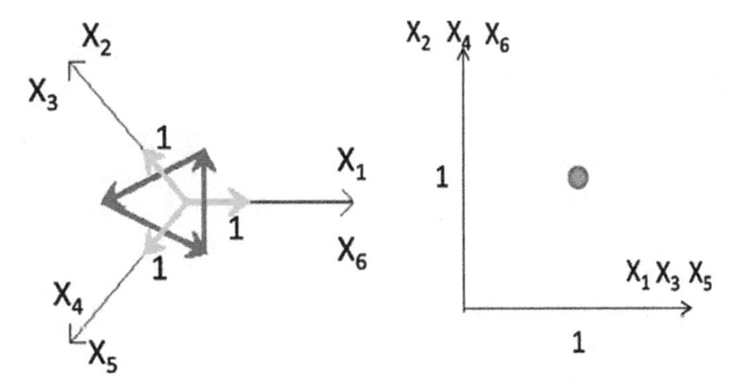

Fig. 2.29 6-D point (1, 1, 1, 1, 1, 1) in two X1–X6 coordinate systems (left—in Radial Collocated Coordinates, right—in Cartesian Collocated Coordinates)

Other versions of this representation are produced when radius r represents x_j and angle α represents x_{j+1}, or vice versa. Alternatively, each pair (x_j, x_{j+1}) can be encoded as is in the *non-orthogonal oblique coordinates* (X_j, X_{j+1}) as it is done in Figs. 2.27 and 2.28a.

In general, CPC Radial Coordinates use $n/2$ 2-D points instead of n 2-D points and dramatically increase the abilities to visualize higher-dimensional data losslessly.

Statement 2.1 (***Half Size***) Any n-D point to be represented in the CPC Radial Coordinates requires $n/2$ 2-D points for even n and $(n + 1)/2$ for odd n.

Proof This statement follows directly from the described Radial CPC representation algorithm.∎

The CPC stars provide a new visualization showing 2-D projections directly. It is especially beneficial for naturally paired data where CPC stars easily interpretable. In comparison with traditional stars, CPC stars:

- have *twice less break points* on the contour of the figure that is significantly decreasing shapes complexity,
- effectively *doubles data dimensions* accessible as Fig. 2.28b shows versus Fig. 2.28c

Thus, CPC stars have important advantages over traditional stars for shape perception. Chapter 6 presents Star coordinates in more detail. That chapter shows that the CPC star representation is quite effective for shape perception for data dimensions up to 200.

2.2.5 *Paired Elliptical Coordinates*

Above Circular Coordinates used curvilinear coordinates. However, that version of Circular Coordinates uses the same number of nodes and edges as Parallel Coordinates. In contrast, paired coordinates CPC, SPC and APC, built with straight lines for coordinates, use *two times less edges and nodes* with less clutter. Below we present a version of **Elliptical Paired Coordinates** (**EPC**), EPC-H, that also hold this property.

Figure 2.30 shows an example of 4-D point $P = (0.3, 0.5, 0.5, 0.2)$ in EPC-H as a short green arrow. For comparison, Fig. 2.31 shows the same point P in Radial Coordinates that present it as a graph with 4 nodes. In Fig. 2.30, the blue ellipse C_E holds four coordinate curves X_1–X_4.

The green arrow ($A \rightarrow B$) is constructed by using 4 ellipses of the size of the blue coordinate ellipse. These four ellipses touch the middle vertical line M and move along it to get different 4-D points. Below we explain this process. The thin red ellipse on the right is defined by the three requirements: to go through $x_1 = 0.3$, to

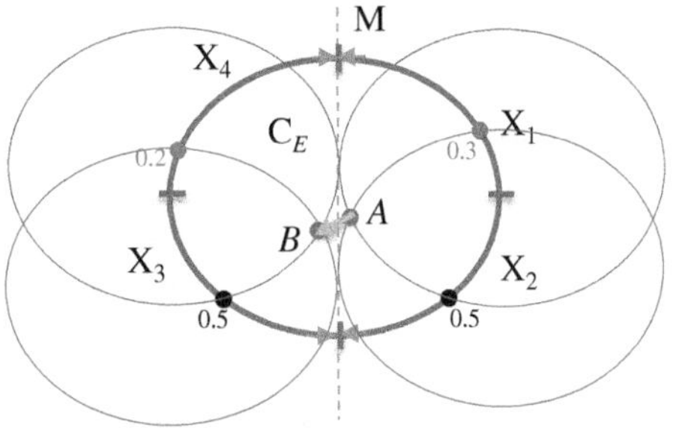

Fig. 2.30 4-D point $P = (0.3, 0.5, 0.5, 0.2)$ in 4-D Elliptic Paired Coordinates, EPC-H as a green arrow. Red mark separate coordinates in the Coordinate ellipse

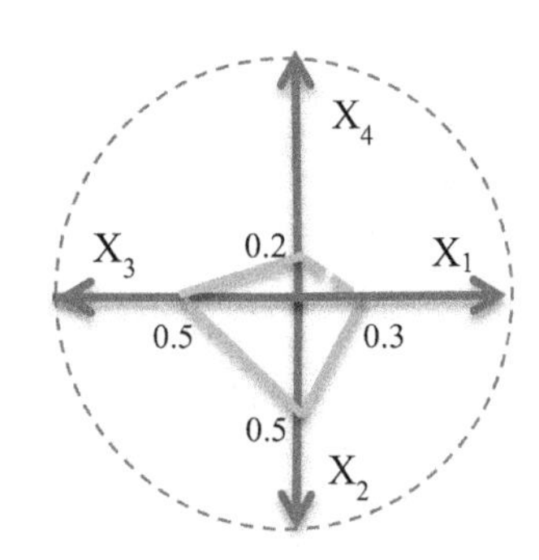

Fig. 2.31 4-D point $P = (0.3, 0.5, 0.5, 0.2)$ in Radial Coordinates

touch line M and be of the size of the blue coordinate ellipse C_E. Visually it is done by copying C_E and shifting the copy to the right to touch line M and then sliding it along M to reach $x_1 = 0.3$ on coordinate X_1.

This process shows that only one red ellipse satisfy these requirements. Then the same process is conducted for $x_2 = 0.5$ with the thin blue ellipse. The point A where these red and blue ellipses cross each other in C_E is the point that represents pair $(x_1, x_2) = (0.3, 0.5)$. Next this process is repeated for $x_3 = 0.5$ and $x_4 = 0.2$ to produce the crossing point B of two respective ellipses on the left that represents pair $(x_3, x_4) = (0.5, 0.2)$. Then the two crossing points A and B are connected to form the green arrow that serves as visualization of 4-D point $P = (0.3, 0.5, 0.5, 0.2)$.

This process is reversible allowing restoring the 4-D point P from the $(A \rightarrow B)$ arrow. The process starts from sliding the thin red ellipse on the right to cross point A. Then the point where this ellipse crosses C_E ellipse gives the point on X_1 coordinate, which is $x_1 = 0.3$. The same process is conducted for the blue circle on the right to find its crossing with X_2 coordinate to find $x_2 = 0.5$. Next it is repeated for point B and thin red and blue ellipses on the left.

2.2.6 Open and Closed Paired Crown Coordinates

This sections describes a method to represent a given n-D point **w** = $(w_1,w_2,\ldots,w_n)$ as an open or closed contour without using radial location of coordinates exploited in Sect. 2.2.4. We will call this class of coordinates as **Paired Crown Coordinates (PWC)** and denote its open and closed contour versions as OPWC, CPWC, respectively.

The process of constructing PWC includes two algorithms:

- *Coordinate Layout* (CL) algorithm to locate n coordinates in 2-D plane, and
- *Point Mapping* (PM) algorithm that maps (locates) a given n-D point on PWC.

Below we describe these algorithms for the even n. If n is odd then the value of coordinate X_n in-D point **w** is repeated to get $n + 1$ coordinates.

The *Coordinate Layout* algorithm:

(1) Normalizes all coordinates $X_1,X_2,\ldots,X_n$ to [0, 1] interval;
(2) Collocates all odd coordinates $X_1,X_3,\ldots,X_{n\text{-}1}$ on a "*crown*" that can be any *convex closed contour*. In Fig. 2.32 we use a square and a circle; We denote the set of all collocated odd coordinates as X coordinate.
(3) Locates all even coordinates $X_2,X_4,\ldots,X_n$ orthogonal to the crown starting from the points on the crown. The location of these points is not static, but *dynamic*. Each even coordinate X_{j+1} starts at the value w_j of the odd coordinate X_j on the crown. We denote the set of all even coordinates as Y coordinate.

The *Point Mapping* (PM) Algorithm:

(1) Normalizes all w_i from **w** to [0, 1] (represents **w** in normalized coordinates);
(2) Pairs normalized values of **w**: (w_1, w_2), (w_3, w_4),…,(w_{n-1}, w_n);
(3) Maps a pair (w_j, w_{j+1}) from **w** to the location of value w_{j+1} on coordinate X_{j+1}, e.g., locates pair (w_1, w_2) on the location of value w_2 on coordinate X_2 and (w_3, w_4) on the location of value w_4 on coordinate X_4.

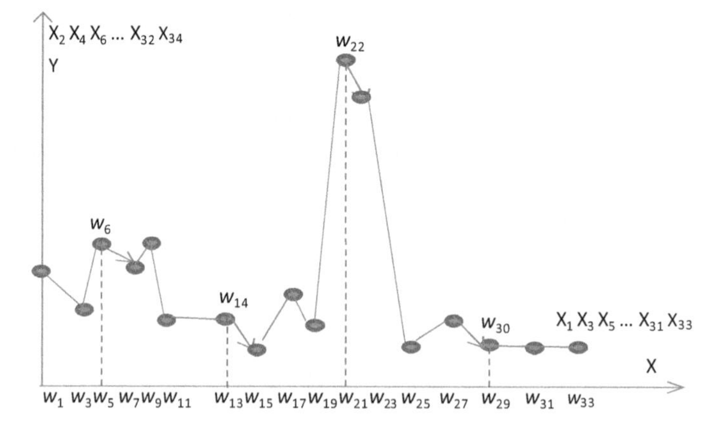

Fig. 2.32 34-D point in Collocated Paired Coordinates

These algorithms are illustrated below for 34-D point **w** with values of all its odd coordinates ordered in ascending order, $w_1 \leq w_3 \leq \ldots \leq w_{n-1}$. First, Fig. 2.32 shows this point CPC, where X and Y stand for sets of odd and even coordinates, respectively. Then Fig. 2.33a shows this point in PWC with the crown as a square and in Fig. 2.33b with the crown as a circle. It is noticeable that the axis X from Fig. 2.32 is "bended" and presented as a square in Fig. 2.33a and as a circle in Fig. 2.33b. Some locations on the square and the circle are labeled from 0 to 8, where 0 and 8 are at the same location (upper left corner) as a result of bending in Fig. 2.33a.

In Fig. 2.33a, the value w_5 (labeled with 1) is a starting point of the normal to the square that forms the coordinate X_6. The length of this norm is the value of w_6.

The location of w_6 represents the pair (w_5, w_6). Both w_5 and w_6 can be restored. To get a closed contour (CPWC version) the location of last pair labeled by 8 in Fig. 2.33a, is connected by a directed edge to the location of first pair. Otherwise, we will have an open contour (OPWC version).

While both these pairs start from the same point labeled by 8, the normals are different. For pair (w_{33}, w_{34}) the normal is for left side of the square and for pair (w_1, w_2) it is for the top of the square. In contrast, the circle in Fig. 2.33b has only *one normal* at all locations.

The bending does not connect locations of w_{34} and w_1, but keeps a fixed gap between them, shown by sector between two red dotted arrows in Fig. 2.33b. The name of these coordinates as crown coordinates is coming from resemblance to the *solar crown* and its protuberances.

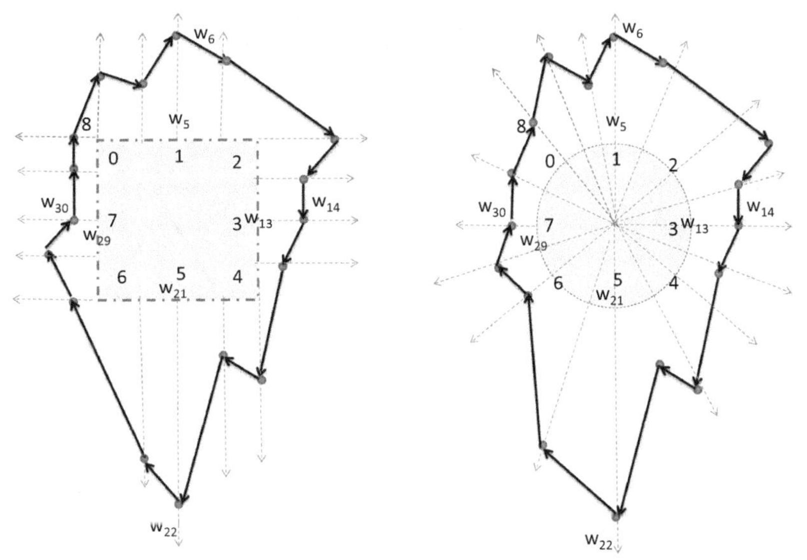

(a) Mapping odd coordinates to a square. (b) Mapping odd coordinates to a circle.

Fig. 2.33 34-D point from Fig. 2.32

The representations of 34-D point **w** in Paired Crown Coordinates in Fig. 2.33 uses 17 nodes and edges, i.e., the *half* of needed to visualize **w** in Parallel Coordinates. The advantage of closed contours in Fig. 2.33 relative to the open contour in CPC in Fig. 2.32 is exactly in its *closeness* that is consistent with Gestalt laws of shape perception. Chapter 6 presents the perception aspects in more detail.

The advantage of these closed contours relative to traditional stars and CPC stars is that they do not *occlude the points near the center*, because all points are located outside of the crown.

These closed contours also *improve visibility of some coordinate values* relative to the polar mapping of CPC Stars and regular stars, because in higher dimension the sectors in CPC Stars and regular stars become smaller. Note that to have a closed contour we need at least 6 dimensional **w**, because the triangle is the simplest closed contour. It needs three pairs of coordinate values (w_1, w_2), (w_3, w_4), and (w_5, w_6) to visualize its three nodes.

Above we described PWC for n-D point **w** with ordered values of its odd coordinates. It ensured that no edge goes backward. Therefore, the PWC graph is planar without self-crossing as it is described below. Figure 2.34 shows another option of the crown as a non-convex closed contour.

Above we assumed that odd coordinates of **w** are ordered. This requirement can be substituted by the requirement to order these coordinates first. Then PWC will be applied to the ordered **w**.

Below we present formal statements showing that this process guarantees that lines of the graph will not cross and will not go backward. These statements are based on the following algorithm called **the odd ordering algorithm:**

Step 1. Represent all nodes of n-D point **x** as a sequence of *pairs* $(x_1, x_2), (x_3, x_4), \ldots, (x_{n-1}, x_n)$ e.g., (0, 0), (1, 0), (0, 1), (1, 1), (0, 0).

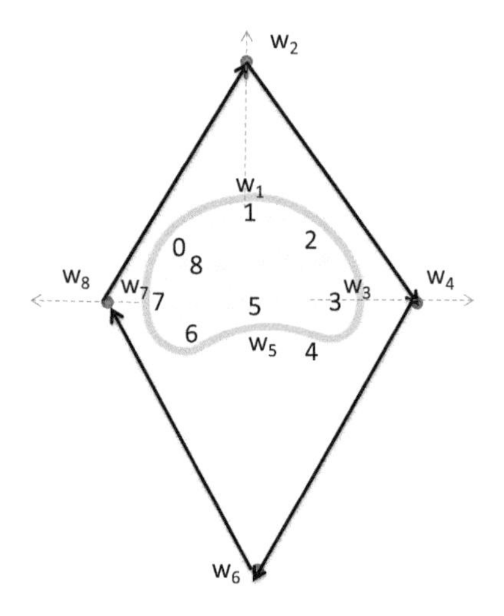

Fig. 2.34 Crown as non-convex closed contour

Step 2. *Order* pairs using their first element in ascending order, e.g., (0, 0), (0, 0), (0, 1), (1, 0), (1, 1), i.e., (x_1, x_2), (x_9, x_{10}), (x_5, x_6), (x_3, x_4), (x_7, x_8).
Step 3. Denote these ordered pairs as (w_1, w_2), (w_3, w_4), (w_5, w_6), (w_7, w_8), (w_9, w_{10}).

These pairs correspond to **w** used in the PWC process above.

Statement 2.2 (*Planarity*) The odd ordering algorithm produces a planar CPC graph such that each edge is located on the right from the previous edges or on the same vertical line as the previous edge.

Proof Let (w_i, w_{i+1}) and (w_{i+2}, w_{i+3}) be two consecutive nodes of a PWC graph for an n-D point **w** after applying the odd ordering algorithm to **w**. Thus, $w_i \leq w_{i+2}$, i.e., each next edge of the CPC graph is located on the right from the previous edge or on the same vertical line as the current node. This location of the next node does not allow the next edge and further edges to cross the current edge and previous edges that are all on the left from the next edge. If $w_i = w_{i+2}$ then the next edge is on the same vertical line as the previous one.■

The next algorithm is the **Complete odd ordering algorithm** where after odd ordering, all pairs with equal first element are ordered in ascending order relative to its *second element*, e.g., (0, 0), (0, 0), (0, 1), (1, 0), (1, 1), i.e., $(x_1, x_2, x_9, x_{10}, x_5, x_6, x_3, x_4, x_7, x_8)$.

The **Complete ordering algorithm** orders *all coordinates* of **x** in the ascending order, e.g., for (0, 0, 1, 0, 0, 1, 1, 1, 0, 0) the order is $(x_1, x_2, x_4, x_5, x_9, x_{10}, x_3, x_6, x_7, x_8) = (0,0,0,0,0,0,1,1,1,1)$.

Statement 2.3 The planarity statement 2.2 is true for both the *complete odd* ordering, and the *complete ordering* algorithm.

Proof Both orderings satisfy the requirements of odd ordering required for the planarity theorem.■

Discussion. Consider another n-D point **q** with at least one odd *i* such that $q_i > q_{i+2}$, i.e., violates the order from **w**. The graph for **q** goes backward from the node for pair (q_i, q_{i+1}) to the node for pair (q_{i+2}, q_{i+3}) and potentially can be self-crossing (non-planar).

Some datasets may have many such points. We analyzed several real datasets. Some datasets have the same orderings practically for all n-D points. Some datasets have large subsets with the same orderings especially when strict ordering is relaxed by allowing violation of the ordering within some *threshold*.

For the datasets that are extremely diverse in ordering it is better to use other GLCs such as Radial Coordinates that have no self-crossings and others that we present in this chapter later. In Chap. 4 we provide methods to simplify visual patterns that are helpful to avoid crossing too.

2.2.7 Clutter Suppressing in Paired Coordinates

Lines in Parallel, Radial, and Paired Coordinates can produce complex shapes especially with larger dimensions that lead to higher density of features. Figure 2.28a shows this for a 192-D point in the traditional Radial Coordinates. This creates clutter where it is difficult to distinguish features of a single n-D point visually. We will call it a **single point clutter**.

Another type of clutter is a result of displaying together several high-dimensional points, which we will call **multiple points clutter**. This clutter issue is well known for popular lossless Parallel and Radial Coordinates (Fanea et al. 2005). It is desirable to have methods for decreasing clutter for GLCs in general.

Side-by-side visualization One of such methods is displaying each n-D point separately as it is commonly done with Chernoff's faces: *side-by-side*. However, only few hundreds of n-D points can be analyzed using the side-by-side method.

While this method does not scale to thousands of n-D points, there are many real world tasks that deal with hundreds not thousands of n-D points as datasets at UCI Machine Learning repository (Lichman 2013) illustrates.

The important issue with this method is switching gaze from one graph to another one. It takes time, requires memorizing the first graph before looking at another one, which complicates the comparison of graphs.

Overlay One of the solutions for this issue is considering one graph as a base, and overlaying other graphs with it one after another. The color of the overlaid graph will differ from the color of the base graph. The sections of two graphs that are practically identical can be highlighted or be shown in a third color. The analysts can indicate interactively that two graphs are similar and potentially from the same class by using graphical user interface.

From Stars to SPC Stars One of the ways to decrease the single graph clutter is moving from traditional Radial (Star) Coordinates to CPC Stars. This increases the scalability of the side-by-side method as Chap. 6 shows.

Envelops and **convex hulls** of CPC graphs is one of the ways to mitigate perceptual difficulties of both single point and multiple points clutter. Envelopes and convex hulls are simpler than CPC graphs.

Users can compare features of envelopes and convex hulls such as orientation, size and symmetries easier than features of CPC graphs. See Fig. 2.35 where the

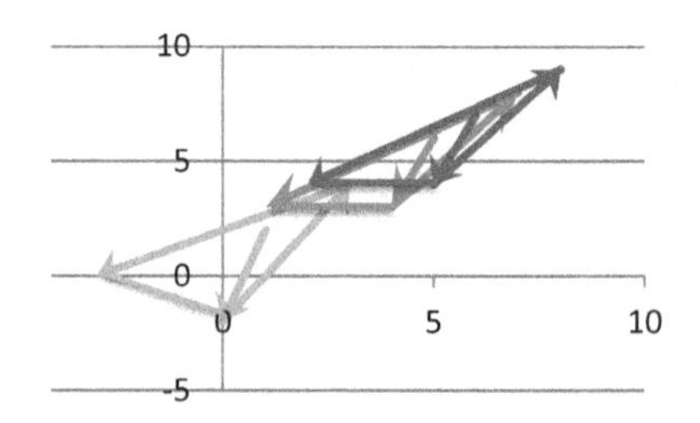

Fig. 2.35 Convex hulls of correlated data from Table 2.3 in CPC

graphs from Fig. 2.20a are converted to their convex hulls that are simple triangles, which can be easily compared visually. Here, the features of each envelope, such as the orientation and size of triangles indicate some relationships, and can be extracted, presented and analyzed mathematically.

In the same way, the envelopes of multiple n-D points can be compared. However, due to generalization of graphs envelopes allow discovering only a *limited* number of statistical properties of data classes such as dominant orientation.

Paired Coordinates presented above provide new visualizations, which show n-D points in a way which could be especially beneficial for the naturally paired data, and could be easily interpretable. All Paired Coordinates representations have **two times fewer break points** than the traditional Stars and Parallel Coordinates on the contour of the figure. This significantly decreases the complexity of the forms. It effectively doubles the representable data dimensions for CPC stars as Fig. 2.28b illustrates and Chap. 6 presents in detail.

The expansion of GLCs for dimensions n up to 1000 is as follows: grouping coordinates x_i of $\mathbf{x}$ by 100–150 and representing them by separate or collocated colored GLC graphs, and/or mapping some x_i values into colors. Lossy reduction of n can be applied after visual analysis of these lossless displays, which can reveal the least informative attributes that can be removed. Another reduction is based on a priori domain knowledge.

2.3 Discussion on Reversible and Non-reversible Visualization Methods

Partial GLC. Lossy non-reversible versions of any GLC can be produced by drawing only a part of the graph $\mathbf{x}^*$ of each n-D point $\mathbf{x}$. We will call them **partial GLC**. For *dynamic coordinates* illustrated in Sect. 2.1.5 in Fig. 2.14a, Kandogan (2000) suggested the extreme case of partial representation—drawing only the last node of the graph $\mathbf{x}^*$. The advantage of it is in *less clutter,* but the disadvantage is in *information loss.*

In general, the *partial dynamic GLCs* have an advantage over *partial static GLCs* because the location of each x_i depends on the location of all previous x_k, $k < i$. In this way, the location of x_n contains some information about all x_i. We consider the issue of dimension reduction for dynamic GLC-L in Chap. 7 in detail.

In contrast, in CPC a node of graph $\mathbf{x}^*$ contains only information about a given pair (x_i, x_{i+1}). Therefore, for partial static GLCs showing most informative coordinates is a preferred way for dimension reduction.

While the focus of this book is on reversible GLCs, there are many other visualization methods and some of them are reversible too. The survey of other visualization methods can be found in (Chan 2006) and several other publications listed in the References. These methods can be combined with GLCs and partial GLCs to produce **hybrid methods**.

Cartesian and Polar versions of the lossless tradition and CPC Star coordinates are defined above. The words "stars" and "radial" are present in several alternative visualization techniques such as the Radial Visualization (RadViz) (Hoffman and Grinstein 2002), and the Star Coordinates (Kandogan 2000, 2001). The Radvis visualization is not reversible but lossy, when representing each n-D data point by a single 2-D point.

Kandogan's Star Coordinates as they are used in (Kandogan 2000, 2001) are non-reversible too. However, the full graph used to get this visualization is reversible as Fig. 2.14a in Sect. 2.1.5 illustrates. The GLC-L and all dynamic GLCs can be viewed as generalization of it.

RadViz and Star Coordinates from (Kandogan 2000, 2001) show clouds of 2-D points and allow getting some attributes of this clouds (sizes, elongation, and localization in data space). These clouds are meaningful mostly for compact classes and if they do not occlude each other.

Thus, the point-based approach has significant limitations being oriented to visual classification and clustering tasks with *relatively simple compact data classes*. While these visualization are lossy and, respectively, incomplete, they has important positive properties such as low occlusion and representing some integral information about the all attributes of an n-D point.

Due to absence of internal structure of a 2-D point, in contrast with a 2-D graph, the ability to extract deep structural information from point clouds is quite limited. In essence, such abraded visual representation prevents deep visual analytics from the very beginning of visual data exploration.

GLC contains well-known Parallel and Radial (Star) coordinates along with new ones that generalize them by locating coordinates in any direction that differ from parallel or radial locations and in any topology (connected or disjoint). For any GLC it is possible to find an n-D data point with *simple representation* in that GLC.

An analyst can do this in two steps: (1) draw a simple figure in a given GLC and then (2) record numeric values of coordinates of that figure. The straight horizontal line in Fig. 2.3c and pentagons in Figs. 2.6 and 2.7 in Sect. 2.1 give examples of these steps.

For a set of n-D points, the concept of best GLC also depends on both dataset and on the user's task. For the classification tasks, the goal is finding the simple representation of the dataset that makes classification easier visually. Discovering such simple parametrized SPC visualization are presented in Chap. 5.

In GLC approach, we attempt to use maximally the unique capabilities of the human vision system to extract the deep structural n-D information. The GLC approach opens opportunity for detecting essentially *nonlinear, non-compact structures* in the n-D data space, and understanding their properties better than by using non-reversible methods. Non-reversible methods simplify the user's visual task, but can remove deep structural information before it can be discovered.

References

Ahonen-Rainio, P., Kraak, M.: Towards multivariate visualization of metadata describing geographic information. In: Dykes J. MacEachren A. Kraak M.J. (eds.) Exploring Geovisualization, pp. 611–626. Elsevier (2005)

Chan WW.: A survey on multivariate data visualization. Department of Computer Science and Engineering. Hong Kong University of Science and Technology. 2006 Jun; **8**(6):1–29. http://people.stat.sc.edu/hansont/stat730/multivis-report-winnie.pdf

Fanea, E., Carpendale, S., Isenberg, T.: An Interactive 3D Integration of Parallel Coordinates and Star Glyphs, In: Proceedings of the 2005 IEEE Symposium on Information Visualization, IEEE Computer Society, Washington, DC, USA, 20. https://doi.org/10.1109/INFOVIS.2005.5

Hoffman PE, Grinstein GG.: A survey of visualizations for high-dimensional data mining. Information visualization in data mining and knowledge discovery. 47–82 (2002)

Kandogan E.: Star coordinates: a multi-dimensional visualization technique with uniform treatment of dimensions. In: Proceedings of the IEEE Information Visualization Symposium 2000 (vol. 650, p. 22)

Kandogan E.: Visualizing multi-dimensional clusters, trends, and outliers using star coordinates. In: Proceedings of the seventh ACM SIGKDD international conference on Knowledge Discovery and Data Mining 2001 Aug 26 (pp. 107–116). ACM

Klippel, A., Hardisty, F., Weaver, C.: Star plots: How shape characteristics influence classification tasks. Cartography Geogr Inf Sci **36**(2), 149–163 (2009)

Lichman, M.: UCI Machine Learning Repository (http://archive.ics.uci.edu/ml). Irvine, CA: University of California, School of Information and Computer Science, 2013

Chapter 3
Theoretical and Mathematical Basis of GLC

The secret of getting ahead is getting started.
Mark Twain

This chapter mathematically defines concepts that form various General Line Coordinates (GLCs). It provides relevant algorithms and statements that describe mathematical properties of GLCs and relations that GLCs represent. The theoretical basis of a GLC is considered in connection with the Johnson-Lindenstrauss Lemma.

3.1 Graphs in General Line Coordinates

Below we give a more formal description of General Line Coordinates in the vector algebra terms. GLC axes can be drawn in 2-D in a variety of ways shown in Chap. 2. The locating and drawing of axes must be accompanied by an algorithm for constructing a 2-D graph $\mathbf{x}^*$ for each n-D point $\mathbf{x} = (x_1, x_2, \ldots, x_n)$ in these located coordinates. We start by presenting concepts needed for defining *located coordinates* and then we outline several algorithms for constructing graphs and present them in detail more rigorously.

Below scalars are denoted as *low case italic* letters such as *x, y, u, w, a, b, c* with or without indices. The n-D points are denoted as **bold** low case letters such as **x**, **y**, **w**, **a**, **b,** or *italic upper case* letters such as *A*, *B*, *C*. Respectively *graphs* that represent in 2-D or 3-D an n-D points are denoted such as $\mathbf{x}^*$, $\mathbf{y}^*$, $\mathbf{w}^*$, $\mathbf{a}^*$, $\mathbf{b}^*$, A^*, B^*, C^*. Coordinates are denoted as upper case letters such as X, W, U, or with indices X_i, W_i, U_i. Symbol ■ will indicate the end of the proof of the statements.

World Coordinates $W_1, W_2, \ldots, W_n$ are n-D coordinates of the given task, i.e., attributes of n-D objects such as mass, length and so on.

Viewport coordinates are coordinates within 2-D screen window where GLC are drawn. We use here the term viewport as it is used in Computer Graphics. We denote these 2-D coordinates as U_1, U_2. *Each coordinate* W_i is defined by its interval of values $[o, e]$, where o is the *origin* value and e is the *end value.*

B. Kovalerchuk, *Visual Knowledge Discovery and Machine Learning*,
Intelligent Systems Reference Library 144,
https://doi.org/10.1007/978-3-319-73040-0_3

Normalized coordinate X = [0, 1] is a coordinate W = [*o*, *e*], where *o* (origin) is mapped to $x = 0$ and *e* is mapped to $x = 1$.

Reversed normalized coordinate is a coordinate X, where *e is mapped to* $x = 0$, and origin *o* to $x = 1$.

Negated normalized coordinate $\bar{X}$ is a coordinate where $x \in X$ is mapped to $1 - x$, on $\bar{X}$. In contrast with the reversed normalized coordinate, it does not change the location of the origin *o* and end point *e*.

Located coordinate is a coordinate located in the viewpoint coordinates (U_1, U_2) where both the origin and the endpoint are mapped to some 2-D points $O = (o_1, o_2)$ and $E = (e_1, e_2)$ in (U_1, U_2).

In the located coordinates *O* and *E* are not scalars, but 2-D points. In other words, a located coordinate is given by a triple $<O, E, L>$, where *L* is a line (straight or curvilinear) between *O* and *E*. More formally the located coordinate can be defined using the notation below.

T is the *length* of line *L;* and

D(*A*, *B*) is the *distance* on the curve L from 2-D point *A* to 2-D point *B*.

A *parametrization function* for ***L*** is mapping M: $\boldsymbol{L} \rightarrow [0,T]$ with $M(O) = 0$, $M(E) = T$, and $M(A) = D(O,A)$ for all $A \in L$, where $D(O, A)$ is the distance on the line from the origin to point *A*.

A **located coordinate** X is a triple $<O, E, L>$ parameterized by some function *M*.

The number *x* from interval [0, *T*] is called a *value on the coordinate* X.

A located **coordinate** X is called a *located linear coordinate* if its line *L* is a linear segment in 2-D. and it is called *located curvilinear* if *L* is curve. Thus, we distinguish *linear* and *curvilinear coordinates* in (U_1, U_2).

A *located vector* **x** in coordinates (U_1, U_2) for a scalar value *x* is an ordered pair of 2-D points $\mathbf{q}_1 = <q_{11}, q_{12}>$, $\mathbf{q}_2 = <q_{21}, q_{22}>$, such that $\|\mathbf{q}_1 - \mathbf{q}_2\| = |x|$.

In Algorithm 1 defined below, each n-D point $\mathbf{x} = (x_1, x_2,\ldots,x_n)$ is mapped to the *n* located vectors $\mathbf{x}_1, \mathbf{x}_2,\ldots,\mathbf{x}_n$.

A linear located coordinate X from point *O* to point *E* in (U_1, U_2) is equivalent to a *set of located vectors* from the origin of (U_1, U_2) to points $\{O + x(E - O)\}$, $x \in [0, 1]$ where *x* is the value of the coordinate X.

Collocated coordinates X_i and X_j are coordinates such that

$$<O_i, E_i, L_i> \; = \; <O_j, E_j, L_j>$$

Horizontal Collocated coordinates X_i and X_j are coordinates such that

$$<O_i, E_i, L> \; = \; <O_j, E_j, L>, O_i = O_j = (o_{i1}, o_{i2}), E_i = E_j = (e_{i1}, o_{i2}).$$

Vertical Collocated coordinates X_i and X_j are coordinates such that

$$<O_i, E_i, L> \; = \; <O_j, E_j, L>, O_i = O_j = (o_{i1}, o_{i2}), E_i = E_j = (o_{i1}, e_{i2}).$$

Here the first coordinate of the end points E_i and E_j differ from end points in horizontal collocated coordinates.

Radial coordinates X_i and X_j are coordinates such that $O_i = O_j$ and $E_i \neq E_j$.

Now we will outline several **representation mapping algorithms** for an n-D point $\mathbf{x} = (x_1, x_2,\ldots,x_n)$ to the set of *located coordinates* $\{X_i\}$ in 2-D. These algorithms are *mappings* that produces a graph in 2-D,

$$F(\mathbf{x}) = \mathbf{x}^*.$$

Therefore, we will call them *graph construction algorithms*. Below several of these mappings F will be presented.

Next, we define the **L^P distance between directed graphs** $\mathbf{x}^*$ and $\mathbf{y}^*$ located in 2-D with equal number of nodes k.

$$D^*(\mathbf{x}^*, \mathbf{y}^*) = (\sum_{i=1}^{k} ||\boldsymbol{n}_{xi} - \boldsymbol{n}_{yi}||^p)^{\frac{1}{p}}$$

where $\boldsymbol{n}_{xi} = (u_{1xi}, u_{2xi})$ and $\boldsymbol{n}_{yi} = (u_{1yi}, u_{2yi})$ are ith nodes of $\mathbf{x}^*$ and $\mathbf{y}^*$ in 2-D (U_1, U_2) coordinates.

The **L^Pdistances between nodes** is defined in 2-D as,

$$\left\|\boldsymbol{n}_{xi} - \boldsymbol{n}_{yi}\right\| = \left(\left|u_{1xi} - u_{1yi}\right|^p + \left|u_{2xi} - u_{2yi}\right|^p\right)^{1/p}$$

The Euclidian distance between graphs and nodes is L^2 distance when $p = 2$ and the sum of absolute values is L^1 distance with $p = 1$.

Mapping F: $\{\mathbf{x}\}\rightarrow\{\mathbf{x}^*\}$ from a set of n-D points $\{\mathbf{x}\}$ to a set of directed graphs $\{\mathbf{x}^*\}$in 2-D is called an ***L*-mapping** if:

(1) F is 1:1 bijective mapping, and
(2) F preserves appropriate L^P distance $D(\mathbf{x}, \mathbf{y})$ between any n-D points $\mathbf{x}$ and $\mathbf{y}$ in 2-D, $D(\mathbf{x}, \mathbf{y}) = D^*(\mathbf{x}^*, \mathbf{y}^*)$.

For instance, D can be Euclidian distance between n-D points

$$D(\mathbf{x}, \mathbf{y}) = ||\mathbf{x} - \mathbf{y}|| = (\sum_{i=1}^{n} (x_i - y_i)^2)^{1/2}$$

and D^* can be Euclidian distance between graphs.

A pair <$\{X_i\}_{i=1:n}$, F>, that consists of a set of located coordinates $\{X_i\}_{i=1:n}$ and a *graph construction algorithm* F is called n-D **General Line Coordinates** (**GLC**).

General Line Coordinates <$\{X_i\}_{i=1:n}$, F> are called **L-GLC** if F is an L-mapping. Thus, only located coordinates $\{X_i\}_{i=1:n}$ with mapping of n-D points to 2-D graphs F that preserve n-D distance and reversible are L-GLC.

Statement 3.1 Parallel Coordinates preserve L^P distances for $p = 1$ and $p = 2$,

$$\mathrm{D}(\mathbf{x}, \mathbf{y}) = \mathrm{D}^*(\mathbf{x}^*, \mathbf{y}^*).$$

Proof In Parallel Coordinates, $u_{1xi} - u_{1yi} = 0$ for any vertical coordinate X_i, because values of x_i and y_i are located on the same vertical line. Also

$$u_{2xi} - u_{2yi} = x_i - y_i$$

for the same reason. ■

Similarly, several other coordinate systems such as CPC and SPC also preserve n-D distance in 2-D graphs.

Statement 3.2 CPC and SPC preserve L^p distances for $p = 1$ and $p = 2$,

$$\mathrm{D}(\mathbf{x}, \mathbf{y}) = \mathrm{D}^*(\mathbf{x}^*, \mathbf{y}^*).$$

Proof In CPC and SPC, each pair (x_i, x_{i+1}) is a node $\boldsymbol{n}_{xi,\ i+1} = (x_i, x_{i+1})$ of the graph x* and each pair (y_i, y_{i+1}) is a node $\boldsymbol{n}_{yi,\ i+1} = (y_i, y_{i+1})$. The p-powered L^p distance between two nodes is

$$\left\|\boldsymbol{n}_{xi,i+1} - \boldsymbol{n}_{yi,i+1}\right\|^p = |x_i - y_i|^p + |x_{i+1} - y_{i+1}|^p$$

The value of $(D^*(\mathbf{x}^*, \mathbf{y}^*))^p$ is the sum of all p-powered distances between nodes which is equal to the p-powered distances between $\mathbf{x}$ and $\mathbf{y}$,

$$\begin{aligned}(D^*(\mathbf{x}^*, \mathbf{y}^*))^p &= \sum\nolimits_{i=1,3,5,\ldots k-1} (|x_i - y_i|^p + |x_{i+1} - y_{i+1}|^p) \\ &= \sum\nolimits_{i=1:k} |x_i - y_i|^p = D^p(\mathbf{x}, \mathbf{y})\end{aligned}$$

This leads to the Statement 3.2. ■

Mapping P:{**w**}→{**u**} from a set of n-D points {**w**} in coordinates $\{W_i\}_{i=1:n}$ to a set of 2-D points in viewport coordinates (U_1.U_2) is called a **P-to-P mapping**.

In contrast with *L*-mapping, known P-to-P mappings such as MDS (Duch et al. 2000) in general do *not preserve the distance* between all n-D points, but only minimize the average difference of distances. Informally, if n-D points **x** and **y** are close to each other, then the graphs **x*** and **y*** are also close to each other in PC, CPC and SPC. The P-P mappings do not guarantee this. For this reason, the visual means that are the focus of this book are the *General Line Coordinates*.

Graph Construction Algorithms

Algorithm 1 Constructing a graph **x*** in 2-D or 3-D as a collection of directed edges (arrows, vectors). Each edge is located on the respective coordinate X_i starting at the origin of this coordinate and ending at point x_i on X_i. See Fig. 3.1 for an example. We will call this algorithm a *basic GLC graph-constructing algorithm* (GLC-B). This example shows that edges of the graph **x*** can be disconnected.

Algorithm 2 Constructing a graph **x*** by connecting the location of x_i on X_i with the location of x_{i+1} on X_{i+1}, starting from $i = 1$, and ending at $i = n$. See Fig. 3.2. The same connections are implemented for coordinates located in parallel in

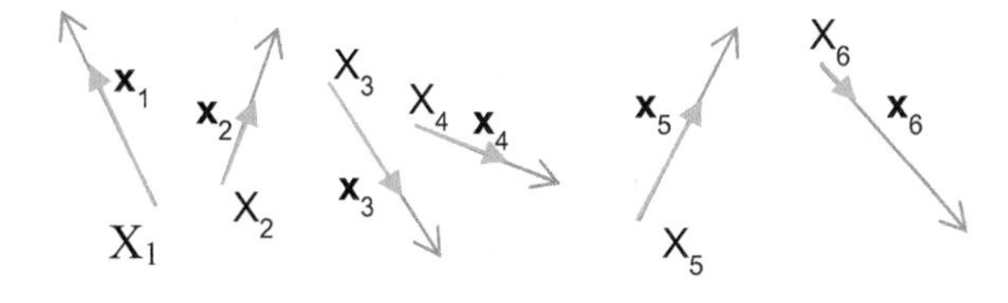

Fig. 3.1 Six coordinates and six vectors that represent a 6-D data point (0.75, 0.5, 0.7, 0.6, 0.7, 0.3)

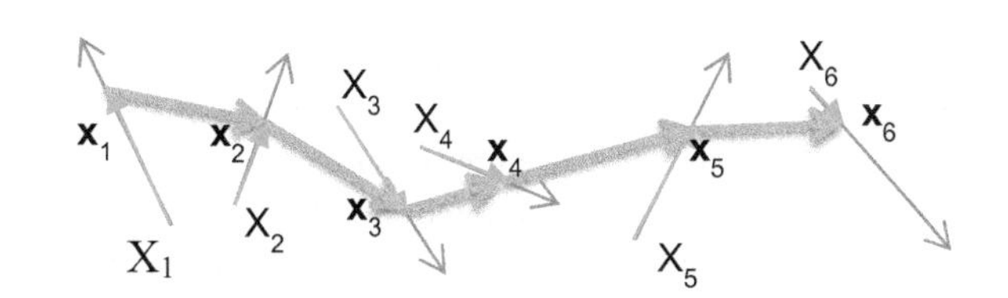

Fig. 3.2 6-D data point (0.75, 0.5, 0.7, 0.6, 0.7, 0.3) in GLC-PC

Parallel Coordinates. Therefore, Algorithm 2 is a generalization to GLC of the algorithm implemented in Parallel Coordinates (PC). Respectively we will call it as *GLC-PC graph constructing algorithm.*

Algorithm 3 Constructing a graph **x*** by the algorithm as illustrated in Fig. 3.3. It moves the start point of each of the vectors $\mathbf{x}_{i+1}$ to the end of vector $\mathbf{x}_i$. This algorithm is a generalization to GLC of the algorithm from (Kandogan 2000, 2001) to what is called there as the Star Coordinates (SC). Respectively we will call it as *GLC-SC1 graph constructing algorithm.*

Algorithm 4 Constructing a graph **x*** by the algorithm that is illustrated in Fig. 3.4. It is a generalization of the CPC algorithm described in Chap. 2. We will call this algorithm the *GLC-CC1 graph constructing algorithm.* In Fig. 3.4 it creates points P_1, P_2 and P_3 in respective pairs of coordinates (X_1, X_2), (X_3, X_4), (X_5, X_6) and connect these points to form a digraph. In contrast with CPC here, it is not required that coordinates are orthogonal and collocated.

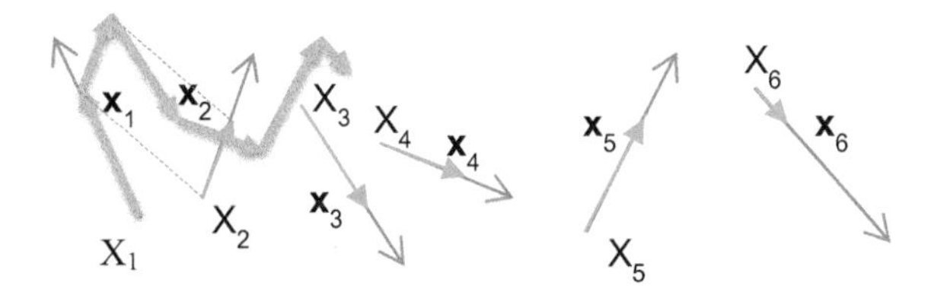

Fig. 3.3 6-D data point (0.75, 0.5, 0.7, 0.6, 0.7, 0.3) in GLC-SC1

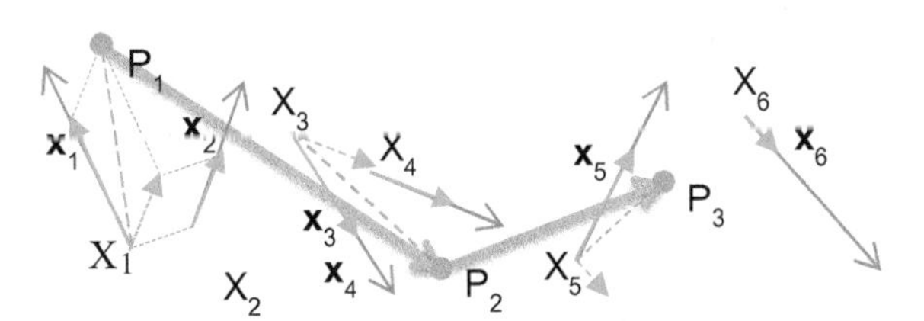

Fig. 3.4 6-D data point (0.75, 0.5, 0.7, 0.6, 0.7, 0.3) in GLC-CC1

Algorithm 5 Constructing a graph **x*** by the algorithm that is illustrated in Fig. 3.5. It is another generalization to GLC of the algorithm for orthogonal Collocated Paired Coordinates presented in Chap. 2 in combination with the idea of Algorithm 3, i.e., moving the next vector to the end of the previous one. We will call this algorithm the *GLC-CC2 graph constructing algorithm*. Here the yellow dotted line, constructed as a sum of vectors $\mathbf{x}_3 + \mathbf{x}_4$, is moved to point P_1. Similarly, the sum of vectors $\mathbf{x}_5 + \mathbf{x}_6$ is moved to point P_2.

Algorithm 6 Constructing a graph **x*** by the algorithm as illustrated in Fig. 3.6. It moves the *end* of the vector $\mathbf{x}_2$ to the *end* of vector $\mathbf{x}_1$ and for each other vectors $\mathbf{x}_{i+1}$ it moves its start point to the *end* point of vector $\mathbf{x}_i$, e.g., the start point of $\mathbf{x}_3$ is moved to the *end* point of vector $\mathbf{x}_2$. This algorithm is a generalization to GLC of the algorithm implemented in the Star Coordinates (SC) (Kandogan 2000, 2001) and is a modification of Algorithm 3. Respectively, we will call it as *GLC-SC2 graph constructing algorithm*.

Figures 3.1, 3.2, 3.3, 3.4, 3.5 and 3.6 show that the algorithms above require a different number of points and edges in the graph **x*** for lossless representation of an n-D point **x**:

- Algorithm 1 requires 12 points and 6 lines;
- Algorithm 2 requires 6 points and 5 lines;
- Algorithms 3 and 6 require 7 points and 6 lines;
- Algorithms 4 and 5 require 3 points and 2 lines.

In general, Algorithm 4 (GLC-CC1) requires two times less points and lines than Algorithms 1–3. This is a *fundamental advantage* of GLC-CC algorithm from human cognitive viewpoint, because it simplifies pattern discovery by a naked eye.

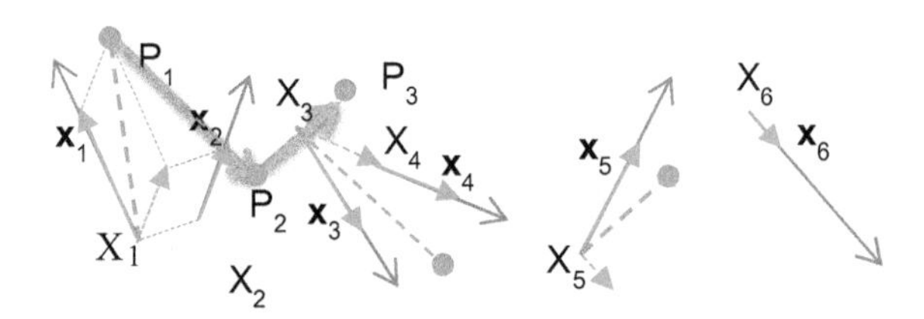

Fig. 3.5 6-D data point (0.75, 0.5, 0.7, 0.6, 0.7, 0.3) in GLC-CC2

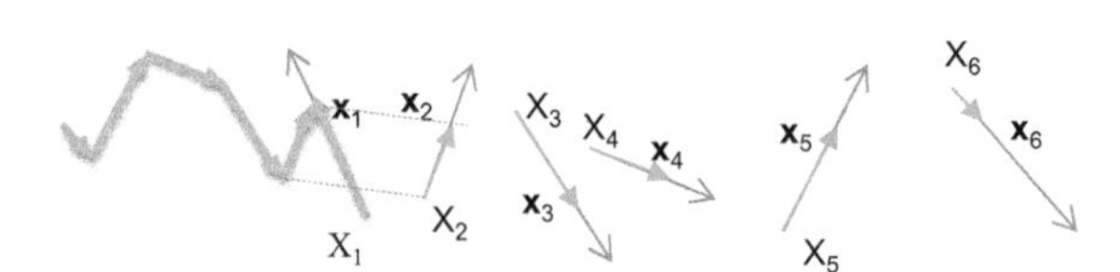

Fig. 3.6 6-D data point (0.75, 0.5, 0.7, 0.6, 0.7, 0.3) in GLC-SC2

3.2 Steps and Properties of Graph Construction Algorithms

Below we present steps and properties of some of the Algorithms 1–4 more formally and rigorously.

Algorithm 1 ***Basic GLC-B Graph Construction Algorithm***

Step 1 Create n located linear coordinates in 2-D coordinates (U_1, U_2).
Step 2 Select an n-D point, e.g., (7, 5, 6, 5, 6, 2).
Step 3 For each i ($i = 1{:}n$) locate value x_i in the coordinate X_i (see Fig. 3.1 for an example), and define n vectors $\mathbf{x}_i$ of length x_i from the origin of X_i that we denote as O_i.

Algorithm 2 ***GLC-PC Graph Construction Algorithm***

Step 1 Apply basic algorithm GLC-B to construct n located linear coordinates and vectors $\mathbf{x}_1$, $\mathbf{x}_2,\ldots,\mathbf{x}_n$ for a given n-D point $\mathbf{x}$.
Step 2 Assign $P_1 = O_1 + \mathbf{x}_1$, $P_2 = O_2 + \mathbf{x}_2,\ldots,P_n = O_n + \mathbf{x}_n$.
Step 3 Connect points P_i to form a graph: $P_1 \rightarrow P_2 \rightarrow \cdots \rightarrow P_{i-1} \rightarrow P_i \cdots \rightarrow P_n$

Statement 3.3 (***n points lossless representation***) If all coordinates X_i do not overlap then GLC-PC algorithm provides bijective 1:1 mapping of any n-D point $\mathbf{x}$ to 2-D directed graph $\mathbf{x}^*$.

Proof Non-overlap ensures getting n unique 2-D points $\mathbf{x}_1$, $\mathbf{x}_2,\ldots,\mathbf{x}_n$ for each n-D point $\mathbf{x}$. Next, GLC-PC reproduces the order of values (x_1, $x_2,\ldots,x_n$) by directed edges in the graph $\mathbf{x}^*$ that allows to restore x_i having 2-D vectors $\mathbf{x}_1$, $\mathbf{x}_2,\ldots,\mathbf{x}_n$. ∎

Algorithm 3 ***GLC-SC1 Graph Construction Algorithm***

Step 1 Apply basic algorithm GLC-B to construct n located linear coordinates and vectors $\mathbf{x}_1$, $\mathbf{x}_2,\ldots,\mathbf{x}_n$ for a given n-D point $\mathbf{x}$.
Step 2 Assign $P_1 = O_1 + \mathbf{x}_1$, $P_2 = P_1 + \mathbf{x}_2,\ldots,P_n = P_{n-1} + \mathbf{x}_n$.
Step 3 Connect points P_i by straight arrows to form a graph:

$$P_1 \rightarrow P_2 \rightarrow \cdots \rightarrow P_{i-1} \rightarrow P_i \rightarrow \cdots \rightarrow P_n.$$

Statement 3.4 (***n points lossless representation***) If all coordinates X_i do not overlap then GLC-PC and GLC-SC1 algorithms provide bijective 1:1 mapping of any n-D point $\mathbf{x}$ to 2-D directed graph $\mathbf{x}^*$.

Proof Non-overlap ensures getting n unique 2-D points $\mathbf{x}_1$, $\mathbf{x}_2,\ldots,\mathbf{x}_n$ for each n-D point $\mathbf{x}$. Next, GLC-PC and GLC-SC1 reproduce the order of values (x_1, $x_2,\ldots,x_n$) by directed edges in the graph $\mathbf{x}^*$. ∎

Also for overlapping GLC-PC and GLC-SC1 coordinates, it is possible to get bijective mapping by labeling edges of the graph and/or by making edges curvilinear for n-D points that come to the overlap area to disambiguate them. In-Line Coordinates defined in Chap. 2 illustrate this with Bezier curves used for edges.

Algorithm 4 ***GLC-CC1 Graph Construction Algorithm***

Step 1 Apply basic algorithm GLC-B to construct n located coordinates and vectors $\mathbf{x}_1, \mathbf{x}_2,\ldots,\mathbf{x}_n$ for a given n-D point $\mathbf{x}$.

Step 2 Compute the sum of vectors $\mathbf{x}_1$ and $\mathbf{x}_2$, $\mathbf{x}_{12} = \mathbf{x}_1 + \mathbf{x}_2$, and then compute the point $P_1 = O_1 + \mathbf{x}_{12}$. Next compute the sum of vectors $\mathbf{x}_3$ and $\mathbf{x}_4$, $\mathbf{x}_{34} = \mathbf{x}_3 + \mathbf{x}_4$ and the point $P_2 = O_3 + \mathbf{x}_{34}$. Repeat this process by computing $P_3 = O_5 + \mathbf{x}_{56}$ and for all next i, $P_i = O_{2i-1} + \mathbf{x}_{2i-1,\,2i}$. For even n, the last point is $P_{n/2} = O_{n-1} + \mathbf{x}_{n-1,\,n}$ (see Fig. 3.4), for odd n, the last $\mathbf{x}_{n-1,\,n} = \mathbf{x}_n$ and the last point $P_{(n+1)/2} = O_n + \mathbf{x}_n$. We denote the last point as P_m which is $P_{n/2}$ for even n and $P_{(n+1)/2}$ for odd n.

Step 3 Build a directed graph by connecting points $\{P\}$:

$$P_1 \rightarrow P_2 \rightarrow \cdots \rightarrow P_{i-1} \rightarrow P_i \cdots \rightarrow P_m.$$

This graph can be closed by adding edge $P_n \rightarrow P_1$.

Statement 3.5 (***n/2 points lossless representation***) If coordinates X_i, and X_{i+1} are not collinear in each pair (X_i, X_{i+1}) then GLC-CC1 algorithm provides bijective 1:1 mapping of any n-D point $\mathbf{x}$ to 2-D directed graph $\mathbf{x}^*$ with $\lceil n/2 \rceil$ nodes and $\lceil n/2 \rceil - 1$ edges.

Proof Non-collinearity allows back projection of each graph node (point P_i) to coordinates $\mathbf{x}_{2i-1}$ and $\mathbf{x}_{2i}$ that are used for constructing it, $P_i = O_{2i-1} + \mathbf{x}_{2i-1,\,2i}$. Next, GLC-CC1 algorithm reproduces the order of values $(x_1, x_2,\ldots,x_n)$ by directed edges from P_i to P_{i+1} for all i. ∎

Algorithm 5 ***GLC-CC2 Graph Construction Algorithm***

Step 1 Apply basic algorithm GLC-B to construct n located coordinates and vectors $\mathbf{x}_1\mathbf{x}_2,\ldots,\mathbf{x}_n$ for a given n-D point $\mathbf{x}$.

Step 2 Compute the sum of vectors $\mathbf{x}_1$ and $\mathbf{x}_2$, $\mathbf{x}_{12} = \mathbf{x}_1 + \mathbf{x}_2$ and then compute the point $P_1 = O_1 + \mathbf{x}_{12}$. Next compute the sum of vectors $\mathbf{x}_3$ and $\mathbf{x}_4$, $\mathbf{x}_{34} = \mathbf{x}_3 + \mathbf{x}_4$ and the point $P_2 = P_1 + \mathbf{x}_{34}$. Repeat this process by computing $P_3 = P_2 + \mathbf{x}_{56}$ and for all next i, $P_i = P_{i-1} + \mathbf{x}_{2i-1,\,2i}$. For even n the last point is $P_{n/2} = P_{n/2-1} + \mathbf{x}_{n-1,\,n}$ (see Fig. 3.5), for odd n, $\mathbf{x}_{n-1,\,n} = \mathbf{x}_n$ and the last point is $P_{(n+1)/2} = P_{(n+1)/2-1} + \mathbf{x}_n$.

Step 3 Build a directed graph by connecting points $\{P\}$:

$$P_1 \rightarrow P_2 \rightarrow \cdots \rightarrow P_{i-1} \rightarrow P_i \rightarrow \cdots \rightarrow P_n.$$

This graph can be closed by adding edge $P_n \rightarrow P_1$.

Statement 3.6 (***n/2 points lossless representation***) If coordinates X_i, and X_{i+1} are not collinear in each pair (X_i, X_{i+1}) then GLC-CC2 algorithm provides bijective 1:1 mapping of any n-D point $\mathbf{x}$ to 2-D directed graph $\mathbf{x}^*$ with $\lceil n/2 \rceil$ nodes and $\lceil n/2 \rceil - 1$ edges.

Proof For a non-collinear pair of coordinates, the point P_1 allows us to restore x_1 value by projecting P_1 to coordinate X_1 as shown in Fig. 3.5. Formally it can be computed by representing the coordinate X_1 as a vector $\mathbf{X}_1$, and using a dot product $(P_1 - O_1) \bullet \mathbf{X}_1$ of $\mathbf{X}_1$ with vector $(P_1 - O_1)$. This gives us value x_1 and vector $\mathbf{x}_1$. Next, the property

$$P_1 = O_1 + \mathbf{x}_{12} = O_1 + \mathbf{x}_1 + \mathbf{x}_2$$

allows us to compute $\mathbf{x}_2 = P_1 - O_1 - \mathbf{x}_1$.

The same can be done by a dot product $(P_1 - O_1) \bullet \mathbf{X}_2$. In the same way by projecting point P_2 to X_3 we get $\mathbf{x}_3$ and then using $P_2 = O_3 + \mathbf{x}_{34} = O_3 + \mathbf{x}_3 + \mathbf{x}_4$ we restore $\mathbf{x}_4 = P_2 - O_1 - \mathbf{x}_3$. These steps are continued for all points P_i until all x_i are restored. The property of $\lceil n/2 \rceil$ nodes and $\lceil n/2 \rceil - 1$ edges follows directly from the process of constructing a single 2-D point $P_i = P_{i-1} + \mathbf{x}_{2i-1,\,2i}$ for each pair $(\mathbf{x}_{2i-1}, \mathbf{x}_{2i})$. ■

Algorithm 6 ***GLC-SC2 Graph Construction Algorithm***

Step 1 Apply basic algorithm GLC-B to construct n located linear coordinates and vectors $\mathbf{x}_1, \mathbf{x}_2,\ldots,\mathbf{x}_n$ for a given n-D point $\mathbf{x}$.

Step 2 Assign $P_1 = O_1 + \mathbf{x}_1$, $P_2 = P_1 - \mathbf{x}_2,\ldots,P_n = P_{n-1} - \mathbf{x}_n$.

Step 3 Connect points P_i by straight arrows to form a graph with arrows going to P_2:

$$P_1 \rightarrow P_2 \leftarrow \cdots \leftarrow P_{i-1} \leftarrow P_i \leftarrow \cdots \leftarrow P_n.$$

Statement 3.7 (***n points lossless representation***) If all coordinates X_i do not overlap then GLC-SC2 algorithm provides bijective 1:1 mapping of any n-D point $\mathbf{x}$ to 2-D directed graph $\mathbf{x}^*$.

Proof Non-overlap ensures getting n unique 2-D points $\mathbf{x}_1, \mathbf{x}_2,\ldots,\mathbf{x}_n$ for each n-D point $\mathbf{x}$. Next, the order of values $(x_1, x_2,\ldots,x_n)$ is restorable from graph $\mathbf{x}^*$ constructed in GLC-SC2. ■

Thus, algorithms GLC-CC1 and GLC-CC2 use about a half of the nodes used in GLC-PC, GLC-SC1 and GLC-SC2 as Figs. 3.2, 3.3, 3.4, 3.5 and 3.6 illustrate this property.

Statement 3.8 GLC-CC1 preserves L^p distances for $p = 1$, $D(\mathbf{x}, \mathbf{y}) = D^*(\mathbf{x}^*, \mathbf{y}^*)$.

Proof In GLC-CC1, each pair (x_i, x_{i+1}) in $\mathbf{x}$ is mapped to node $n_{xi,\,i+1} = (O_i + \mathbf{x}_i + \mathbf{x}_{i+1})$ of the graph x* and each pair (y_i, y_{i+1}) in $\mathbf{y}$ is mapped to node $n_{yi,\,i+1} = (O_i + \mathbf{y}_i + \mathbf{y}_{i+1})$ of the graph $\mathbf{y}^*$. The L^1 distance between these nodes is

$$\begin{aligned}
&\left\|n_{xi,i+1} - n_{yi,i+1}\right\| = \left\|(O_i + \mathbf{x}_i + \mathbf{x}_{i+1}) - (O_i + \mathbf{y}_i + \mathbf{y}_{i+1})\right\| = \\
&\left\|\mathbf{x}_i + \mathbf{x}_{i+1} - \mathbf{y}_i - \mathbf{y}_{i+1}\right\| = \left\|(\mathbf{x}_i - \mathbf{y}_i) + (\mathbf{x}_{i+1} - \mathbf{y}_{i+1}\right\| = \\
&\left\|\mathbf{x}_i - \mathbf{y}_i\right\| + \left\|\mathbf{x}_{i+1} - \mathbf{y}_{i+1}\right\| = |x_i - y_i| + |x_{i+1} - y_{i+1}|
\end{aligned}$$

The last property is derived for even n from the fact that for each i vectors $\mathbf{x}_i$ and $\mathbf{y}_i$ are located on the same coordinate X_i, i.e., their difference is a their scalar difference due to construction process of vectors $\mathbf{x}_i$ and $\mathbf{y}_i$ in GLC-CC1.

The total L^1 distance $D^*(\mathbf{x}^*, \mathbf{y}^*)$ is the sum of L^1 distances between all nodes of $\mathbf{x}^*$ and $\mathbf{y}^*$, which is equal to the distance between $\mathbf{x}$ and $\mathbf{y}$ in GLC-CC1 due to the property of this distance derived above,

$$D^*(\mathbf{x}^*, \mathbf{y}^*) = \sum\nolimits_{i=1,3,5,\ldots n-1} \|n_{xi,i+1} - n_{yi,i+1}\| = \sum\nolimits_{i=1,3,5,\ldots n-1} (|x_i - y_i| + |x_{i+1} - y_{i+1}|) = \sum\nolimits_{i=1:n} |x_i - y_i| = D(\mathbf{x}, \mathbf{y}).$$

For odd n we duplicate the last coordinate to get this statement. ∎

3.3 Fixed Single Point Approach

3.3.1 *Single Point Algorithm*

So far, we have shown a cognitive advantage of both GLC-CC representations, i.e., its *twice-smaller footprint in 2-D*, relative to GLC-PC and GLC-SC. This leads to *much smaller occlusion* when multiple n-D data are represented in 2-D. Below we show its other *advantage*—the ability to represent losslessly any given n-D point $(x_1, x_2, \ldots, x_n)$ as a *single 2-D point instead of a graph*. The algorithm to produce this representation is the Single Point (SP) algorithm

Steps of the **Single Point (SP) Algorithm**

Step 1 Select an arbitrary 2-D point $A = (a_1, a_2)$ on the plane. This point will be called the *anchor 2-D point*. Then select the n-D point $\mathbf{x} = (x_1, x_2, \ldots, x_n)$ that will be called the *base n-D point* (or *n-D anchor point*). Next, select a set of positive constants $c_1, c_2, \ldots, c_n$ that will be used as lengths of coordinates $X_1, X_2, \ldots, X_n$.

Step 2 Compute 2-D points $O_1 = (a_1 - x_1, a_2 - x_2)$ and $E_1 = (a_1 - x_1 + c_1, a_2 - x_2)$. Coordinate line X_1 is defined as the located vector $\mathbf{X}_1 = (O_1, E_1)$.

Step 3 Define the points O_1 and O_2, $O_2 = O_1$, and $E_2 = (a_1 - x_1, a_2 - x_2 + c_2)$. Coordinate line X_2 is defined as a vector $\mathbf{X}_2 = (O_2, E_2)$.

Step 4 Repeat the steps 2 and 3 for all other coordinates to build the coordinate system $X_1, X_2, \ldots, X_n$.

This algorithm creates a system of **Parameterized Shifted Paired Coordinates (PSPC)**, where each next pair of coordinates is drawn in the shifted Cartesian coordinates. These coordinates are defined by *parameters* which are respective components of a *base n-D point* $\mathbf{x}$ and *2-D anchor point* A. See Fig. 3.7.

Statement 3.9 In the coordinate system $X_1, X_2, \ldots, X_n$ constructed by the Single Point algorithm with the given base n-D point $\mathbf{x} = (x_1, x_2, \ldots, x_n)$ and the anchor 2-D

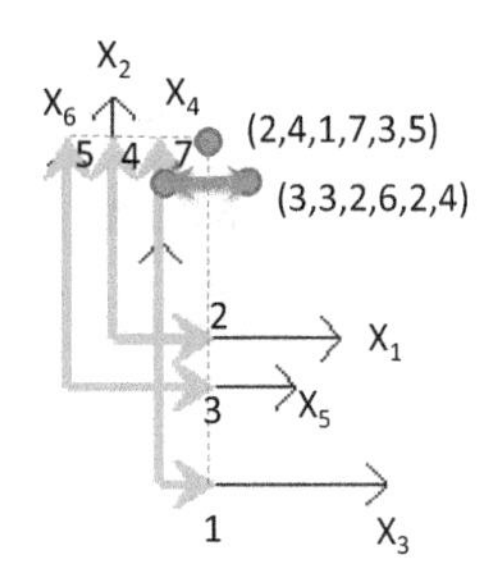

Fig. 3.7 6-D points (3, 3, 2, 6, 2, 4) and (2, 4, 1, 7, 3, 5) in X_1–X_6 coordinate system build using point (2, 4, 1, 7, 3, 5) as an anchor

point A, the n-D point **x** is mapped one-to-one to a single 2-D point A by GLC-CC algorithm.

Proof Consider coordinate X_1 and a point located on X_1 at the distance x_1 from O_1. According to Step 2 of SP algorithm $O_1 = (a_1 - x_1, a_2 - x_2)$. Thus, it is the point $(a_1 - x_1 + x_1,\ a_2 - x_2) = (a_1,\ a_2 - x_2)$. It is projection of pair (x_1, x_2) to X_1 coordinate.

Similarly consider coordinate X_2 and a point located on X_2 at the distance x_2 from O_2. According to Step 2 of SP algorithm $O_2 = (a_1 - x_1, a_2 - x_2)$. Thus, it is the point $(a_1 - x_1, a_2 - x_2 + x_2) = (a_1 - x_1, a_2)$. It is a projection of pair (x_1, x_2) to X_2 coordinate. Therefore, pair (x_1, x_2) is represented in X_1, X_2 coordinate system as (a_1, a_2). In the same way, the pair (x_3, x_4) is also mapped to the point (a_1, a_2). The repeat of this reasoning for all other pairs (x_i, x_{i+1}) will match them to the same point (a_1, a_2). This concludes the proof. ■ Figure 3.7 illustrates this proof for a 6-D point (2, 4, 1, 7, 3, 5).

3.3.2 *Statements Based on Single Point Algorithm*

Another advantage of the combination of GLC-CC and SP algorithms is that all n-D points of an n-D hypercube around a given base n-D point $\mathbf{x} = (x_1, x_2,\ldots,x_n)$ are mapped to graphs that are located within a square defined by the square algorithm presented below. In other words informally, *n-D locality* is converted *to 2-D locality* and vice versa. Here an n-D point **y** is close to the base n-D point **x** if and only if the graph **y*** of **y** is close to 2-D anchor point A.

Steps of the Square algorithm for Parameterized SPC

Step 1 Construct a hypercube H with the center at the base point $\mathbf{x} = (x_1, x_2,\ldots,x_n)$ and distance d to its faces. Respectively, 2^n nodes $\mathbf{N}$ of this hypercube are $(x_1 + \alpha d,\ x_2 + \alpha d,\ldots,x_n + \alpha d)$, where $\alpha = 1$ or $\alpha = -1$ depending on the node, e.g., $(x_1 + d,\ x_2 + d,\ldots,x_n + d)$, $(x_1 - d,\ x_2 - d,\ldots,x_n - d)$, $(x_1 + d,\ x_2 - d,\ldots,x_n - d)$.

Step 2 Construct a square S around point (a_1, a_2) with corners: $(a_1 + d,\ a_2 + d)$, $(a_1 + d,\ a_2 - d)$, $(a_1 - d,\ a_2 + d)$, $(a_1 - d,\ a_2 - d)$

Statement 3.10 (*locality statement*) All graphs that represent nodes ***N*** of n-D hypercube *H* are within square S.

Proof Consider the n-D node $(x_1 + d, x_2 + d, \ldots, x_n + d)$ of *H* where *d* is added to all coordinates of the n-D point **x**. This node is mapped to the 2-D point $(a_1 + d, a_2 + d)$ which is a corner of the square *S*. Similarly, the node $(x_1 - d, x_2 - d, \ldots, x_n - d)$ of *H* where *d* is subtracted from all coordinates of **x** is mapped to the 2-D point $(a_1 - d, a_2 - d)$ which is another corner of the square *S*. Similarly, the n-D node of the hypercube that contains pairs $(x_1 + d, x_2 - d)$, $(x_3 + d, x_4 - d), \ldots, (x_i + d, x_{i+1} - d), \ldots, (x_{n-1} + d, x_n - d)$, i.e., with positive *d* for the odd coordinates ($X_1, X_3, \ldots$) and negative *d* for the even coordinates ($X_2, X_4, \ldots$) is mapped to the 2-D point $(a_1 + d, a_2 - d)$. Similarly, a node with alternation of positive and negative *d* in all such pairs $(x_i - d, x_{i+1} + d)$ will be mapped to $(a_1 - d, a_2 + d)$. Both these points are also corners of the square S.

If an n-D node of *H* includes two pairs such as $(x_i + d, x_{i+1} + d)$ and $(x_j + d, x_{j+1} - d)$ then it is mapped to the graph that contains two 2-D nodes $(a_1 + d, a_2 + d)$ and $(a_1 + d, a_2 - d)$ that are corners of the square *S*. Similarly, if an n-D node of *H* includes two other pairs $(x_k - d, x_{k+1} + d)$ and $(x_m - d, x_{m+1} - d)$ it is mapped to the graph that contains two 2-D nodes $(a_1 - d, a_2 + d)$ and $(a_1 - d, a_2 - d)$ that are two other corners of the square S. At most a hypercube's n-D node has all these four types of pairs that can be present several times, and respectively all of them will be mapped to four corners of the 2-D square S. Respectively, all edges of this graph will be within the square *S*. Any other n-D point **y** of the hypercube *H* has at least one coordinate that is less than this coordinate for some node Q of this hypercube.

For example, let $y_1 < q_1$ and $y_i = q_i$ for all other *i*, then all pairs (y_i, y_{i+1}), but the first pair (y_1, y_2) will be mapped to the corners of the square S. The first pair (y_1, y_2) will be mapped to the 2-D point, which is inside of the square *S* because $y_1 < q_1$. This concludes the proof. ∎

Figures 3.8 and 3.9 illustrate this statement and its proof.

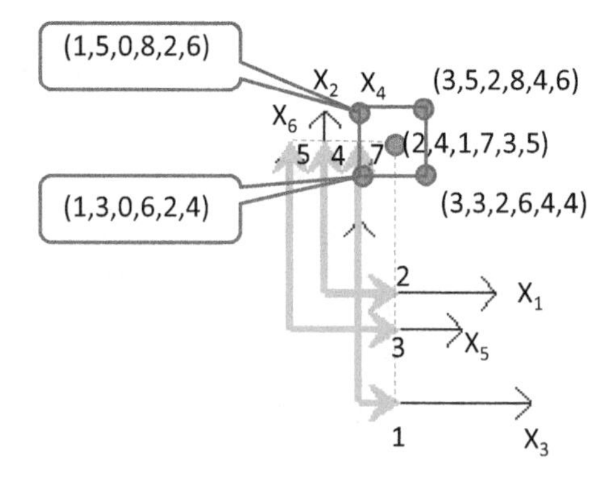

Fig. 3.8 Data in parameterized shifted paired coordinates. Blue dots are corners of the square S that contains all graphs of all n-D points of hypercube *H* for 6-D base point (2, 4, 1, 7, 3, 5) with distance 1 from this base point

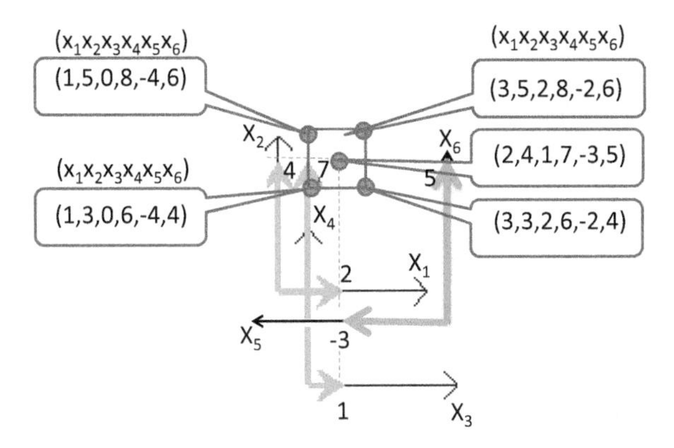

Fig. 3.9 Data in Parameterized Shifted Paired Coordinates. Blue dots are corners of the square S that contains all graphs of all n-D points of hypercube *H* for 6-D base point (2, 4, 1, 7, −3, 5) with distance 1 from this base point

Both the Collocated Paired Coordinates and the Parameterized Shifted Paired Coordinates are *reversible (lossless),* and represent a *similar n-D point as similar 2-D graphs*, i.e., 2-D nodes of similar n-D points are *located closely* as Figs. 3.10 and 3.11 illustrate.

Figure 3.10 shows an example of 4-D data of two classes in Collocated Paired Coordinates in blue and green ellipses. Figure 3.11 shows data from Fig. 3.10 in the Parameterized Shifted Paired Coordinates with 4-D point (3, 13, 13, 2) from the green class as the base point for parameterized shift.

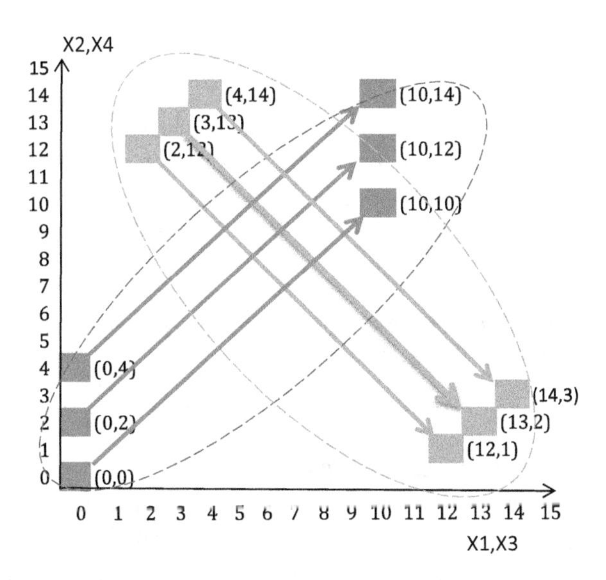

Fig. 3.10 4-D data of two classes in Collocated Paired Coordinates shown in blue and green ellipses

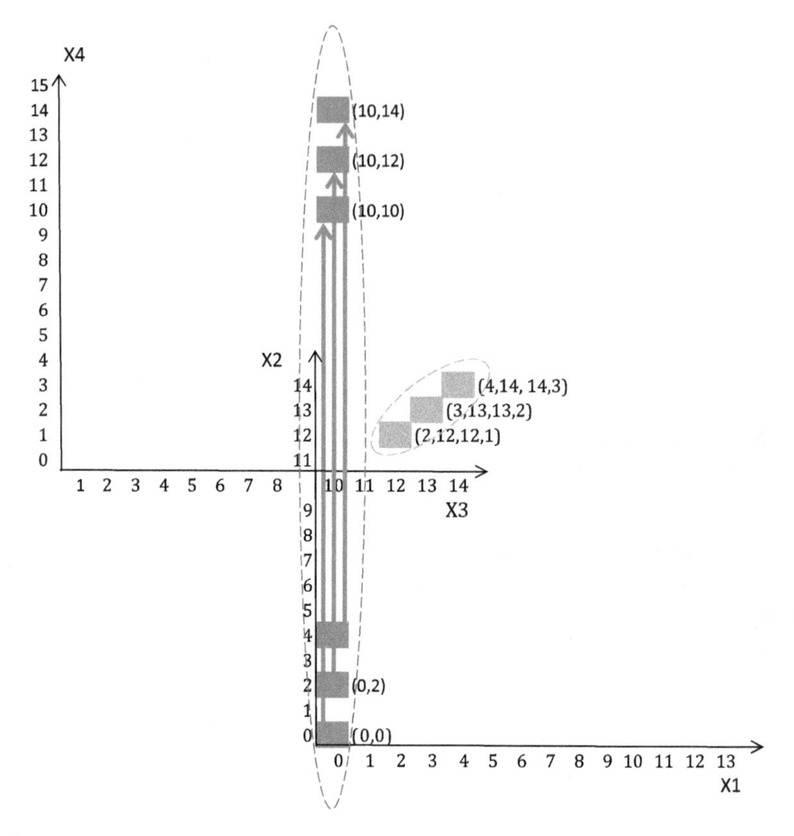

Fig. 3.11 4-D data of two classes in Parameterized Shifted Paired Coordinates

Both Figs. 3.10 and 3.11 show the separation of two classes, but in Fig. 3.11, the separation between these blue and green classes is much simpler than in Fig. 3.10. This is a demonstration of the promising **advantages** of parameterized shifted coordinates to **simplify visual patterns** of n-D data in 2-D in tasks such as clustering and supervised classification. This gives the direction for future studies to solve a major challenge. This challenge is finding the conditions where this empirical observation can be converted into the provable property of simpler and less overlapped 2-D representation of non-intersecting hyper-ellipses, hyper-rectangles, and other shapes in n-D.

3.3.3 Generalization of a Fixed Point to GLC

Other general line coordinates, not only the shifted paired coordinates, also allow the creating of a single 2-D point from an n-D point **x** reversible and losslessly. These GLCs include Shifted Bush and Parallel Coordinates, Partially Collocated

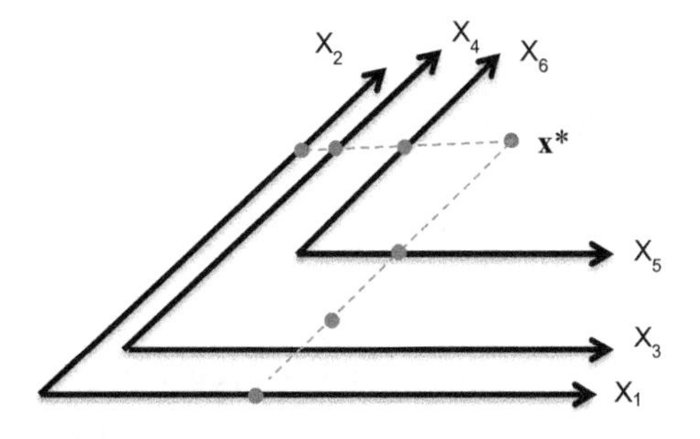

Fig. 3.12 Representation of 6-D point **x** = (3, 4, 2, 3, 1, 2) as a single 2-D point **x*** in non-orthogonal shifted coordinates

and In-Line Coordinates, which is illustrated in Chap. 2. Figure 3.12 provides additional illustrations for the *non-orthogonal shifted paired coordinates.*

Statement 3.11 Any n-D general line coordinates (including the connected or disconnected, the orthogonal or non-orthogonal) constructed by Algorithm 2 with Algorithm 1 *can* be shifted to represent a given n-D point as a single 2-D point losslessly.

Proof Consider a 2-D point A, coordinate axes $X_1 - X_n$ and values $(x_1, x_2,\ldots,x_n)$ of an n-D point **x** that are drawn in some 2-D coordinate system (U_1, U_2) in accordance with the Algorithm 1. Let $u(O_{i1}, O_{i2})$, $u(E_{i1}, E_{i2})$ be respectively coordinates of the origin and the end of axis X_i in (U_1, U_2) coordinate system, and pair (u_{i1}, u_{i2}) be a location in (U_1, U_2) where vector $\mathbf{x}_i$ ends on the axis X_i, i.e., represents the value x_i.

Coordinates of point A are given as (a_1, a_2) in coordinates (U_1, U_2). Consider a vector $\mathbf{v}_i = (a_1 - u_{i1}, a_2 - u_{i2})$ that represents the difference between points A and (u_{i1}, u_{i2}) that represents value x_i. Shifting axis X_i by this difference will bring the end point of vector $\boldsymbol{x}_i$ from (u_{i1}, u_{i2}) to A:

$$(a_1 - u_{i1}, a_2 - u_{i2}) + (u_{i1}, u_{i2}) = (a_1, a_2).$$

Doing this for all axes, X_i will move all x_i to A. ∎

Statement 3.12 Any n-D general line coordinates (including the connected or disconnected, orthogonal or non-orthogonal) constructed by Algorithm 4 with Algorithm 1 can be shifted to represent a given n-D point as a single 2-D point losslessly.

Proof Below we use the same notation and approach as in the proof for Algorithms 1 and 2. The difference is that instead of shifting end points of vectors x_i we shift points $P_i = (p_{i1}, p_{i2})$ to $A = (a_1, a_2)$. This is done by shifting a whole pair of coordinates that forms P_i by vector $\mathbf{w}_i = (a_1 - p_{i1}, a_2 - p_{i2})$. This will bring point P_i to A: $(p_{i1}, p_{i2}) + (a_1 - p_{i1}, a_2 - p_{i2}) = (a_1, a_2)$. ∎

Statement 3.13 Any n-D general line coordinates (including connected or disconnected, orthogonal or non-orthogonal) constructed by any of Algorithms 3, 5 and 6 with Algorithm 1 *cannot* be shifted to represent a given n-D point as a single 2-D point losslessly.

Proof In Algorithms 3, 5, and 6 in contrast with Algorithms 2 and 4 the length of each edge of the graph **x*** of the n-D point **x** is fixed and cannot be changed under shifting.

This prevents the collapsing of any two nodes of **x*** to a single node that is needed for getting a single 2-D point representing **x**. Below we use the same notation and approach as in the proof for Algorithm 2 with Algorithm 1. ∎

3.4 Theoretical Limits to Preserve n-D Distances in 2-D: Johnson-Lindenstrauss Lemma

The *curse of dimensionality* challenge was originally associated with the optimization problems (Bellman 1957) where the number of alternative to be explored is growing exponentially with growth of dimension (Curse of Dimensionality 2010).

The curse of dimensionality challenge in visual representation of n-D data for machine learning is related to the fundamental problem of *preserving n-D distanced in 2-D*. The source of this problem is the *drastic differences in the neighborhood capacities*.

In the discrete case in n-D, the center **c** of the hypercube H_n has 2^n nodes of hypercube around this n-D point ("neighbors") with equal distances $D(\mathbf{c}, \mathbf{x})$ from this center. In 2-D, the square H_2 has just $2^2 = 4$ nodes, but H_{10} in 10-D has $2^{10} = 1024$ nodes.

Mapping those 1024 10-D points to four 2-D points means that in average 1024/4 = 256 10-D points will be mapped to a single 2-D point. Thus, multiple non-zero distances, $D_{10}(\mathbf{x}, \mathbf{u}) > 0$, between such 256 points **x**, **u** in 10-D are nullified to the zero distance $D_2(\mathbf{x}, \mathbf{u}) = 0$ between them in 2-D.

In other words, the 2-D space does not have enough neighbors with equal distances to represent all n-D neighbors of a given n-D point with equal distances in n-D space. This leads to the significant **corruption** of n-D distances in 2-D for datasets that have multiple n-D points with equal or close to equal distances from a given n-D point. This is a situation for the mean point of an n-D Gaussian distribution, because multiple points are concentrated around the mean in this distribution.

Note that mapping nodes of n-D hypercube H_n to 2-D points to the circle preserves the distance from nodes to the center of the hypercube, but corrupts the distances between these n-D points themselves. These difficulties are reflected in the *Johnson-Lindenstrauss lemma* that implies that only *a small number of n-D points can be represented with preserving distances with small deviations* in k-D when $k < n$ as we show below.

Lemma (Johnson and Lindenstrauss 1984) Given $0 < \varepsilon < 1$, a set X of m points in R^n, and a number $k > 8\ln(m)/\varepsilon^2$, there is a linear map $f: R^n \to R^k$ such that

$$(1 - \varepsilon)\|u - v\|^2 \leq \|f(u) - f(v)\|^2 \leq (1 + \varepsilon)\|u - v\|^2$$

for all $u, v \in X$.

In other words, this lemma sets up a relation between n, k and m when the distance can be preserved with some allowable error ε.

The version of the Johnson-Lindenstrauss lemma (Dasgupta and Gupta 2003) allows one to define the possible dimensions $k < n$ such that for any set of m points in R^n there is a mapping f: $R^n \rightarrow R^k$ with "similar" distances in R^n and R^k between mapped points. This similarity is expressed in terms of error $0 < \varepsilon < 1$.

For $\varepsilon = 0$ these distances are equal. For $\varepsilon = 1$ the distances in R^k are less or equal to $\sqrt{2}S$, where S is the distance in R^n. This means that distance s in R^k will be in the interval [0, 1.42S].

In other words, the distances will not be more than 142% of the original distance, i.e., it will not be much exaggerated. However, it can dramatically diminish to 0. The exact formulation of this version of the Johnson-Lindenstrauss lemma is as follows.

Theorem 1 (Dasgupta and Gupta 2003, Theorem 2.1) *For any* $0 < \varepsilon < 1$ *and any integer n, let k be a positive integer such that*

$$k \geq 4(\varepsilon^2/2 - \varepsilon^3/3)^{-1} \ln n \quad (3.1)$$

then for any set V of m points in R^k *there is a mapping f:* $R^n \rightarrow R^k$ *such that for all* $u, v \in V$

$$(1-\varepsilon)\|u-v\|^2 \leq \|f(u)-f(v)\|^2 \leq (1+\varepsilon)\|u-v\|^2 \quad (3.2)$$

It is also shown in (Dasgupta and Gupta 2003) that this mapping can be found in randomized polynomial time. A formula (3.3) is presented in (Frankl and Maehara 1988) stating that k dimensions are sufficient, where

$$k = \left\lceil 9(\varepsilon^2 - 2\varepsilon^3/3)^{-1} \ln n \right\rceil + 1 \quad (3.3)$$

Table 3.1 presents the values of the number of points in high-dimensional space and dimension k required to keep error in R^k within about 31% of the actual distance in the R^n ($\sqrt{\varepsilon} = \sqrt{0.1} = 0.316228$) using the formulas (3.1)–(3.3).

It shows that to keep distance errors within about 30% for just 10 arbitrary high-dimensional points, we need over 1900 dimensions, and over 4500 dimensions for 300 arbitrary points.

This is hard to accomplish in many tasks. Figure 3.13 illustrates data from Table 3.1. The bounds computed for other values of ε such as 0.3, 0.5, 0.8, 1.0 and the number of samples up to 10^6 can be found in (Pedregosa et al. 2011).

3.5 Visual Representation of n-D Relations in GLC

This section describes relations in n-D spaces that can be represented by different GLCs including PC, CPC, SPC and GLC-L that are defined below. In Paired GLC such as CPC and SPC each directed edge of the graph directly visualizes relations

Table 3.1 Dimensions to support ±31% of error (ε = 0.1)

Number of arbitrary points in high-dimensional space	Sufficient dimension with formula (3.1)	Sufficient dimension with formula (3.2)	Insufficient dimension with formula (3.3)
10	1974	2145	1842
20	2568	2791	2397
30	2915	3168	2721
40	3162	3436	2951
50	3353	3644	3130
60	3509	3813	3275
70	3642	3957	3399
80	3756	4081	3506
90	3857	4191	3600
100	3947	4289	3684
200	4541	4934	4239
300	4889	5312	4563

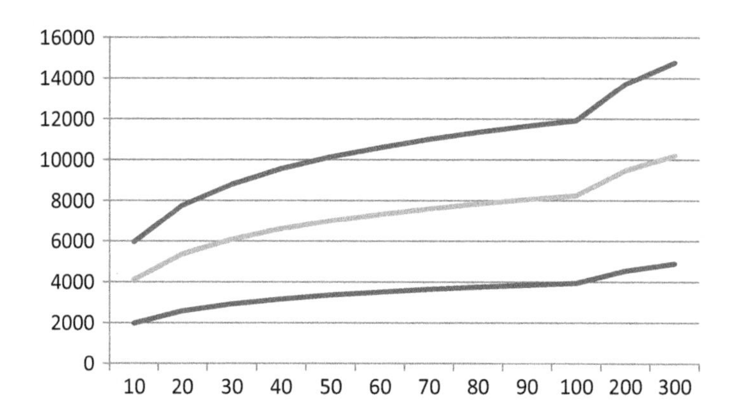

Fig. 3.13 Dimensions required supporting ±31% of error ε

of four dimensions. In contrast, each edge of the graph in PC directly visualized only relations between two adjacent coordinates. For instance, for all coordinates scaled to [0.1], in 4-D CPC, an edge going up and to the right indicates relation $(x_3 > x_1)$ and $(x_4 > x_2)$. In contrast, in PC the edge going to the same direction (up and right) indicates only relation $(x_2 > x_1)$. In CPC if this edge is in the first quadrant below its diagonal then in addition it also expresses the relation $(x_2 < x_1)$ and $(x_4 < x_3)$.

3.5.1 Hyper-cubes and Clustering in CPC

Consider an n-D *hypercube*, that is a set of n-D points {**w**} that are in the vicinity of an n-D point **a** within distance r from **a** for each dimension X_i, {**w**: $\forall$ i d_i(**w**, **a**) $\leq$ r}. We denote this set {**w**} of n-D points as **w**(**a**, r).

Statement 3.14 For any n-D point **w** within the n-D hypercube **w**(**a**, r), a CPC graph **w*** for **w** is a CPC graph **a*** for **a** shifted to no more than the distance r in each dimension.

Proof Let **a** = $(a_1, a_2,\ldots,a_n)$. Then the farthest points **w** of the hypercube are **w** = $(a_1 \pm r, a_2 \pm r_2,\ldots,a_n \pm r)$. If these points are shifted relative to **a** no more than r then all other points are shifted within r too. Therefore, the statement is true for these points. In CPC **w** is shown as a directed graph **w*** that connects pairs

$$(w_1, w_2) \rightarrow (w_3, w_4) \rightarrow \cdots \rightarrow (w_{n-1}, w_n).$$

Consider for all i, $w_i = a_i + r$ then $(w_1, w_2) = (a_1 + r, a_2 + r)$ of **w** is shifted relative to the first pair (a_1, a_2) of **a** with the shift (r, r). Similarly the pair (w_i, w_{i+1}) of **w** is shifted relative to the pair (a_i, a_{i+1}) of **a** with the same shift (r, r). Thus, to produce the graph **w*** for **w** we need to add shift r in X coordinate and shift r on Y coordinate to the graph for **a**. Similarly shifts $(-r, r)$ or $(r, -r)$ will happen when some w_i will be equal to $a_i - r$ keeping the max of distance equal to r. ∎

Now consider two classes of n-D vectors that are within two different hypercubes with property Q: centers **a** and **b** of W(**a**, r) and W(**b**, r) are with the distance greater than $2r$ in each dimension X_i, that is D_i(**a**, **b**) > $2r$ and have $a_i < b_j$. for all i and j.

Statement 3.15 Nodes of graphs of CPC representations of all n-D points from hyper-cubes W(**a**, r) and W(**b**, r) with property Q do not overlap.

Proof First, maxD_i(**w**$_k$, **w**$_s$) = $2r$ and max D_i(**w**$_k$, **a**) = r when both **w**$_k$ and **w**$_s$ are within W(**a**, r). The same properties maxD_i(**w**$_k$, **w**$_s$) = $2r$ and maxD_i(**w**$_k$, **b**) = r are true when both **w**$_k$ and **w**$_s$ are within W(**b**, r). In n-D if hypercubes W(**a**, r) and W(**b**, r) overlap then a dimension X_i must exist such that D_i(**a**, **b**) $\leq$ $2r$ because it must be less than or equal to the sum of max distances D_i(**a**, **w**) = r and D_i(**b**, **w**) = r within each hyper-cube that is $r + r$. This would contradics property Q that requires for all X_i that D_i(**a**, **b**) > $2r$. This distance is sufficient for graphs of n-D points from these hyper-cubes not overlap because for all i and j $a_i < b_j$., i.e., the lowest node of the graph **b*** will be above the highest node of **a*** with that distance. ∎

A similar statement can be formulated for the hyper-spheres. For some situations that do not satisfy property Q we still can make two hyper-cubes non-overlapping by reversing some coordinates.

Example Let in 4-D **a** = (1, 4, 4, 1), **b** = (4, 1, 1, 4), and r = 1. The hyper cubes W (**a**, r) and W(**b**, r) do not overlap in 4-D, but in CPC in 2-D, graph **a*** is the arrow

(1, 4) → (4, 1) and graph b* is the arrow (4, 1) → (1, 4), i.e., it is the same line with the opposite direction. Assume that values of each coordinate X_i are in [0, 5] interval. Next, we reverse coordinates X_2 and X_3 making reversed coordinates X_2' and X_3' with values $x_2' = 5 - x$ and $x_3' = 5 - x$ for each n-D point **x**, where 5 is the max of X_2 and X_3 in this example. This produces **a**′ = (1, 1, 1, 1), and **b**′ = (4, 4, 4, 4) with graphs (1, 1) → (1, 1) and (4, 4) → (4, 4) that do not overlap.

The benefit of Statement 3.15 is that if two classes are within such non-overlapping hyper-cubes then it will be visible in CPC visualization without knowing in advance centers and lengths of these hyper-cubes. In contrast, analytical discovery of this separation would require search for centers and lengths of appropriate hypercubes.

For analytical clustering, we need to seed n-D points to start it and the clustering result is sensitive to selection of these points. Not knowing the centers and lengths of hypercubes, we need to search through many of them. In addition, the selection of clustering objective function itself is quite subjective.

The visualization of the clusters found analytically is a way to confirm them and to check their meaning. The challenges for pure analytical methods for discovering more complex data structures such as non-central tubes, piecewise tubes and overlapping structures grow due to the complexity of these data. Thus hybrid methods that combine them with visual methods are promising.

3.5.2 Comparison of Linear Dependencies in PC, CPC and SPC

In PC a 2-D line L: $x_j = mx_i + b$ is visualized by an *infinite set of lines* (Inselberg 2009) (Figs. 3.14 and 3.15). In CPC and SPC the same line is represented in a classical Cartesian form as a *single line* (Figs. 3.16 and 3.17), because CPC and SPC consist of a set of pairs of classical Cartesian coordinates that are collocated or shifted. In PC there is a point L^ that does not show the value of x_j having x_i. One must draw a line to coordinate X_j via points x_i and L^ to get x_j.

A different line must be drawn for each other x_i value. If all these lines are drawn together, they will completely cover a large segment between coordinates X_i and X_j and none of the line will be visible creating an extreme case of full occlusion (see the grey area in Fig. 3.14).

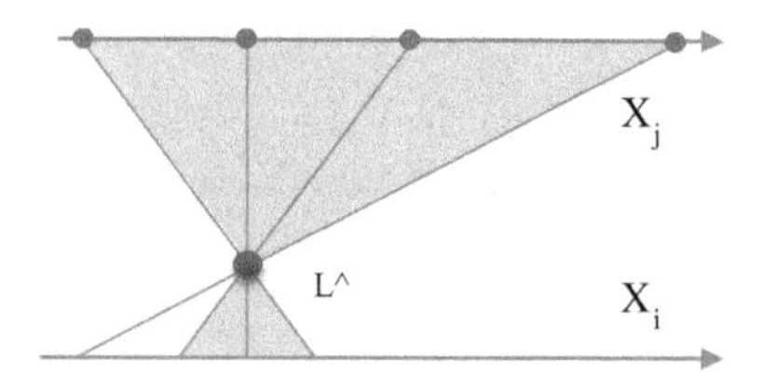

Fig. 3.14 Line L: $x_j = mx_i + b$ for $m < 0$ and point L^ that represent L in Parallel Coordinates

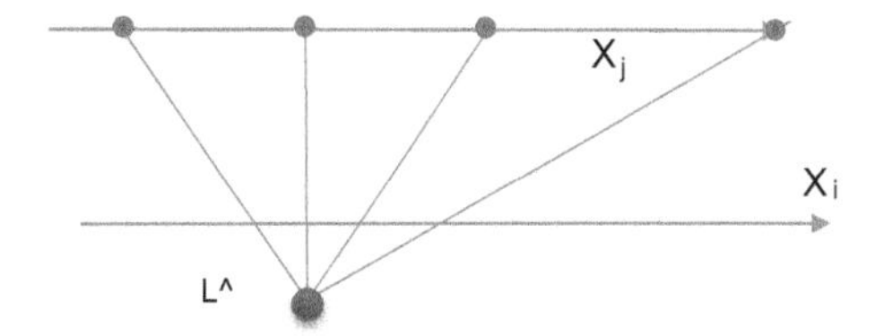

Fig. 3.15 Line *L:* $x_j = mx_i + b$ for $m > 0$ and point *L^* that represents *L* in Parallel Coordinates

In contrast in the classical Cartesian Coordinates, the same single line is used for all *x*. This classical metaphor is familiar to everyone and learning a new metaphor is not required. Also while point *L^* fully represents line *L:* $y = kx + m$, the 2-D point $H^\wedge = (k, m)$ fully represents L also in a compact way in CPC in a pair of coordinates such as (X_1, X_2). The visual process of getting *y* from a given *x* is drawing line *L* using points (0, *m*), (1, $m - k$), then drawing a line from *x* on X_1 to *L*, and projecting the crossing point to X_2 to get *y* (see Fig. 3.17).

In SPC, we have two cases for representing line *L*. The first case is for consecutive pairs such as (X_1, X_2), and (X_3, X_4), where *L* is visualized in the classical Cartesian form discussed above. The second case is for odd pairs such as (X_1, X_3), (X_3, X_5), (X_5, X_7), where *L:* $x_j = mx_i + b$, $j = i + 2$.

In SPC these coordinates are parallel, but shifted, thus the classical Cartesian visualization is not applicable. When X_j coordinate is shifted the point *L^* can keep its locations in SPC if one or both coordinates are rescaled to accommodate the shift. Alternatively, *L^* is shifted if coordinates are not rescaled. To see it compare Fig. 3.14 with Fig. 3.18 where X_j is shifted. Thus, the property that *L^* fully represents a 2-D line *L* in PC holds for SPC.

Now consider situations for Collocated Paired Coordinates. For consecutive pairs (X_i, X_{i+1}) for odd *i* it is the same as for SPC considered above, but for collocated pairs (X_i, X_{i+2}) with a common origin for pairs of coordinates such as

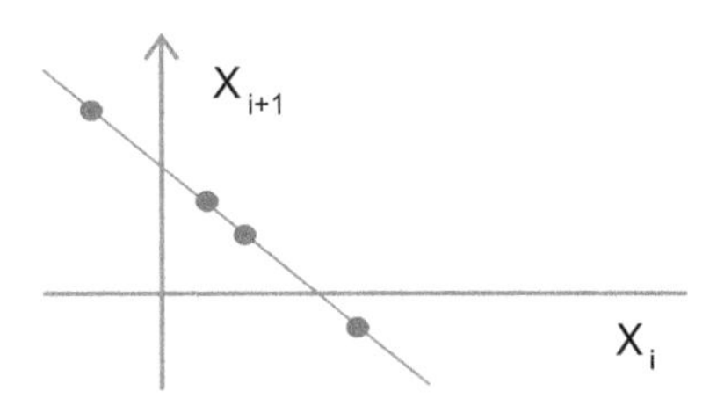

Fig. 3.16 Line *L:* $x_j = mx_i + b$ for $m < 0$ in Cartesian visualization used in CPC and SPC for pairs of coordinates (X_i, X_{i+1})

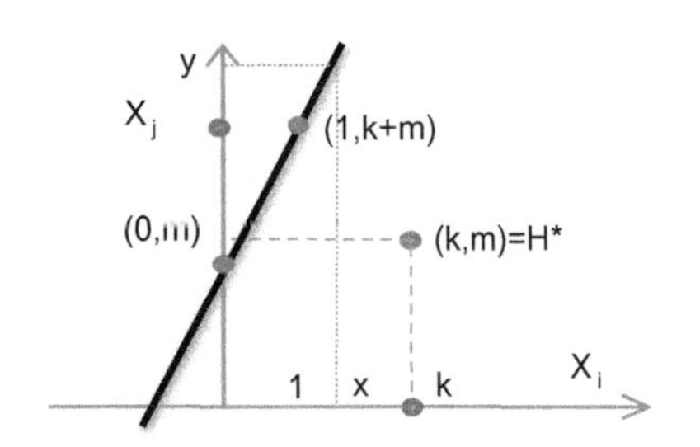

Fig. 3.17 Line *L:* $x_j = mx_i + b$ for $m > 0$ in Cartesian visualization used in CPC and SPC for pairs of coordinates (X_i, X_{i+1}) with point *H^* that completely represents *L*

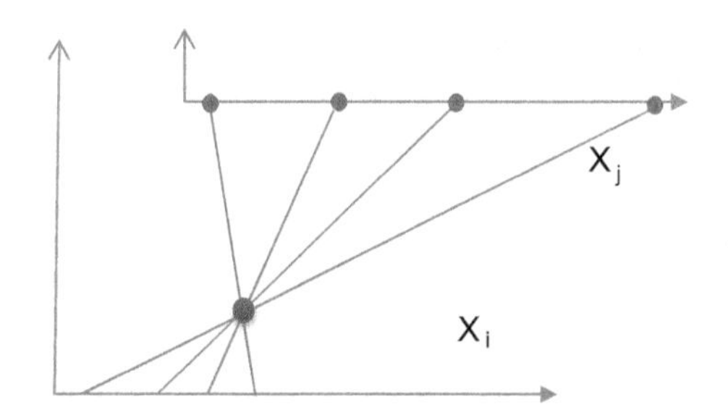

Fig. 3.18 line *L:* $x_j = mx_i + b$ and point L^ that represents L in SPC where both i and j are odd or even

(X_1, X_3) and (X_2, X_4) the situation is different and another solution is required. This solution is shown in Figs. 3.15, 3.16 and 3.17. First a distance d is set up and coordinate X_{i+2} is moved up to this distance parallel to X_i. In this way we get a situation of finding point L^ as it is done in PC in Fig. 3.14. This stage is shown in Fig. 3.19, where blue dotted lines show how x_j is identified on X_j by getting line via x_i and L^. Next, these lines are reflected back to X_i relative to the middle line that is at height $d/2$ for a given d (see Fig. 3.19) to get value of x_j on coordinate X_j that is collocated with coordinate X_i.

This process is equivalent to building a triangle with two equal sides. Thus, the X_j can be removed from its temporary location (see Fig. 3.20). In the final algorithm the triangular property allows avoiding moving X_j to the temporary location

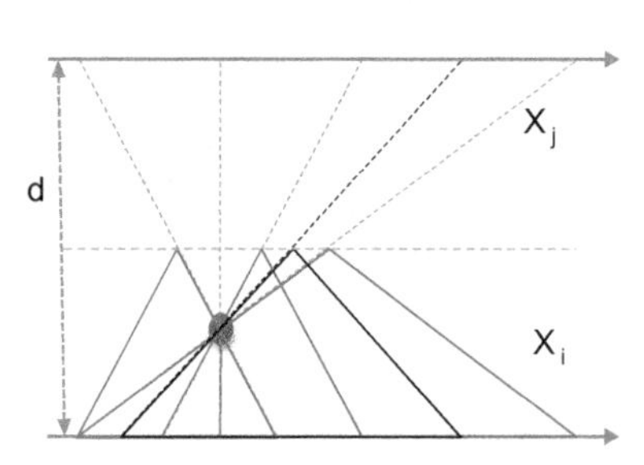

Fig. 3.19 Line *L:* $x_j = mx_i + b$ in CPC constructed by reflecting points relative to the middle line at height $d/2$ for a given d

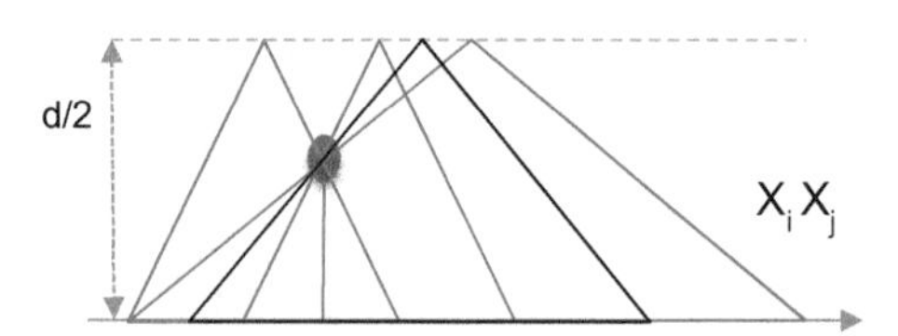

Fig. 3.20 Figure 3.19 with removed X_j from its temporary location

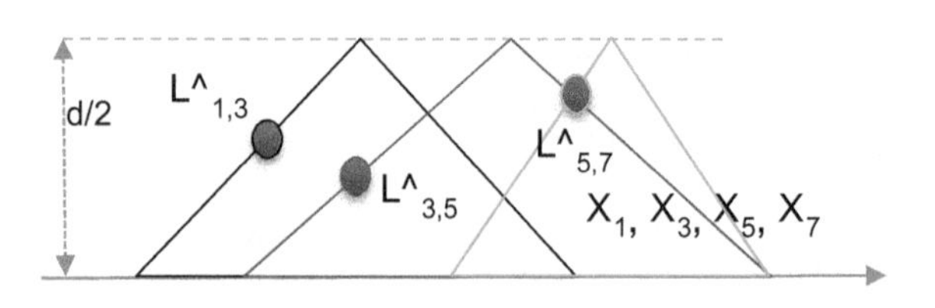

Fig. 3.21 Multiple points L^ for 8-D data for pairs (X_1, X_3), (X_3, X_5) and (X_5, X_7) in CPC

substituting it by building triangles. Finally, Fig. 3.21 shows multiple points L^ for 8-D data for pairs (X_1, X_3), (X_3, X_5) and (X_5, X_7) in CPC.

3.5.3 Visualization of n-D Linear Functions and Operators in CPC, SPC and PC

At the mathematical level, n-D linear structure is a parameterized straight line $Y = A + t\mathbf{v}$ in n-D space, where $A = (a_1, a_2,\ldots,a_n)$ and $Y = (y_1, y_2,\ldots,y_n)$ are n-D points and $\mathbf{v} = (v_1, v_2,\ldots,v_n)$ is an n-D vector, $y_i = a_i + tv_i$, $i = 1,\ldots,n$, and scalar t is changing in some interval, e.g., [0, 1].

The PC visualization in Figs. 3.14 and 3.15 is not actual visualization of the n-D straight line: $Y = A + t\mathbf{v}$. It is visualization of a single *projection* of that n-D line to a pair of coordinates (X_i, X_j) (Inselberg 2009). To fully represent the n-D straight line: $A + t\mathbf{v}$, Inselberg (2009) uses *a set of projections* of that n-D line to pairs of coordinates (X_i, X_2), (X_2, X_3),…,(X_{n-1}, X_n) using the property that a line in R^n is the intersection of $n - 1$ non-parallel hyperplanes. Those hyperplanes can be built from those projections, e.g., $x_2 = mx_1 + b$ can be converted to the n-D hyperplane

$$x_2 = mx_1 + b + 0x_3 + 0x_4 + \cdots + 0x_n,$$

i.e., any n-D point $\mathbf{x} = (x_1, x_2, x_3, x_4,\ldots,x_n)$ such that $x_2 = mx_1 + b$ will be on that hyperplane when all other x_i for $i > 2$ can take any values due to their zero coefficients.

Conceptually the idea of this visualization is related to the idea of the scatter plot matrix that represents a set of n-D data in their projections to all pairs of coordinates (X_i, X_j). PC present projections only for adjacent pairs of coordinates, while it is sufficient for restoration of the n-D line it does not show the n-D line itself. This situation is also similar to showing 2-D projections of a 3-D object instead of that 3-D object.

To represent several parallel lines $\{L_k: x_{i+1} = mx_i + b_k\}$ for a given pair (X_i, X_{i+1}) in PC a set of points L_k^ is used that are located on the same vertical line (Inselberg 2009). In CPC and SPC the parallel lines are represented by a set of H_i^ points also located on the same vertical line. Alternatively, CPC and SPC have a classical and familiar visual representation of one line shifted above or below another one. As with a single line L, an attempt in PC to draw all points of two parallel lines L_{k1} and L_{k2} will end up with a completely covered "black" segment between coordinates X_1 and X_2. The generalization of this visualization for n-D line L^n: $A + t\mathbf{v}$ in PC requires to show sets of L^ points for all consecutive pairs (X_i, X_{i+1}) (Inselberg 2009).

Linear Operators. The situation with parallel n-D lines also represents another type of linear relation in n-D, where n-D line L^n_2 is a linear function of another n-D line $L^n_1 : L^n_2 = L^n_1 + u\mathbf{h}$, where $\mathbf{h}$ is a given n-D vector and u is a scalar coefficient. This is equivalent to defining a set of *linear operators* $T_u(\mathbf{x})$ in n-D space for different u. At the level of individual $\mathbf{x} = (x_1, x_2,\ldots,x_n)$, we have $L^n_2(\mathbf{x}) = T_u(\mathbf{x}) = L^n_1(\mathbf{x}) + u\mathbf{h}$, where $\mathbf{h}$ represents a shift vector of the n-D line. In the notation where $L^n_1 = A_1 + t\mathbf{v}$ and $L^n_2 = A_2 + t\mathbf{v}$ we have $L^n_2 = A_2 + t\mathbf{v} = A_1 + t\mathbf{v} + u\mathbf{h} = (A_1 + u\mathbf{h}) + t\mathbf{v}$. Thus $A_2 = (A_1 + u\mathbf{h})$, i.e., n-D point A_1 moves to direction $\mathbf{h}$ for the distance $|u\mathbf{h}|$. In other words, the lines L^n_1 and L^n_2 go from points A_1 and A_2 in the same direction given by n-D vector $\mathbf{v}$.

Under such a linear operator for given $\mathbf{x} = A + t\mathbf{v}$, t, $\mathbf{h} = (\mathrm{h}_1, \mathrm{h}_2,\ldots,\mathrm{h}_n)$ and u, a graph $\mathbf{x}^*$ for $\mathbf{x}$ in PC moves to $(a_1 + uh_1 + tv_1, a_2 + uh_2 + tv_2,\ldots,a_n + uh_n + tv_n)$. If in vector $\mathbf{v}$ all its coordinates v_i are equal, then visually in PC the graph $\mathbf{x}^*$ for $\mathbf{x}$ is shifted by that value. This is the case in Fig. 3.22b.

For CPC and SPC consider paired spaces $(X_1 \times X_2)$, $(X_3 \times X_4)$,…, $(X_{n-1} \times X_n)$. In the space $(X_1 \times X_2)$, point (a_1, a_2) is moving to the direction $\mathbf{v}_{1,2} = (v_1, v_2)$ with the parameter t: $(y_1, y_2) = (a_1, a_2) + t(v_1, v_2)$. Similarly in $(X_i \times X_{i+1})$ point (a_i, a_{i+1}) is moving to the direction $\mathbf{v}_{i,i+1}=(v_i, v_{i+1})$ with the same parameter

$$t : (y_i, y_{i+1}) = (a_i, a_{i+1}) + t(v_i, v_{i+1}).$$

When all of these spaces are collocated in CPC, an n-D point A is represented as a set of 2-D points connected by arrows forming a graph A^*. Applying $t\mathbf{v}$ to A will produce n-D point Y. The 2-D representation of this n-D point is graph Y^* obtained

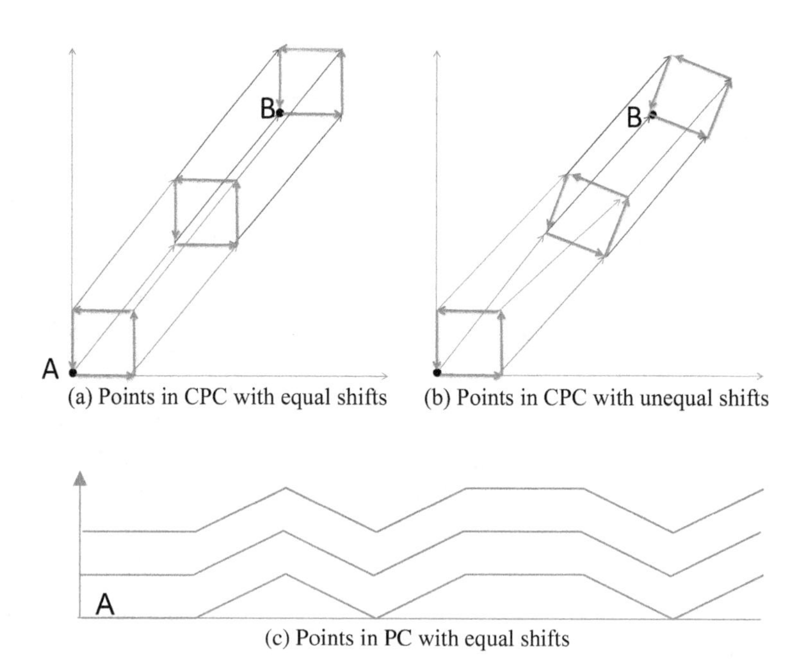

Fig. 3.22 Linearly related 8-D points in CPC (**a**), (**b**) and PC (**c**)

from graph A^* by moving each of its 2-D points (a_i, a_{i+1}) to the direction $t\mathbf{v}_{i,\ i+1}$. For different 2-D points (a_i, a_{i+1}) these directions are different.

A linear segment in n-D is a set of points $\{W\colon W = A + t\mathbf{v},\ t \in [\mathrm{a, b}]\}$.

Statement 3.16 An n-D linear segment is represented as a set of shifted graphs in CPC.

Proof Two arbitrary n-D vectors $W_1 = A + t_1\mathbf{v}$ and $W_2 = A + t_2\mathbf{v}$ from this linear segment with $\mathbf{v} = (v_1, v_2, \ldots, v_n)$ are shifted from n-D point A in the $\mathbf{v}$ direction for different distances of these shifts in accordance with t_1 and t_2, respectively. These shifts are translated to shifts of nodes of graphs W_1^* and W_2^*. For any W_1 and W_2 their first 2-D points (w_{11}, w_{12}) and (w_{21}, w_{22}) are shifted in the directions of (v_1, v_2) on 2-D relative to the first 2-D point of A which is (a_1, a_2). Similarly, the second points are shifted in the direction (v_3, v_4). Thus, directions of the linear shifts can differ for different points/nodes of the same graph. ∎

In Fig. 3.22, the 8-D structure consists of three 8-D points. In Fig. 3.22a, c CPC the first 8-D point is $A = (0, 0, 1, 0, 1, 1, 0, 1)$. It forms a square of four 2-D points in 2-D in Fig. 3.22a when the first and last nodes are connected. The 8-D linear transform $\mathbf{v}$ creates four 2-D vectors (v_1, v_2), (v_3, v_4), (v_5, v_6), (v_7, v_8) that transform these four 2-D points in different directions depending on the values of these vectors. These vectors are shown in red in Figs. 3.22a when all four 2-D transformation vectors are the same. A more complex situation when transform vectors have different norms but the same direction is shown in Fig. 3.22b. It produces rotation in 2-D. In such cases, rescaling of coordinates in CPC and SPC allows making all 2-D pairs (v_1, v_2), (v_3, v_4), (v_5, v_6), (v_7, v_8) equal resulting in figures like Fig. 3.22a.

The comparison of Fig. 3.22a, c shows that in PC $\mathbf{v}$ also captures the n-D structure in 2-D, but again uses significantly more lines.

Analytical discovery and visual insight. Consider 8-D linear segment from A to B,

$$[\mathrm{A}, \mathrm{B}] = \{\mathbf{x}_t : \mathbf{x}_t = A + t\mathbf{v}, t \in \mathrm{Z}\}$$

with $A = (0, 0, 1, 0, 1, 1, 0, 1)$ from, $\mathbf{v} = (1, 1, 1, 1, 1, 1, 1, 1)$ visualized partially in Fig. 3.22a.

While the square structure is a visible structure in this figure it is not obvious how it can be discovered analytically. The analytical discovering of relations is done by searching in an assumed class of relations. It is difficult to guess this class. Even when the class is guessed correctly, but the class is very large; the search may not be feasible computationally. The visual insight helps to identify and to narrow this class. Figures 3.23 and 3.24 show the more complex n-D linear structures visualized in CPC and in PC. These 8-D structures consist of three 8-D points.

Figure 3.23a shows the case with opposing 2-D transform vectors directed outside of the square. This leads to expanding the base square similar to zooming out. In the case of Fig. 3.23a these transform vectors are equal; in the case of Fig. 3.23b they have different norms.

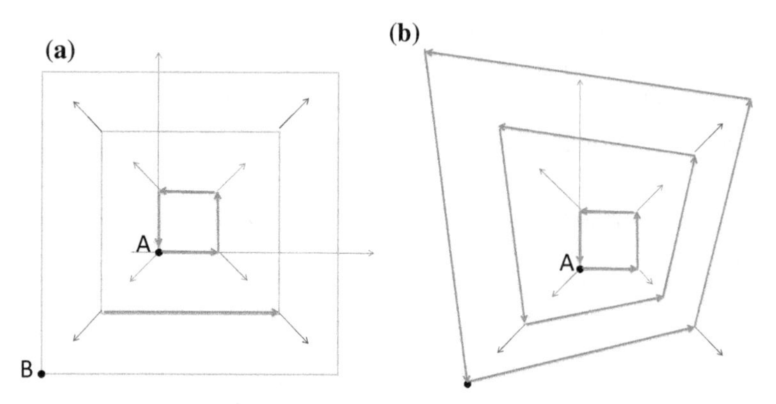

Fig. 3.23 Linearly related 8-D points in CPC with opposing shifts outside of equal value (**a**), and opposing shifts with unequal values (**b**)

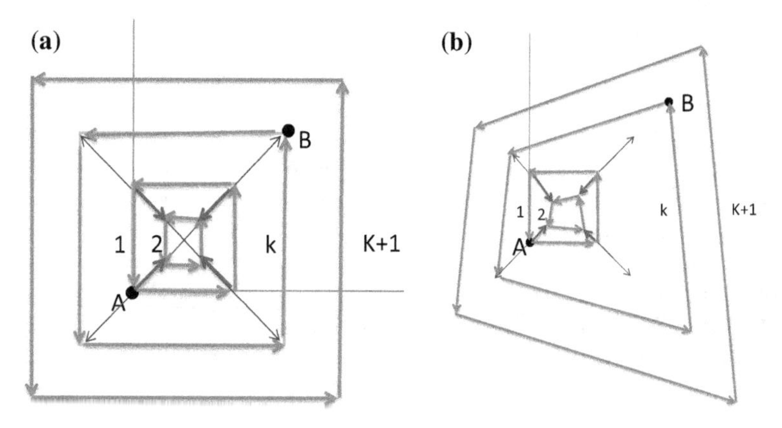

Fig. 3.24 Linearly related 8-D points in CPC with opposing shifts inside of equal value (**a**), and opposing shifts with unequal values (**b**)

An interesting case represents the situation when opposing 2-D transform vectors are directed inside of the square. At first, it decreases the square and then increases it ("swapping", "turning out" 2-D points).

See trajectory of point A to point B in Fig. 3.24a. In the case of Fig. 3.24a, the transform vectors are equal; in the case of Fig. 3.24b they have different norms.

n-D tubes. Consider data of two classes that satisfy two different linear relations: $W = A + t\mathbf{v}$ and $U = B + t\mathbf{q}$. These data are represented in CPC as two sets of graphs shifted in **v** and **q** directions, respectively. If $W = A + t\mathbf{v} + \mathbf{e}$, where **e** is a noise vector, then we have graphs for n-D points W in the "tube" with its width defined by **e**. Later in Sect. 5.2 (Chap. 5), we compare PC and CPC visualizations of n-D *linear relations corrupted by noise* given by vector **e**: $L^n_2 = A_2 + t\mathbf{v} + \mathbf{e}$ for **v** with all equal coordinates v_i. In Chap. 6, tubes with noise are explored in detail.

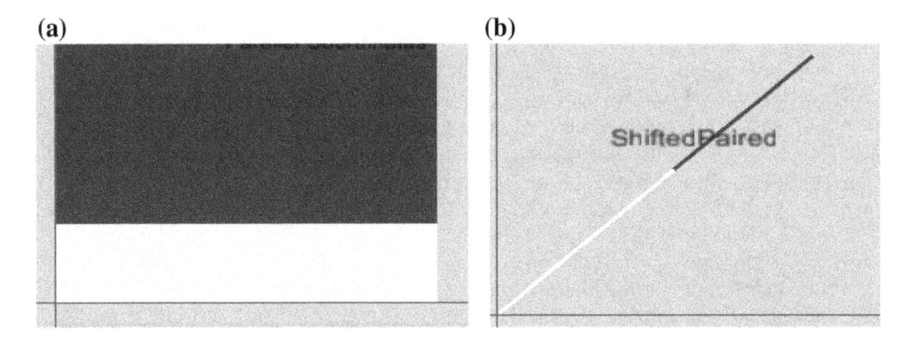

Fig. 3.25 8-D linear segment with 100 8-D points in Parallel Coordinates (**a**) and Shifted Paired Coordinates (**b**). Red box in (**a**) and red line in (**b**) show the same 100 8-D points

Scalability and occlusion of linear dependencies. There is an advantage of CPC and SPC over PC for representing some n-D linear structures. Consider 100 linearly dependent 8-D points $\mathbf{w}_i = A + t_i\mathbf{v}$, where $A = (5, 5, 5, 5, 5, 5, 5, 5)$, $\mathbf{v} = (1, 1, 1, 1, 1, 1, 1, 1)$, and $t_{i+1} = t_i + 0.001$, $t_1 = 0$, $i = 1{:}100$. In PC each of them is a horizontal line. Figure 3.25a shows these 100 lines in PC. They fill the red area forming a complete box without any individual PC line being visible. Also, no room left to show any other n-D data. This is the case of the complete occlusion in PC. In contrast, CPC and SPC provide more meaningful visualizations.

Figure 3.25b shows the same 100 8-D points in SPC forming a red line that corresponds to the traditional visualization of the linear dependence. CPC forms a similar red line, where each of these 8-D points is a single red dot. This a significant scalability issue for PC. In contrast, in this example CPC and SPC left plenty of room for other data to display without overlap and occlusion.

Norms. In CPC we can compute norms $|\mathbf{v}_{i,\,i+1}|$ of all $\mathbf{v}_{i,\,i+1}$ vectors and find all vectors with the max of these norms. If only a single vector has the max norm value then this is a prevailing direction of the 2-D shape transformation that represents the n-D linear map/operator $t\mathbf{v}$. This transformation can also be described by a multi-linear map of n variables (n-linear map) or more generally by a tensor.

Above we considered n-D linear relation where the output is an n-D point not a scalar. Chapter 7 is devoted to a way to visualize an n-D linear function $F(\mathbf{x}) = y$ where y is a scalar,

$$y = c_1x_1 + c_2x_2 + c_3x_3 + \cdots + c_nx_n + c_{n+1}$$

for machine learning tasks.

References

Bellman, R.: Dynamic Programming. Princeton University Press, New Jersey (1957)
Curse of Dimensionality, Encyclopedia of Machine Learning, pp. 257–258, Springer, Berlin (2010)

Dasgupta, S., Gupta, A.: An elementary proof of a theorem of Johnson and Lindenstrauss. Random Struct. Algorithms **22**(1), 60–65 (2003). https://doi.org/10.1002/rsa.10073

Duch, W., Adamczak R., Grąbczewski K., Grudziński K., Jankowski N., Naud A.: Extraction of Knowledge from Data Using Computational Intelligence methods, Copernicus University, Toruń, Poland (2000). https://www.fizyka.umk.pl/~duch/ref/kdd-tut/Antoine/mds.htm

Frankl, P., Maehara, H.: The Johnson-Lindenstrauss lemma and the sphericity of some graphs. J Comb. Theor. Ser. B **44**(3), 355–362 (1988)

Inselberg, A.: Parallel Coordinates: Visual Multidimensional Geometry and Its Applications, Springer, Berlin (2009)

Johnson, W., Lindenstrauss, J.: Extensions of Lipschitz mappings into a Hilbert space. In Beals, et al. (eds.) Conference in Modern Analysis and Probability (New Haven, Conn., 1982). Contemporary Mathematics. 26. Providence, RI: AMS, pp. 189–206 (1984). https://doi.org/10.1090/conm/026/737400

Kandogan, E.: Star coordinates: a multi-dimensional visualization technique with uniform treatment of dimensions. In: Proceedings of the IEEE Information Visualization Symposium 2000, vol. 650, p. 22 (2000)

Kandogan, E.: Visualizing multi-dimensional clusters, trends, and outliers using star coordinates. In: Proceedings of the Seventh ACM SIGKDD International Conference on Knowledge Discovery and Data Mining 2001 Aug 26, pp. 107–116. ACM (2001)

Pedregosa, et al.: Scikit-learn: Machine Learning in Python, JMLR 12, pp. 2825–2830 (2011). http://scikit-learn.org/stable/modules/random_projection.html

Chapter 4
Adjustable GLCs for Decreasing Occlusion and Pattern Simplification

Simplicity is the ultimate sophistication.

Leonardo da Vinci

Occlusion is one of the major problems for visualization methods in finding the patterns in the n-D data. This chapter describes the methods for decreasing the occlusion, and pattern simplification in different General Line Coordinates by adjusting GLCs to the given data via shifting, relocating, and scaling coordinates. In contrast, in Parallel and Radial Coordinates such adjustments of parameters are more limited. Below these adjustment transformations are applied to the Radial, Parallel, Shifted Paired, Circular and n-Gon Coordinates. Cognitive load can be significantly decreased, when a more complex visualization of the same data is simplified.

4.1 Decreasing Occlusion by Shifting and Disconnecting Radial Coordinates

In *Radial Coordinates,* the different n-D data points *occlude* each other, when their values are close to the common coordinate origin, because that area is small. Figure 4.1 illustrates this occlusion, where it is impossible to see the full difference between the three 8-D points shown as red, green, and blue lines.

The **Unconnected Radial Coordinates (URC)** shown in Fig. 4.2 resolve this occlusion issue by starting all coordinates at the edge of the circle instead of at the common origin, i.e., by *shifting* all of the coordinates to that edge. Thus, more freedom, in locating coordinates, shows its benefits in the decrease of the occlusion in Radial Coordinates.

The same origin-based occlusion takes place in the Cartesian Coordinates, Collocated Cartesian, and Collocated Star Coordinates, because all of them have a common origin of all coordinates. The *way to resolve* this origin-base occlusion is the same as for the Radial Coordinates—shifting the coordinates from the common

B. Kovalerchuk, *Visual Knowledge Discovery and Machine Learning*,
Intelligent Systems Reference Library 144,
https://doi.org/10.1007/978-3-319-73040-0_4

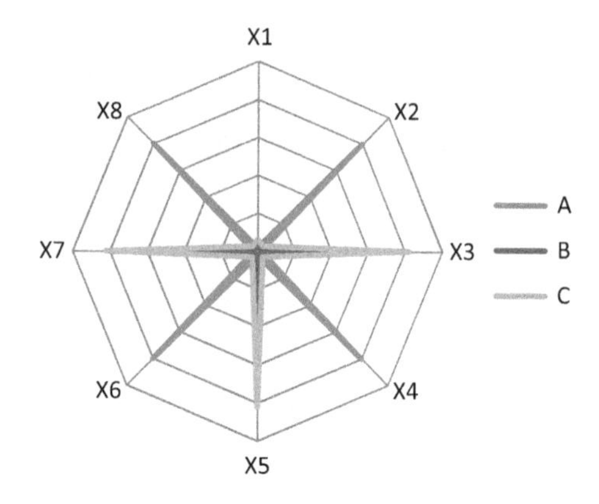

Fig. 4.1 Traditional Radial Coordinates: values of 8-D point A occluded by 8-D points B and C, near the origin, due to the connectedness of coordinates at the origin

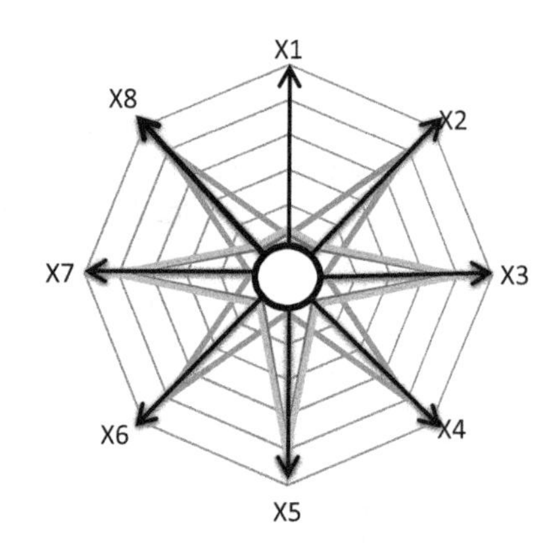

Fig. 4.2 Unconnected radial coordinates: occlusion removal demonstration, values of 8-D points A, B, and C are not occluded

origin along their directions, i.e., making these coordinates *unconnected*. Parallel Coordinates are free from this occlusion problem, due to the absence of the common origin.

4.2 Simplifying Patterns by Relocating and Scaling Parallel Coordinates

4.2.1 Shifting and Tilting Parallel Coordinates

For *Parallel Coordinates* shifts of coordinates allow revealing visual patterns *faster* and make patterns s*impler* by presenting them as preattentive straight lines as we show below. It exploits a well-known property that straight lines are *preattentive features* (Few 2004; Appelbaum and Norcia 2009).

Below we provide a summary of the experimental electroencephalogram (EEG) study in (Appelbaum and Norcia 2009), on straight line preattentive perception. They measured the time to distinguish the vertical and the horizontal lines, in the central region, from the lines in the background (see Fig. 4.3).

Each straight line is represented as a one-dimensional random luminance bar, with a minimum bar width of 6 arc min. In the experiment, the frequency of changes of lines in the center differs from the frequency of change of the background (3 Hz vs. 3.6 Hz). This difference allowed the measurement of time for processing straight lines in the center.

Participants were asked to detect the subtle changes in a set of straight lines (Fig. 4.3b) with a button press. In 20% of the 1.67-second stimulus cycles, the aspect ratio (horizontal to vertical) became elliptical vs. circular at the beginning, as Fig. 4.3b shows. The aspect ratio was monitored and adjusted to maintain the performance at approximately 80% of correct detection.

These authors assessed the amplitude and timing of brain responses in EEG for the sets of straight lines in the center versus the background processing. They concluded: the separation of the straight lines from the background proceeds *pre-attentively,* based on the statistical analysis of the electroencephalogram collected with a whole-head 128-channel Geodesic EEG system.

This result covers many straight lines. In our examples in this chapter, just a few straight lines are used, i.e., much simpler cases, which respectively also should be preattentive, and can be generalized for more lines.

Consider the two 6-D data points $A = (0.3, 0.6, 0.4, 0.8, 0.2, 0.9)$ in blue and $B = (0.35, 0.68, 0.48, 0.85, 0.28, 0.98)$ in orange in Fig. 4.4. Figure 4.4a shows A and B, in the standard Parallel Coordinates, as non-preattentive zig-zag lines.

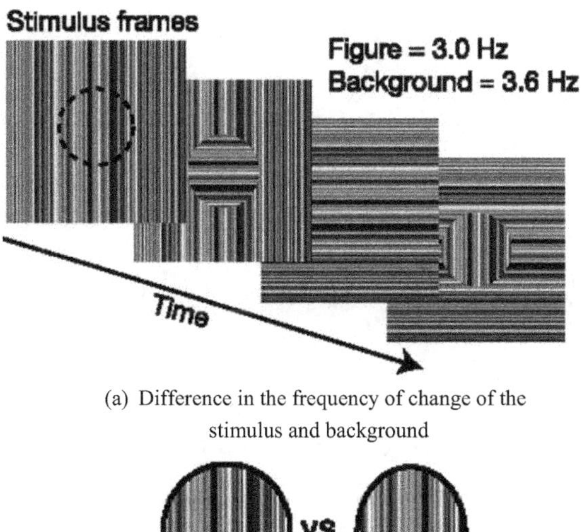

(a) Difference in the frequency of change of the stimulus and background

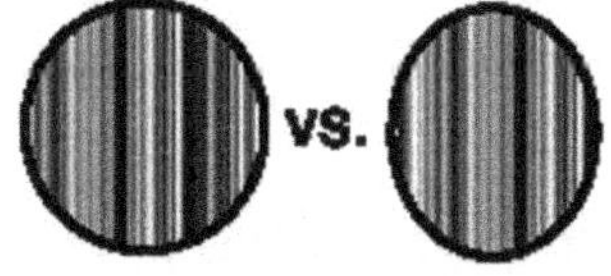

(b) Shape discrimination

Fig. 4.3 Preattentive aspects of the straight horizontal and vertical line processing (Appelbaum and Norcia 2009)

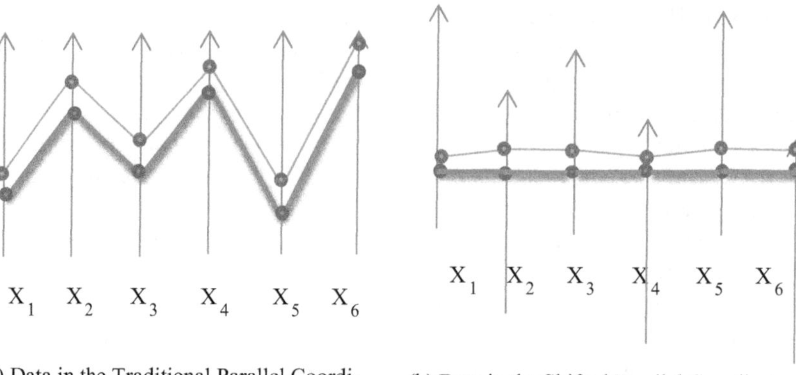

(a) Data in the Traditional Parallel Coordinates – non-preattentive representation.

(b) Data in the Shifted Parallel Coordinates - pre- attentive representation.

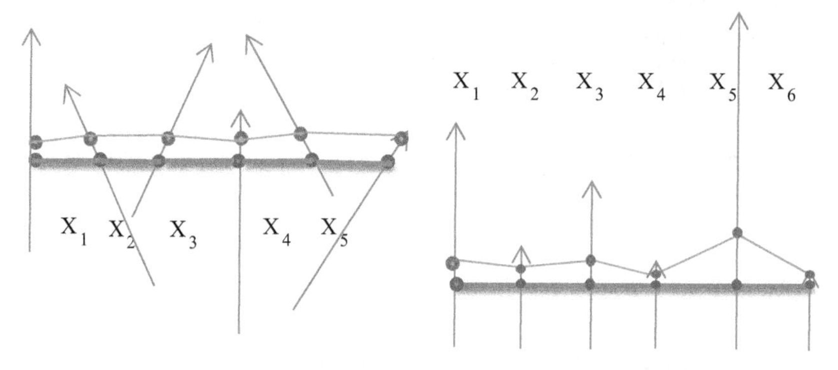

(c) Data in the Shifted General Line Coordinates- preattentive representation.

(d) Data in the scaled Parallel Coordinates – pre- attentive representation.

Fig. 4.4 Non-preattentive versus preattentive visual representations (linearized patterns): the 6-D point $A = (3, 6, 4, 8, 2, 9)$ in blue, and the 6-D point $B = (3.5, 6.8, 4.8, 8.5, 2.8, 9.8)$ in orange in the traditional and the shifted parallel coordinates and GLCs

In contrast, in Fig. 4.4b, c *A* is a preattentive straight line, and *B* is much simpler than in Fig. 4.4a. The lines in Fig. 4.4b, c can be compared and correlated easier. This simplification was achieved by changing the Parallel Coordinates to the **Shifted Coordinates**.

4.2.2 *Shifting and Reordering of Parallel Coordinates*

The Shifted Parallel Coordinates is a visual way to implement the idea of designing a *complex non-linear transform* of the n-D dataspace into another space, where a linear discriminant function, or a hyper-plane, can be built for the n-D data classification. This linearization idea is behind the algorithm based on **rescaling** (Vityaev and Kovalerchuk 2005).

Shifting coordinates, in the Parallel Coordinates, to get a linear representation is similar to applying the *Rescaled Parallel Coordinates* (Theus 2008). Rescaling visually can be done without shifting, but by contracting or expanding the coordinates, as it is shown in Fig. 4.4d. Note that rescaling may change the perception, because the resolution of some coordinates can decrease.

Rescaling in the Parallel Coordinates can be done with minimal visual changes in the coordinates, by changing the number of pixels used for the units in a coordinate. Shrinking the coordinate line of the given length leads to the shrinking of the units of that coordinate (decreasing the number of pixels devoted to the scale unit).

The same decreased number of pixels can be implemented with keeping the length of the coordinate. In this case, the range of the values that the coordinate line carries will be larger. This would require redrawing the dividers of units on the coordinate making them denser.

Figure 4.5 illustrates the simplification of the visual representation of n-D data of the two classes by *shifting and reordering* of the Parallel Coordinates. In Fig. 4.5b, some coordinates are moved up, and some of the others are moved down. The order of the coordinates is also changed. As a result, the first class (red) became preattentive, being represented by horizontal straight lines. In Fig. 4.5b, the second class (green) became simpler too. It is now a set of monotone increasing lines, which is recognizable easier and faster, than the zig-zag lines in Fig. 4.5a.

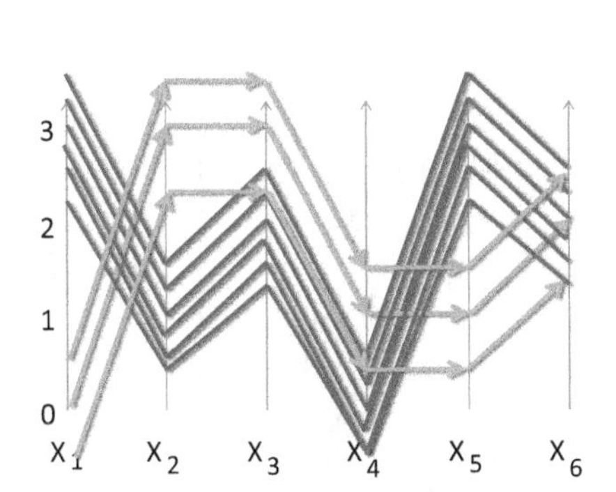

(a) Original visual representation of the two classes in the Parallel Coordinates.

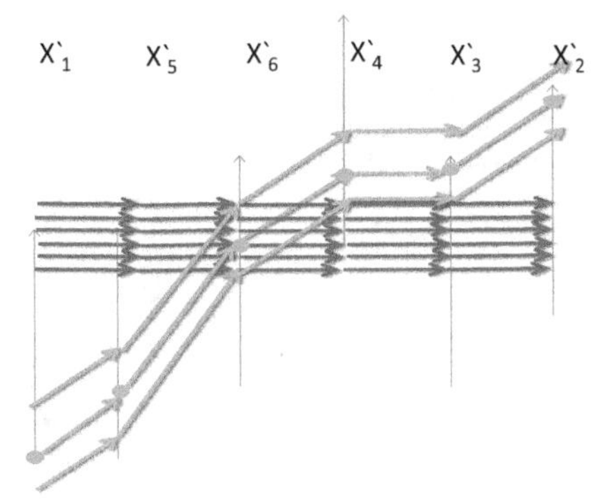

(b) Simplified visual representation after the shifting and reordering of the Parallel Coordinates.

Fig. 4.5 Simplification of the visual representation by the shifting and reordering of the parallel coordinates

4.3 Simplifying Patterns and Decreasing Occlusion by Relocating, Reordering, and Negating Shifted Paired Coordinates

4.3.1 *Negating Shifted Paired Coordinates for Removing Crossings*

Figures 4.6, 4.7, 4.8, 4.9, 4.10, 4.11, 4.12 and 4.13 illustrate the simplification process, for the Shifted Paired Coordinates, in comparison with the Parallel Coordinates. All of these figures show the same four 6-D points: $A = (0.3, 0.6, 0.4, 0.8, 0.2, 0.9)$ (thick blue line), and $C = (0.8, 0.9, 0.4, 0.5, 0.2, 0.3)$ (thick green line), and the two other 6-D points (thin lines) of the blue and green classes.

Points A and C can be viewed as the representative points (centers) of these classes. We start from the representation of these data in the Parallel Coordinates in Fig. 4.6. This representation is quite complex, with 6 points and 5 zigzag lines for every 6-D point, without a clear visual separation between the points of the two classes. In contrast, the representation in SPCs in Fig. 4.7 is simpler with the 3 points, and the two lines for each 6-D point. However the lines of the two classes cross each other in (X_1, X_2), (X_3, X_4) spaces. A simpler pattern would be the one, where the lines do not cross in these spaces. This is implemented, in Fig. 4.8, by substituting the coordinate X_4 to $1-X_4$. While it removes one, it creates another crossing in $(X_3, 1-X_4)$, and (X_5, X_6) spaces. This crossing is resolved, in Fig. 4.9, by substituting the coordinate X_6 by $1-X_6$. Now, the cases of the two classes are separated, but still not preattentive.

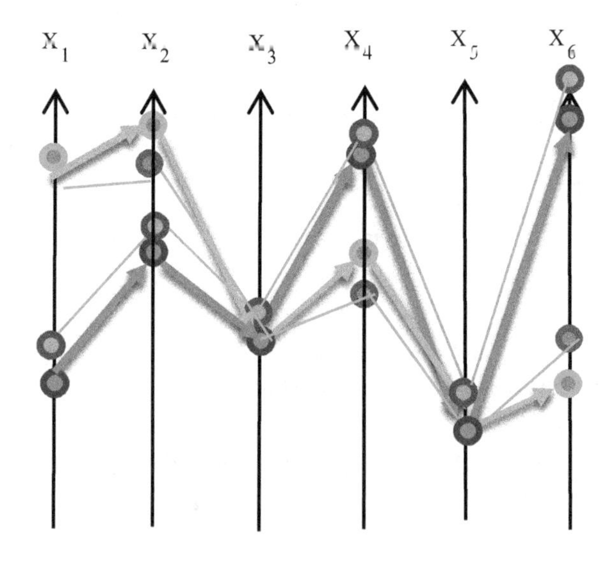

Fig. 4.6 Four 6-D points of the two classes in the parallel coordinates

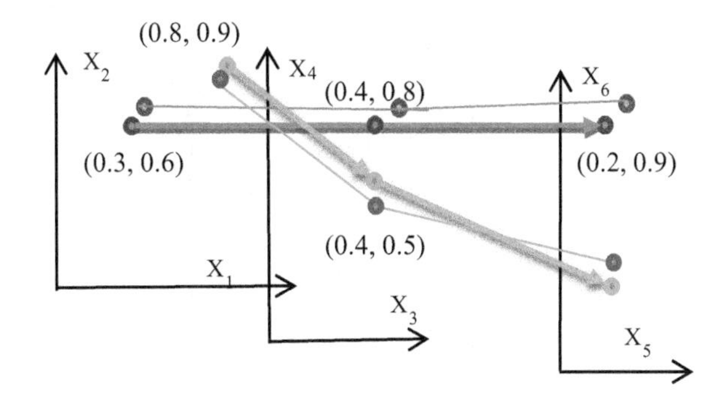

Fig. 4.7 Four 6-D points of the two classes in SPCs with the crossing lines

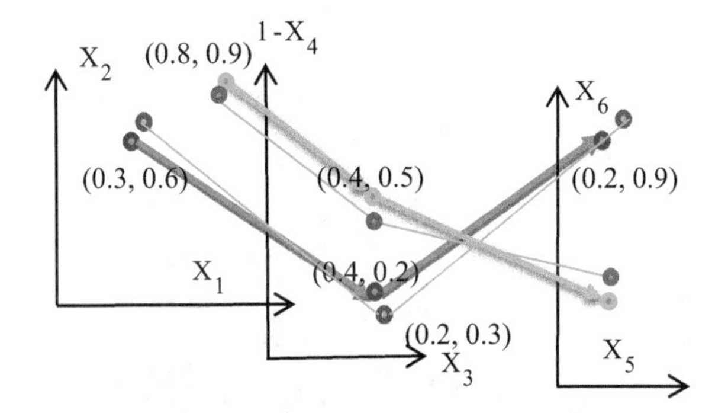

Fig. 4.8 Four 6-D points of the two classes in SPCs with the crossing lines with the coordinate X_4 changed to $1-X_4$, which removes the crossing of lines caused by X_4, where: A' = (0.3, 0.6, 0.4, **0.2**, 0.2, 0.9) (thick blue line), and C' = (0.8, 0.9, 0.4, **0.5**, 0.2, 0.3) (thick green line)

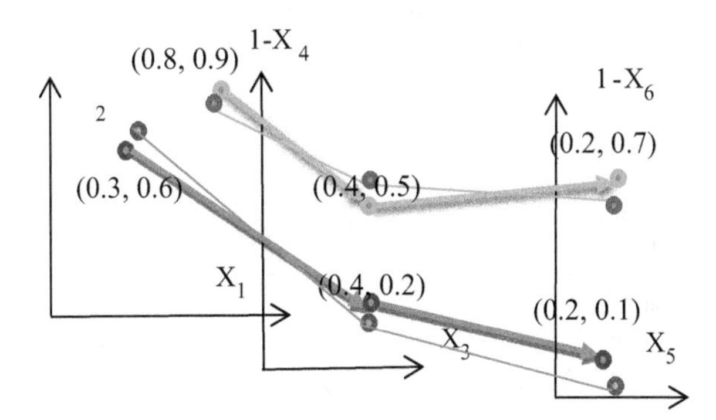

Fig. 4.9 Four 6-D points of the two classes in SPCs with the crossing of the lines caused by X_6 eliminated by changing the coordinate X_6 to $1-X_6$ with A'' = (0.3, 0.6, 0.4, **0.2**, 0.2, **0.1**) (thick blue line), and C'' = (0.8, 0.9, 0.4, **0.5**, 0.2, **0.7**) (thick green line)

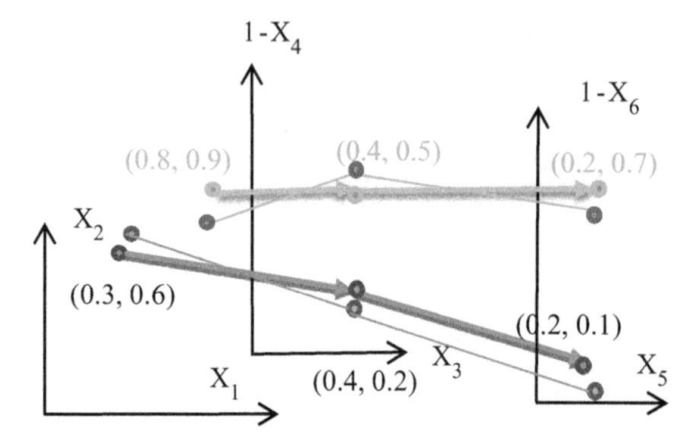

Fig. 4.10 Four 6-D points of the two classes in the shifted paired coordinates, without the crossing lines and with the monotone blue lines, and the horizontal thick green line obtained by shifting the pairs of coordinates (X_1, X_2) and (X_5, $1-X_6$) down, to make the thick green line C horizontal

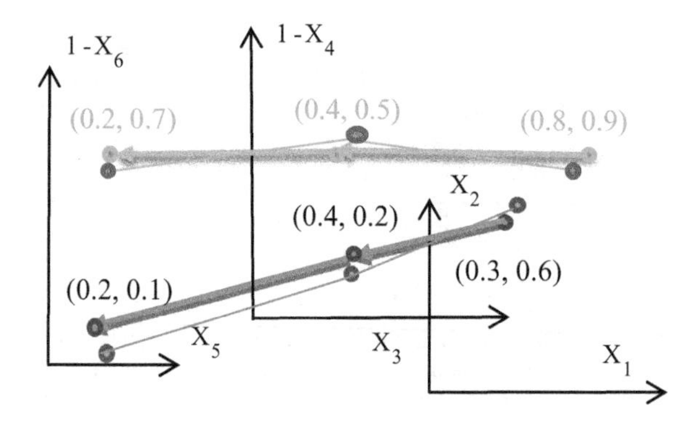

Fig. 4.11 Four 6-D points of two classes in Shifted Paired Coordinates, without the crossing lines and with the horizontal thick green line, and with a straight monotone thick blue line, obtained by shifting the pairs of coordinates (X_1, X_2) and (X_5, $1-X_6$) horizontally, to make the thick blue line a straight line

Fig. 4.12 Four 6-D points of the two classes in the Shifted Paired Coordinates without the crossing lines, and with a straight monotone thick blue line, and a horizontal thick green line, obtained by shifting a pair of coordinates (X_1, X_2) to the right, to collapse *C* into a single arrow

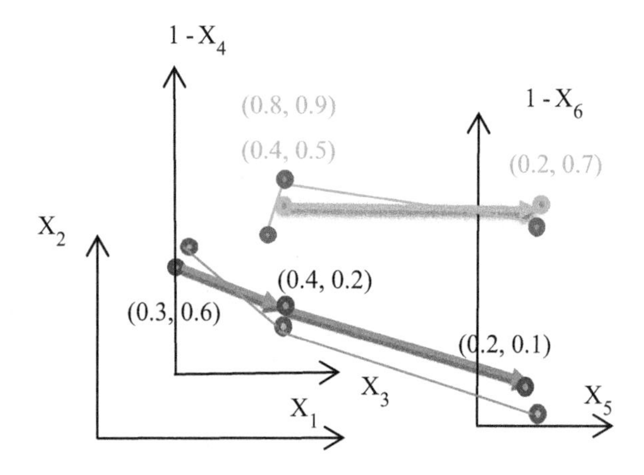

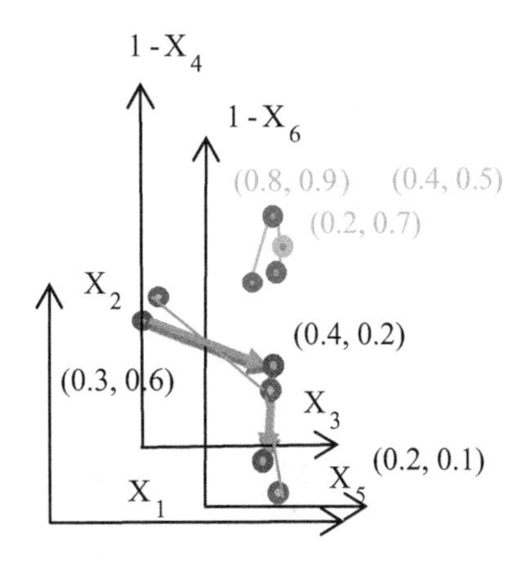

Fig. 4.13 Four 6-D points of the two classes in the shifted paired coordinates without the crossing lines, obtained by shifting pair of coordinates (X_5, $1{-}X_6$) to the left, to collapse the thick green line C into a single point

4.3.2 Relocating Shifted Paired Coordinates for Making the Straight Horizontal Lines

In Fig. 4.10, the green line *C* became a preattentive horizontal straight line by shifting the coordinates (X_1, X_2) and (X_5, $1{-}X_6$) down. Here the blue line is still not preattentive. In Fig. 4.11, it is made more preattentive by shifting pairs of the coordinates (X_1, X_2) and (X_5, $1{-}X_6$) horizontally, to make the thick blue line a straight line.

4.3.3 Relocating Shifted Paired Coordinates for Making a Single 2-D Point

Figure 4.12 shows the next step of simplification, where the green line *C* is collapsed into a *single arrow* between the two points. The first point is obtained by shifting the pair of coordinates (X_1, X_2) to the right, in such way that (x_1, x_2) = (0.2, 0.7) will be at the same position, where (x_3, x_4) = (0.4, 0.5) is located in (X_3, X_4) coordinates.

The final simplification step is presented in Fig. 4.23, where the green line *C* is collapsed into a single preattentive 2-D *point*, shown as the green dot, by shifting a pair of coordinates (X_5, $1{-}X_6$) to the left. This representation remains reversible, i.e., all values of these four 6-D points can be restored, from these graphs.

Figure 4.14 shows another numeric example of the preattentive lossless representation in the Shifted Paired Coordinates. A straight horizontal line that represents the 6-D point *A* = (3, 6, 4, 8, 2, 9) in blue in Fig. 4.14 is preattentive. It also shows the point *B* = (3.5, 6.8, 4.8, 8.5, 2.8, 9.8) in orange from the same class. While *B* is not preattentive, its representation, and comparison with *A*, is simplified, when *A* was made preattentive.

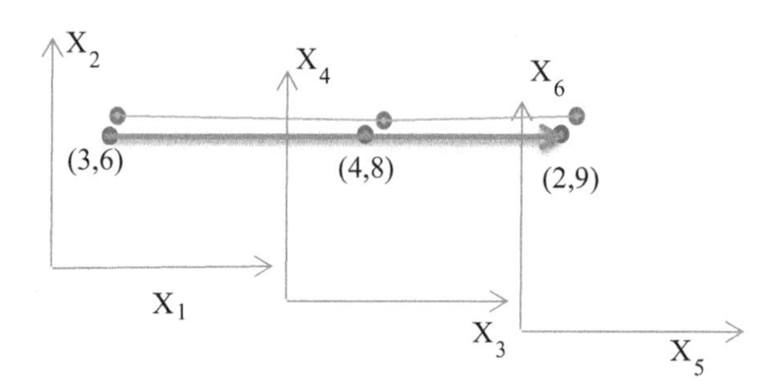

Fig. 4.14 Preattentive lossless visualization of the 6-D point A = (3, 6, 4, 8, 2, 9) in the blue, as a preattentive horizontal straight line, and a simplified visualization of the 6-D point B = (3.5, 6.8, 4.8, 8.5, 2.8, 9.8) in orange in the SPCs

The paired lossless representation of the 6-D point A, in Fig. 4.14, not only is preattentive, but is also two times simpler, than in the Parallel Coordinates having only the three 2-D points vs. the six 2-D points, in Parallel Coordinates, in Fig. 4.4a. Figure 4.15 represents the same 6-D point A, as a single 2-D point, losslessly and preattentively by shifting the coordinates (X_1, X_2) and (X_5, X_6), in the same way, as in Fig. 4.13.

4.4 Simplifying Patterns by Relocating and Scaling Circular and n-Gon Coordinates

Figure 4.16 shows the ways to simplify the visualization of other GLCs using the shifting and rescaling of coordinates. In Fig. 4.16, this approach is applied to the traditional Radial Coordinates, and the new circular and n-Gon coordinates, which were defined in Chap. 2, without these simplifications. For the comparison, the same data are shown in Fig. 4.16a in the Parallel Coordinates, and in Fig. 4.17a in the Radial Coordinates.

In the Circular Coordinates (Fig. 4.16b), the circle is divided into the segments, and each segment encodes a coordinate (e.g., in a normalized scale within [0, 1]), where each x_i of an n-D point **x** is located in the respective coordinate X_i. For instance, x_1 = 0.3, in the Fig. 4.16b, is located at the respective distance from the origin of X_1, along the X_i segment of the circle.

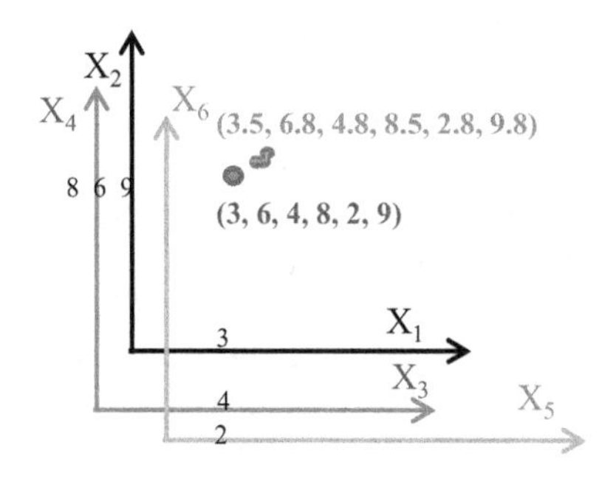

Fig. 4.15 Preattentive lossless visualization of the 6-D point in blue, as a preattentive single 2-D point, and the simplified visualization of the 6-D point B in orange in the SPCs

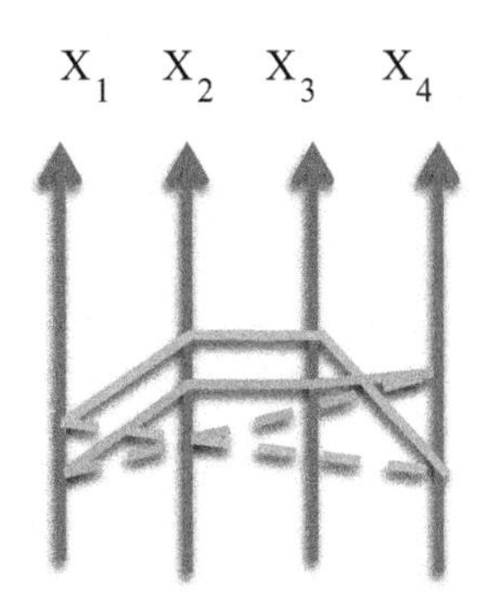

(a) Two 4-D points in Parallel Coordinates

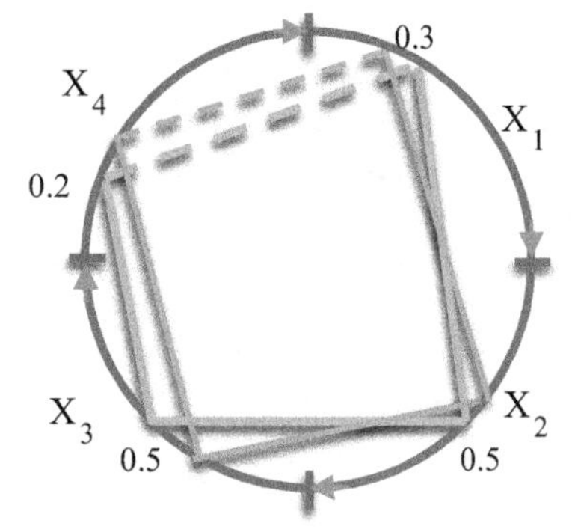

(b) Two 4-D points in Circular Coordinates

Fig. 4.16 4-D points $A = (0.3, 0.5, 0.5, 0.2)$ in green, and $B = (0.2, 0.45, 0.4, 0.4)$ in orange, in parallel and circular coordinates, fully connected

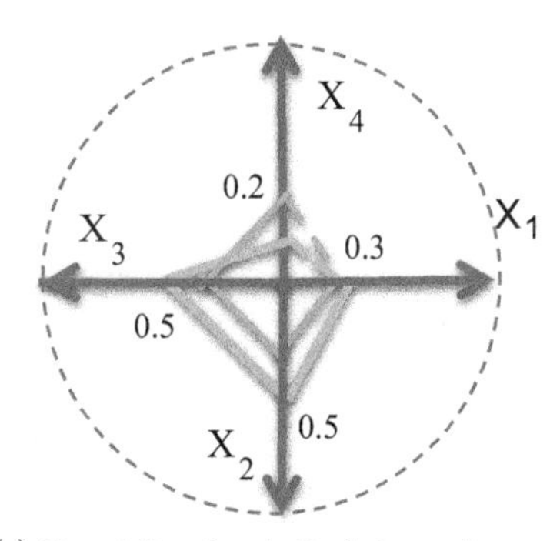

(a) Two 4-D points in Radial coordinates

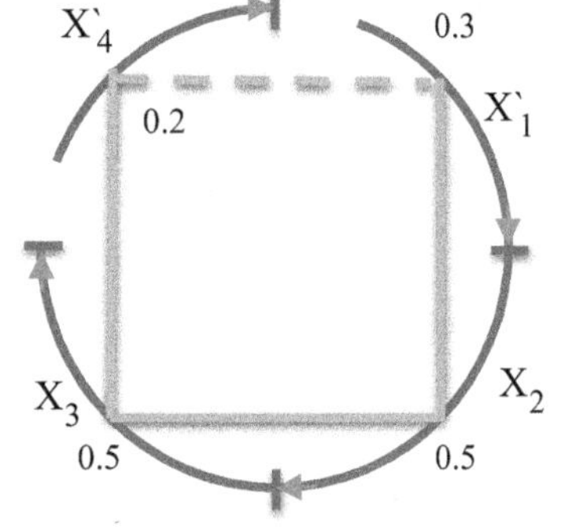

(b) 4-D point A in Contracted Circular Coordinates

Fig. 4.17 Simplification of visualization by relocating and rescaling of the coordinates for 4-D points A in green and B in orange displayed in the radial and circular coordinates partially connected

Next, these points x_i are connected, to form the directed graph, starting from x_1. Connecting x_1 and x_n leads to a closed contour (see the dotted lines in Fig. 4.17b). It is advantageous perceptually, which we will illustrate later in detail. Closed contours are colored to distinguish the n-D points, and their classes. Similarly, the n-Gon (triangle, rectangle, pentagon, and so on) is divided into the segments, and each segment encodes a coordinate.

These points are connected to form a graph (see Fig. 4.19a). Figures 4.16a, b, 4.17a and 4.19a shows that the Circular Coordinates, and the n-Gon coordinates, have the same complexity. They require the same number of lines, as the known Parallel and Radial Coordinates, for the same 4-D point. Therefore, these new coordinates have a potential to be successful, in the same types of applications, where the Parallel and Radial Coordinates have been successful.

The idea of simplification is transforming the original 2-D graph, from the irregular shape, to a *familiar regular symmetric shape* such as a square, pentagon, hexagon, and so on. Figures 4.17b and 4.18a, b show the results of such a transformation, of an irregular rectangle into a square, for the 4-D point *A* in Circular Coordinates; and the Figs. 4.19a, b and 4.20 show this in the n-Gon (Square) Coordinates and Radial Coordinates. All of these transformations are linear. Coordinates shrink under these transformations; respective segments of the circle shrink into subsegments, shown in Figs. 4.17b and 4.18a. In Fig. 4.17b, this resulted in moving $x_1 = 0.3$ down and $x_4 = 0.2$ up to form a horizontal line at the position of 0.5. Moving both x_1 and x_4 to the location of 0.5, on the respective coordinates, ensures getting a square because x_2 and x_3 are already in those 0.5 positions.

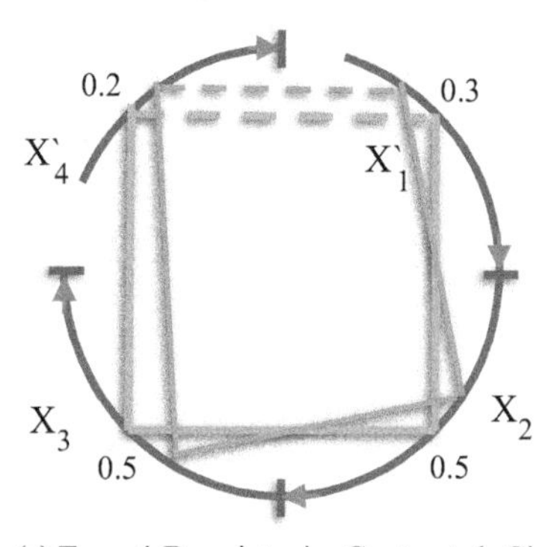

(a) Two 4-D points in Contracted Circular Coordinates.

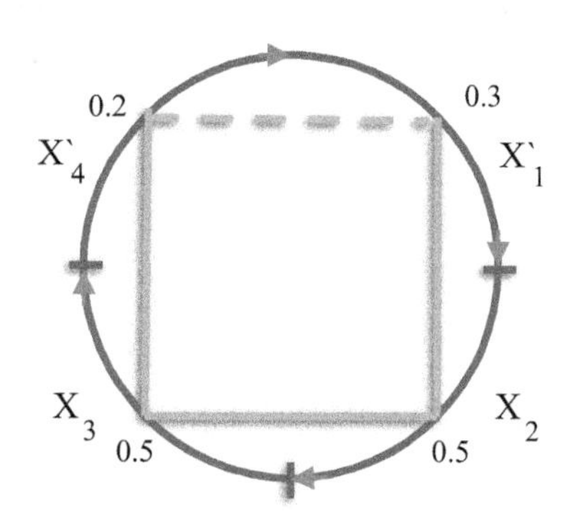

(b) 4-D point A in Disproportional Circular Coordinates.

Fig. 4.18 Simplification of visualization by relocating and rescaling of the coordinates for 4-D points *A* in green and *B* in orange displayed in the circular coordinates fully and partially connected

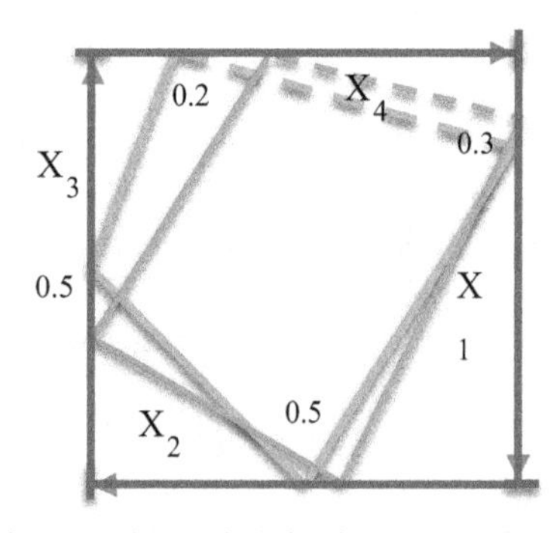

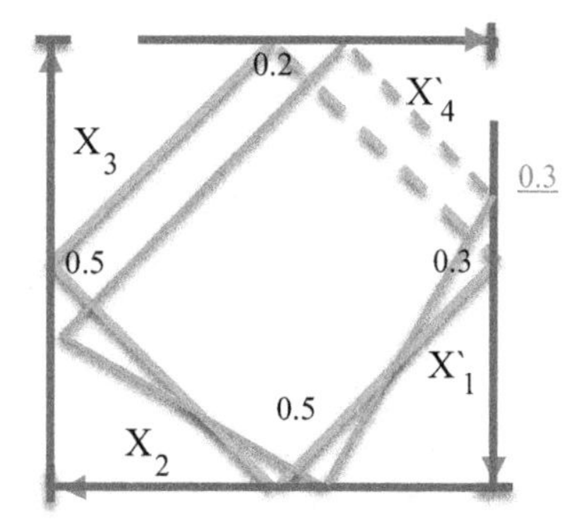

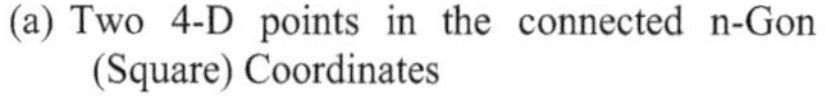

(a) Two 4-D points in the connected n-Gon (Square) Coordinates

(b) Two 4-D points in the Partially Connected n-Gon (Square) Coordinates

Fig. 4.19 Simplification of visualization by relocating and rescaling of the coordinates for 4-D points *A* in green and *B* in orange displayed in the square coordinates, fully and partially connected

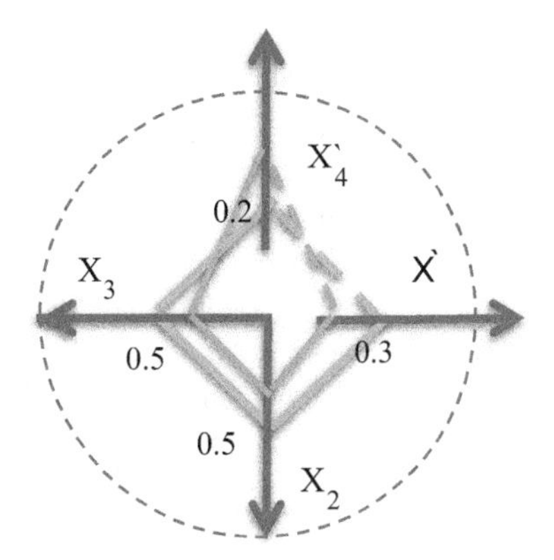

Fig. 4.20 Simplification of visualization by relocating and rescaling the coordinates for the 4-D points *A* in green and *B* in orange partially connected radial coordinates

These linear transformations map X_1 to X'_1 with $0 \rightarrow b$, $0.3 \rightarrow 0.5$, $1.0 \rightarrow 1.0$, and X_4 to X'_4 with $0 \rightarrow d$, $0.2 \rightarrow 0.5$, and $1.0 \rightarrow 1.0$. The linear equations to find these transforms are $y = ax + b$ and $y = cx + d$, where for $y = ax + b$ and X_1, we have the two pairs ($x = 0.3$, $y = 0.5$) and ($x = 1$, $y = 1$). This leads to a system of two linear equations: $0.5 = 0.3a + b$ and $1 = a + b$ with a solution: $a = 0.714$ and $b = 0.286$. Thus, the coordinate X'_1 starts at the location 0.286 on the circle segment X_1. In the polar coordinates, this location is given by a pair (R, α). Similar computations are conducted for $x_4 = 0.2$, on the coordinate X_4. In contrast in Fig. 4.18b, the shape is simplified by a non-linear monotone transform, which maps X_1 to X'_1 with $0 \rightarrow 0$, $0.3 \rightarrow 0.5$, $1.0 \rightarrow 1.0$, and maps X_4 to X'_4 with $0 \rightarrow 0$, $0.2 \rightarrow 0.5$, and $1.0 \rightarrow 1.0$. It can be modeled by a piecewise linear transformation, with two linear parts, which can be found by solving two systems of two linear equations.

The scaling approach without changing the length and location of the Parallel Coordinates was implemented for the Parallel Coordinates in (Theus 2008). Above this idea was generalized for various GLCs, allowing relocation, and resizing of the coordinates in addition to rescaling. The advantages of rescaling were shown in (Theus 2008) for n-D data, for the Tour de France 2005, with a conclusion, that after each axis is aligned at the individual medians, the display clearly reveals the most information.

In terms of the examples, in Figs. 4.16, 4.17, 4.18, 4.19 and 4.20, the role of the linearized graph of n-D point *A*, is played by a set of individual medians in each axis in Theus (2008). Theus (2008) proposed several options for selecting such an n-D point: the mean, the median, a specific case, or a specific value. We will call this n-D point: an *alignment* n-D point. He also noticed that the alignment of coordinates can be controlled without using a particular n-D point by either individually scaling the axes or by using some common scale over all axes with a general conclusion: "Parallel coordinate plots are not very useful "out of the box," i.e., without features like α-blending, and the scaling options".

The same simplification approach as described above can be applied for Circular and n-Gon coordinates for a higher number of dimensions with substituting squares to pentagons, hexagons, and other respective n-Gons. It is likely that the highest dimensions will be comparable with the dimension that is workable for the Radial Coordinates. It is illustrated, in Figs. 4.16, 4.17, 4.18, 4.19 and 4.20, where Figs. 4.17a and 4.20 shows the same two n-D points, in Radial Coordinates, and Partially Connected Radial Coordinates. Circular, n-Gon coordinates, and Radial Coordinates use the same number of 2-D points, around a common center, to represent each n-D point. The difference is in how the points are located. Therefore, easiness for people in discovering clusters, in Circular and n-Gon coordinates, should be similar to that for the Radial Coordinates. In more detail, this issue is considered in Chap. 6, for the dimensions up to $n = 192$.

4.5 Decreasing Occlusion with the Expanding and Shrinking Datasets

4.5.1 Expansion Alternatives

Above in this chapter we simplified the visualization for the four n-D points. The justification for considering the four n-D points is the assumption that these four points are *representative for the larger datasets*. The thick lines are viewed as centers of the two classes, centers of subsets of the class (clusters), or other n-D points of interest. For short, we denote these n-D points as centers $\mathbf{c}_1$ and $\mathbf{c}_2$. The thin lines represent other points of these classes, which are close to the first ones. For the four n-D points, the major issue is the simplification of the pattern, while for the larger number of n-D points, both the decreasing occlusion, and the simplification of the visual pattern, are critical for the success of visual discovery.

Below we present five expansion alternatives E1–E5, for the given vicinities of the n-D *center points* $\mathbf{c}_1$ and $\mathbf{c}_2$, to be able to visualize more data, with less occlusion. The vicinities are defined, by some parameter of clustering. Typically, it is a closeness threshold, which we denote as T. Changing the threshold T, will show the dependence, of the class separation on it.

Alternative expansions of the visualization process presented in the previous sections of this chapter are:

(E1) Keep the center point $\mathbf{c}_1$, and visualize all the n-D points from both classes C_1 and C_2, which are in the given *vicinity* of $\mathbf{c}_1$. This will visualize how *close* the two classes are, and how they overlap.

(E2) Keep the center point $\mathbf{c}_2$ and visualize all the n-D points from both classes C_1 and C_2, which are in the given vicinity of $\mathbf{c}_2$. This will visualize how *close* the two classes are, and how they overlap.

(E3) Keep the center point $\mathbf{c}_1$, and visualize all the n-D points of class C_1 in the given vicinity of that center $\mathbf{c}_1$, and all the n-D points of class 2, which are outside of this vicinity. This visualizes how *far* the n-D points of the two classes are.

(E4) Keep the center point $\mathbf{c}_2$, and visualize all the n-D points of class C_2, in the given vicinity of $\mathbf{c}_2$, and all of the n-D points of class C_1, which are outside of this vicinity. This will visualize how *far* the n-D points of the two classes.

(E5) Keep both the center points $\mathbf{c}_1$ and $\mathbf{c}_2$, and visualize all the n-D points which are in the given vicinities of $\mathbf{c}_1$ and $\mathbf{c}_2$. This will visualize how *close,* and how *far* the n-D points of the two classes are.

The motivation for these *controlled expansions,* of the process, with four points, is that throwing all data, of both classes, into the visualization will typically end up, in a messy highly occluded picture, with the useful pattern hidden. Clustering, before visualizing, is a natural approach to avoid such an output. In Machine Learning, using the clustering before the classification is quite common, even without the visualization, because it simplifies the pattern, and therefore the process of discovering it (Rokach et al. 2005; Cohen et al. 2007).

4.5.2 Rules and Classification Accuracy for Vicinity in E1

The visualizations for E1 allow the generation of a simple classification rule

$$\text{If}\,\mathbf{x} \in V(\mathbf{c}_1, T)\ \text{then}\,\mathbf{x} \in \text{Class 1},$$

i.e., if n-D point $\mathbf{x}$ is in the vicinity $V(\mathbf{c}_1, T)$ of the n-D point with the threshold T, then $\mathbf{x}$ belongs to Class 1. Also for vicinity $V(\mathbf{c}_1, T)$ of $\mathbf{c}_1$, we can get two numbers;

- the total number $N(\mathbf{c}_1, T)$ of n-D points in $V(\mathbf{c}_1, T)$, and
- the number $N(1, \mathbf{c}_1, T)$ of n-D points of class 1 in $V(\mathbf{c}_1, T)$.

This gives us the accuracy of this rule as a ratio: $R = N(1, \mathbf{c}_1, T)/N(\mathbf{c}_1, T)$.

Finding a threshold Below we describe an algorithm for finding a threshold T. Consider an n-D dataset $\mathbf{A}$ of the labeled points of several classes, e.g., of two classes, which we will denote class 1 and class 2. Let also $\mathbf{c}_1, \mathbf{c}_2 \in \mathbf{A}$ be the n-D points of interest, which belong to class 1 and class 2, respectively.

The steps of **the algorithm** that we denote as **GLC-S** are as follows:

Step 1 Selecting an n-D point of interest $\mathbf{c}_1$, from A, by a user (domain expert) or by the algorithm (as the center of some cluster, produced by a clustering algorithm).

Step 2 Checking for elements of other classes in a vicinity of the n-D point $\mathbf{c}_1$.

Step 3 Finding the largest vicinity $V_{\text{kmax}}(\mathbf{c}_1)$ of $\mathbf{c}_1$, where only the elements of class 1 are present. An efficient way for doing this is combining the analytical and visual means.

Step 4 Checking if $V_{\text{kmax}}(\mathbf{c}_1)$ is large enough against the size threshold T, $|V_{\text{kmax}}(\mathbf{c}_1)| > T$. This step is to avoid the potential overfitting.

Step 5 If $|V_{kmax}(\mathbf{c}_1)| > T$ then form a sub-rule

$$\text{If } \mathbf{x} \in V_{kmax}(c_1), \text{then } \mathbf{x} \in \text{class } 1. \tag{4.1}$$

Step 6 Expanding the vicinity V_{kmax} to a larger $V(\mathbf{c}_1, T)$, with a larger threshold T:

$$|c_{11} - x_1| < T \,\&\, |c_{12} - x_2| < T \,\&\ldots\&\, |c_{1n} - x_n| < T \tag{4.2}$$

This vicinity is a hypercube, centered in $\mathbf{c}_1$, with the sides of length $2T$. In other words, an object is contained in the vicinity $V(\mathbf{c}_1, T)$ of $\mathbf{c}_1$, if the difference between each of the dimensions is less than T, for the all coordinates of the object $\mathbf{x}$.

Step 7 Exploring $V(\mathbf{c}_1, T)$, using GLC visualizations, to find the classification patterns in it. If the patterns are not found, for a given threshold, it is interactively changed. If this step is ended, with patterns P_1 and P_2 in $V(\mathbf{c}_1, T)$ and $V(\mathbf{c}_2, T)$, for classes 1 and 2, respectively, with the acceptable errors; these P_1 and P_2 are recorded as rules along with the confusion matrix M, which shows their accuracy.

Step 8 Forming the sub-rules for $V(\mathbf{c}_1, T)$ and $V(\mathbf{c}_2, T)$:

$$\text{If } \mathbf{x} \in V(\mathbf{c}_1, T) \& P_1(\mathbf{x}) \text{ then } \mathbf{x} \in \text{class } 1$$
$$\text{If } \mathbf{x} \in V(\mathbf{c}_2, T) \& P_2(\mathbf{x}) \text{ then } \mathbf{x} \in \text{class } 2$$

Step 9 Looping steps 1–7 for the other n-D points $\mathbf{x}$, which are outside of $V(\mathbf{c}_1, T)$ and $V(\mathbf{c}_2, T)$. Points $\mathbf{x}$ are selected by the user, or by the algorithm, as the centers of other clusters. The scalability of this algorithm, to the large data sets, depends on the number of $V(\mathbf{c}_i, T)$, which needs to be explored.

4.6 Case Studies for the Expansion E1

This section illustrates the expansion E1, on several case studies. In three case studies, only the one n-D point, from class 2, is in the given vicinity $V(\mathbf{c}_1, T)$. In these cases, there is a significant vicinity of point $\mathbf{c}$, where classification is quite correct, in terms of the accuracy ratio R with the rule (4.1).

Case study 4.1 This case study uses the *glass identification data* from the UCI Machine Learning repository (Lichman 2013). These data include the 10-D instances from seven classes. Figure 4.21 shows the results for E1 for these data in CPCs, and in SPCs with the two different mutual locations of paired coordinates denoted as SPC1 (Fig. 4.21b), and SPC2 (Fig. 4.21c). Here $V(\mathbf{c}_1, T)$ contains only a single blue line, from class 2. The rule (4.1) is quite accurate, on these data, because only the one n-D point is misclassified. We denote its graph as L_4, in Fig. 4.21a. This line differs from the lines of class 1, in all three visualizations. Therefore, we

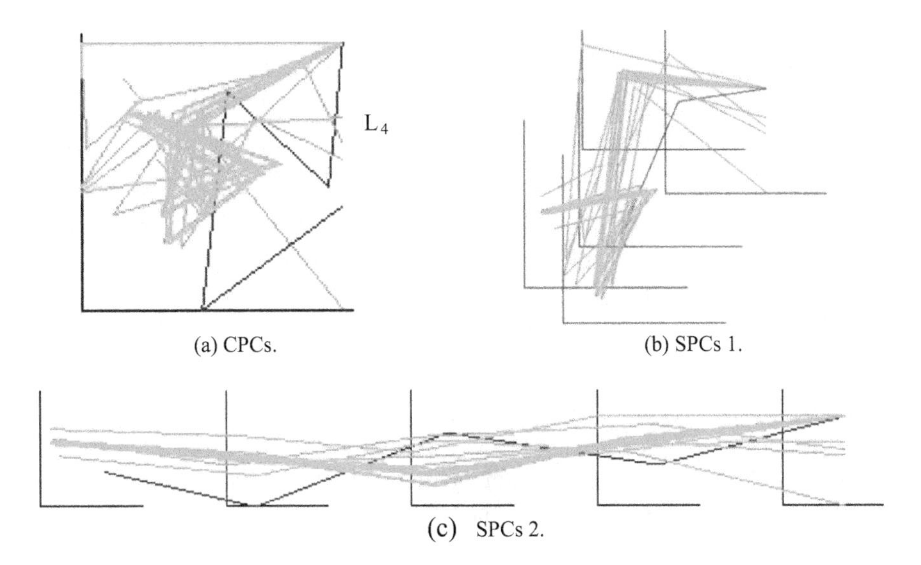

Fig. 4.21 Set $V(1, \mathbf{c}_1, T)$: objects of class 1 (green), and class 2 (blue), which are close to the object $\mathbf{c}_1$ in the Shifted Paired Coordinates, the and Collocated Paired Coordinates

can construct a classification sub-rule for n-D point **a**, which excludes it from class 1, by using a distance threshold from it,

$$\text{If } \mathbf{a} \in V(\mathbf{c}, T) \& |L_5 - \mathbf{a}^*| > e(L_5), \text{ then } \mathbf{a} \in \text{class } 1,$$

where a* is a graph for **a**, and $e(L_4)$ is a distance threshold between L_4 and **a***.

Case study 4.2 This case study uses *Blood transfusion* 4-D data from UCI Machine Learning repository (Lichman 2013). The attributes are: months since last donation, total number of donations, total blood donated, months since first donation, and a binary class variable (1 for donated before the given date in red, in Fig. 4.22, and 0 for not donated, before the given date, in blue in Fig. 4.22).

Figure 4.22 illustrates the GLC-S algorithm for a given **c** from this dataset. It shows these data in CPCs and PSPCs. Here $V(\mathbf{c}, T)$ contains only a single blue line from class 2, denoted as L_5, in Fig. 4.22a, which differs from the lines of the n-D

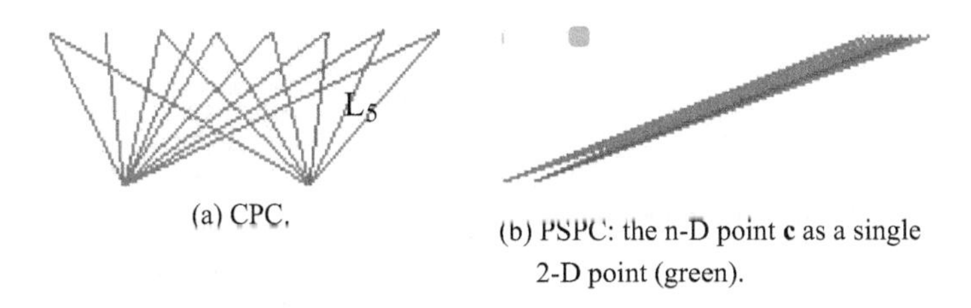

Fig. 4.22 Blood transfusion 4-D data, which are close to the data point **c,** within a given threshold *T,* in CPCs and PSPCs. Points of class 1 are in red, and points of class 2 are in blue

points of class 1, in CPC visualizations, and in PSPCs, in Fig. 4.22b. This allows us to create the classification sub-rule:

$$\text{If } \mathbf{a} \in V(\mathbf{c}, T)\&|L_5 - \mathbf{a}^*| > e(L_5), \text{ then } \mathbf{a} \in \text{class } 1,$$

where **a*** is a graph for the n-D point **a**, and $e(L_5)$ is a distance threshold, between L_5 and **a***. Alternatively, we can ignore L_5, and get a simpler, but less accurate sub-rule: If $\mathbf{a} \in V(\mathbf{a}_k, T)$, then $\mathbf{a} \in$ class 1. The location of the green dot (point **c**), in Fig. 4.22b, allows seeing the closest cases relative to it. For any point **c**, which belongs to a specific class (e.g., class 1), it helps to assess the number of similar points, from the opposite class, and closeness to **c** of the similar points, from the opposite class. CPC and PSPC show that a blue line is similar to red lines. This shows that accurate classification of this blue point is challenging. In this way, a user can visually compare the red lines, which are closest to the blue line. A user can analyze the blue line in connection with two nearest red lines and reveal other features for classification. Interactive change of the threshold T gives a user a way to dynamically assess: how common **c** is, for its own class, and for the opposite class.

Case study 4.3 This case study uses *Ecoli 7-D data* of 8 classes from the UCI Machine Learning Repository (Lichman 2013). Figure 4.23 illustrates the GLC-S algorithm for a given **c,** from this dataset.

It shows these data in the CPCs and the SPCs. Here $V(\mathbf{c}, T)$ contains a set of lines from class 1 (red), which is distinct from lines of class 2 (blue) in the SPCs visualization (Fig. 4.23a). Denote the average line of the lines of class 2, as shown in Fig. 4.23a, as L_6, and construct the classification sub-rule for n-D point **a**,

$$\text{If } \mathbf{a} \in V(\mathbf{c}, T)\ \&\ |L_6 - \mathbf{a}^*| > e(L_6), \text{then } \mathbf{a} \in \text{class } 1,$$

where a* is a graph for **a**, and $e(L_6)$ is a distance threshold between L_6 and **a***.

Case study 4.4 This case study uses *7-D Seeds data* of three classes from the UCI Machine Learning Repository (Lichman 2013).

Figure 4.24 illustrates the GLC-S algorithm, for a given **c,** from this dataset, with 4 attributes used. It shows these data in SPCs. Here $V(\mathbf{c}, T)$ contains a set of lines, from class 1 (red), and only one line from class 2 (blue). We denote this line as L_7

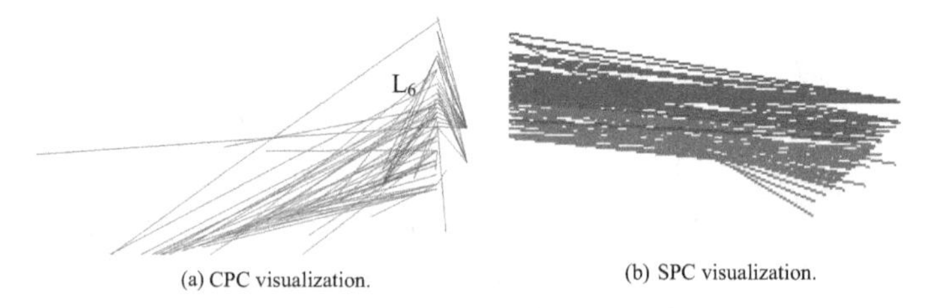

Fig. 4.23 Ecoli 7-D data: ecoli cytoplasm (red), and the ecoli inner membrane (blue) in CPCs and SPCs, which are close to point **c**

Fig. 4.24 4-D subset of seeds dataset in SPC: class 1 (red), and class 2 (blue). The green dot is **c**

(see Fig. 4.24), and construct the classification sub-rule shown below that filters out lines $\mathbf{a}^*$ that are very close to L_7,

$$\text{If } \mathbf{a} \in V(\mathbf{c}, T) \,\&\, |L_7 - \mathrm{a}^*| > \mathrm{e}(L_7), \text{ then } \mathbf{a} \in \text{class } 1,$$

where $\mathbf{a}^*$ is a graph for n-D point $\mathbf{a}$, and $e(L_7)$ is a distance threshold between L_7 and $\mathbf{a}^*$. Alternatively, L_7 can be ignored with a simpler, but slightly less accurate sub-rule as follows: If $\mathbf{a} \in V(c, T)$, then $\mathbf{a} \in$ class 1.

Case study 4.5 This case study is based on the user knowledge modeling dataset from the UCI Machine Learning Repository (Lichman 2013). This dataset contains the 258 5-D records of the four classes. The fifth attribute is repeated to make the sixth attribute, because the SPC require the even number of attributes. First, the data were split into the 172 training cases and the 86 validation cases (2/3 and 1/3). Then the centers $\mathbf{c}_1$–$\mathbf{c}_4$ of classes on the training data are computed in two steps: (1) computing average 6-D points $\mathbf{e}_1$–$\mathbf{e}_4$ of each class, and (2) finding centers $\mathbf{c}_i$ as 6-D points, with the smallest Euclidian distance, to the respective $\mathbf{e}_i$.

Figure 4.25 shows the colored SPC graphs of these four points $\mathbf{c}_i$, and the colored nodes of all the graphs, from the vicinity sets (hypercubes) $V(\mathbf{c}_i, T_i)$. The edges of the graphs are omitted to decrease clutter. Here $T_1 = 0.22$, $T_2 = 0.25$, $T_3 = 0.2$, and $T_4 = 0.15$. As this figure shows, the 6-D points in each $V(\mathbf{c}_i, T_i)$ belong only to the training data of the respective class without any point from other classes. In other words, the 6-D hyper-cubes defined by pairs $(\mathbf{c}_i, T_i)$ separates these

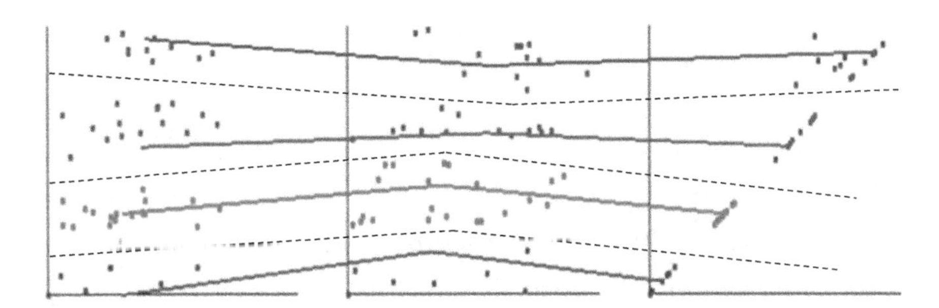

Fig. 4.25 Phase 1: 6-D training data subset of knowledge modeling data in SPC. Centers of classes are colored lines. Black dotted lines are discrimination lines. Dots are nodes of graphs

training data with 100% accuracy. Moreover, this separation allows constructing the simple and accurate discrimination functions in SPC, shown as dotted black lines in Fig. 4.25. Figure 4.26 shows the centers $\mathbf{c}_1$–$\mathbf{c}_4$, and the black discrimination lines from Fig. 4.25, along with all the validation 6-D points, which are located within the hypercubes, found on the training data. These 6-D points are shown, by the nodes of their graphs, in the respective colors of their classes. As Fig. 4.26, shows all these points in each hypercube, belong to the respective class, without any point from other classes. This means 100% accuracy, on the validation cases.

Each next phase repeats the same process, for unclassified data, remaining from the previous phases. This includes computing the new centers $\mathbf{c}_1$–$\mathbf{c}_4$ and the new vicinity hyper-cubes $V(\mathbf{c}_i, T_i)$. Figures 4.27, 4.28 and 4.29 shows the results for the phases 2 and 3. Table 4.1 summarizes the results numerically. All training and validation cases are correctly classified in the phases 1-3. Three training cases and two validation cases remain unclassified (see Fig. 4.30). Thus, the classification rule R_V: for this dataset is R_V: If $\mathbf{x} \in V_1(\mathbf{c}_i, T_i) \cup V_2(\mathbf{c}_i, T_i) \cup V_2(\mathbf{c}_i, T_i)$ then $\mathbf{x} \in$ *class* $\boldsymbol{i}$, where $V_1(\mathbf{c}_i, T_i)$, $V_2(\mathbf{c}_i, T_i)$, $V_2(\mathbf{c}_i, T_i)$ are hypercubes found in the phases 1–3.

The five remaining cases $\mathbf{a}_k$, k = 1:5, which are outside of these hyper-cubes, can be classified, by memorizing them, without making the rule significantly overfitted, because they are only 1.9% of the dataset, with a new rule R_a to augment R_v:

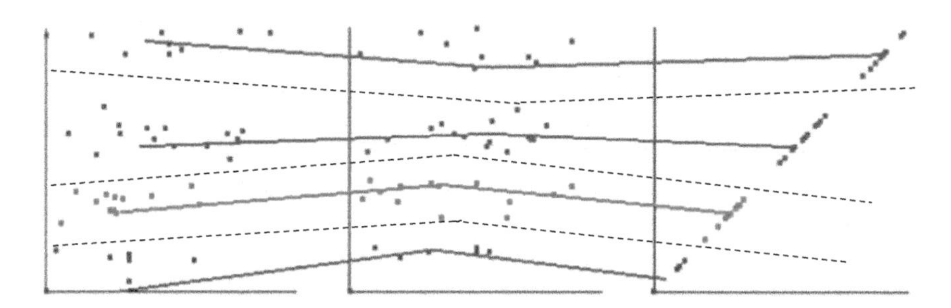

Fig. 4.26 Phase 1: 6-D validation data subset of knowledge modeling data in SPC. Centers of classes (colored lines), and discrimination dotted lines are from training data. Dots are nodes of graphs

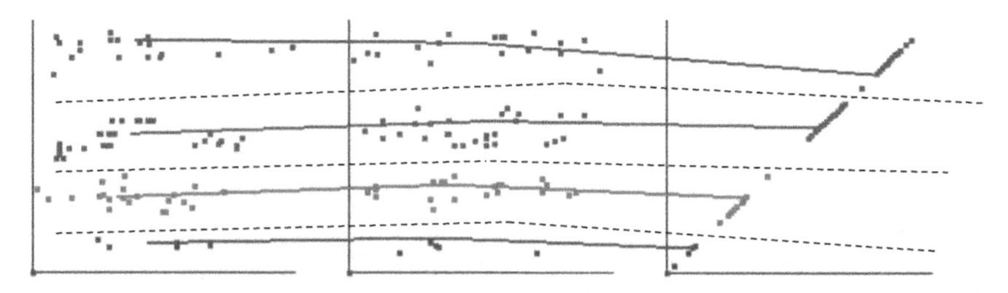

Fig. 4.27 Phase 2: 6-D training data subset of knowledge modeling data in SPC. Centers of classes are colored lines. Black dotted lines are discrimination lines. Dots are nodes of graphs

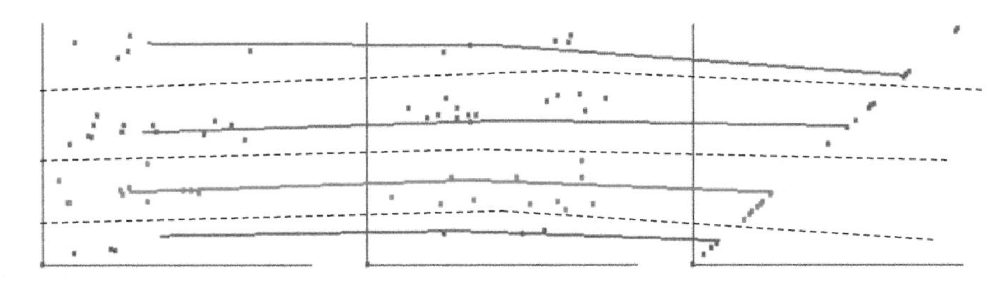

Fig. 4.28 Phase 2: 6-D validation data subset of knowledge modeling data in SPC. Centers of classes (colored lines) and discrimination dotted lines are from training data. Dots are nodes of graphs

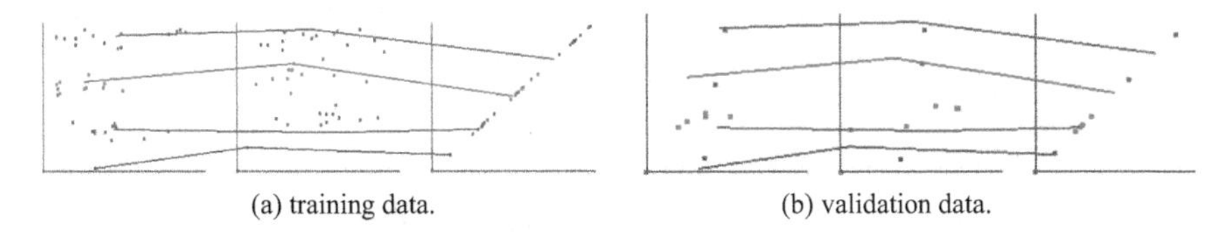

(a) training data. (b) validation data.

Fig. 4.29 Phase 3: 6-D training and validation subsets of knowledge modeling data in SPC. Centers of classes are colored lines. Black dotted lines are discrimination lines. Dots are nodes of graphs

Table 4.1 All phases of visual classification with expanding and shrinking the user knowledge dataset

	Class 1	Class 2	Class 3	Class 4	Total	Data%	Accuracy %
Total training	48	54	57	13	172	66.67	100
Total validation	15	29	31	11	86	33.33	100
Phase 1 training	12	22	16	6	56	21.71	100
Phase 1 validation	9	13	16	6	44	17.05	100
Phase 2 training	19	21	28	5	73	28.29	100
Phase 2 validation	5	11	14	3	33	12.79	100
Phase 3 training	14	11	11	2	38	14.73	100
Phase 3 validation	1	5	1	2	9	3.4	100
Remaining training data	3	0	0	0	3	1.1	
Remaining validation data	0	2	0	0	2	0.78%	

If $(\mathbf{x} = \mathbf{a}_1 \vee \mathbf{x} = \mathbf{a}_2 \vee \mathbf{x} = \mathbf{a}_3)$ then $\mathbf{x} \in$ *class* 1, else if $(\mathbf{x} = \mathbf{a}_4 \vee \mathbf{x} = \mathbf{a}_5)$, then $\mathbf{x} \in$ *class* 3.

Case study 4.6 This case study uses *Wisconsin Breast Cancer* (WBC) dataset from the UCI Machine Learning Repository (Lichman 2013). It contains 444 benign cases and 239 malignant cases as full 9-D records. The 9th attribute is repeated to

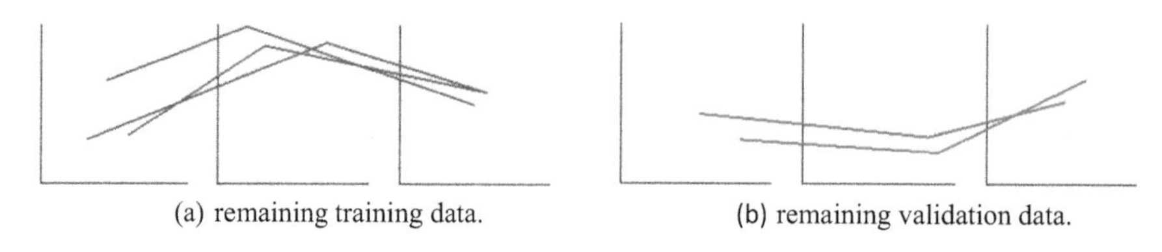

(a) remaining training data. (b) remaining validation data.

Fig. 4.30 Remaining 6-D training and validation subsets of knowledge modeling data in SPC

make the 10th attribute. The data were split to training and testing cases (70%:30%). The original plotting of data in SPC is cluttered (see Fig. 4.31). The search for rectangular rules decsribed below allowed decreasing clutter in SPC and discovering efficient rules (see Fig. 4.32). These rules have accuracy over 90% and cover majority of the cases of the given class. The rectangular rules include the rules of the following structure for a 10-D point $\mathbf{x} = (x_1, x_2,\ldots, x_n)$:

$$\text{If}\left(x_i, x_j\right) \in R_1 \text{ then } \mathbf{x} \in \text{class C.}$$

$$\text{If}\left(x_i, x_j\right) \in R_1 \,\&\, (x_k, x_m) \notin R_2 \,\&\, (x_s, x_t) \notin R_3 \text{ then } \mathbf{x} \in \text{class C}$$

where R_1–R_3 are rectangles in projections of **x** in respective paired of coordinates (X_i, X_j), (X_k, X_m) and (X_s, X_t).

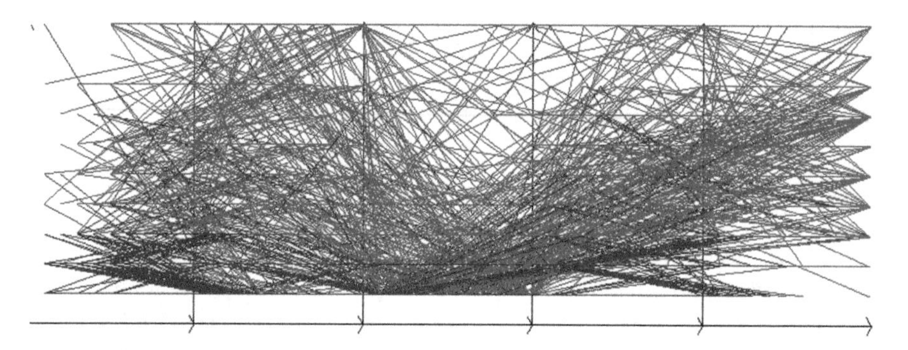

Fig. 4.31 9-D WBC data in SPC with high level of clutter

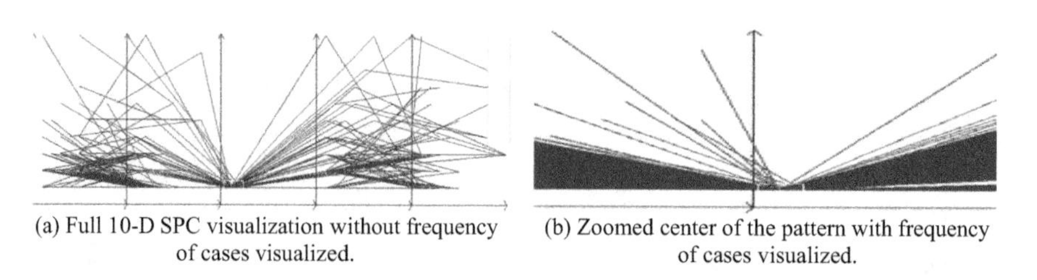

(a) Full 10-D SPC visualization without frequency of cases visualized. (b) Zoomed center of the pattern with frequency of cases visualized.

Fig. 4.32 Subset of WBC data with dominant red class cases found by a rectangular rule

4.7 Discussion

Parallel Coordinates are used, for over 25 years, and many applications had shown their benefits. However, there are plenty of observations in the literature which help to specify the better areas of the application of, and the improvement for the PCs. GLCs are a new technique, which is going through the same process. There are datasets, where the different GLCs reveal the patterns better, by changing the directions and mutual location, as this chapter shows in multiple cases. Also there are other datasets, where PCs reveal the patterns better than some other GLCs.

References

Appelbaum, L.G., Norcia, A.M.: Attentive and pre-attentive aspects of figural processing. J. Vis. **9** (11), 18 (2009)

Cohen, S., Rokach, L., Maimon, O.: Decision-tree instance-space decomposition with grouped gain-ratio. Inform. Sci. **177**(17), 3592–3612 (2007)

Few, S.: Tapping the power of visual perception. Vis. Bus. Intell. Newsl. **39**, 41–42 (2004)

Lichman, M.: UCI Machine Learning Repository [http://archive.ics.uci.edu/ml]. University of California, School of Information and Computer Science, Irvine, CA (2013)

Rokach, L., Maimon, O., Arad, O.: Improving supervised learning by sample decomposition. J. Comput. Intell. Appl. **5**(1), 37–54 (2005)

Theus M.: High-dimensional data visualization. In: Handbook of data visualization, pp. 151–178. Springer (2008)

Vityaev, E., Kovalerchuk, B.: Visual data mining with simultaneous rescaling. In: Kovalerchuk, B., Schwing, J. (eds.) Visual and spatial analysis: advances in data mining, reasoning, and problem solving, pp. 371–386. Springer Science & Business Media, Dordrecht (2005)

Chapter 5
GLC Case Studies

A pretty experiment is in itself often more valuable than twenty formulae extracted from our minds.
No amount of experimentation can ever prove me right; a single experiment can prove me wrong.

Albert Einstein

This chapter provides several successful case studies on the use of GLC for knowledge discovery and supervised learning mostly from the data from the University of California Irvine Machine Learning repository (Lichman 2013). The real world tasks include Health monitoring, Iris data classification, Challenger disaster and others. These cases studies involve several GLC methods, two-layer visual representation, and comparison with alternative methods such as Parallel Coordinates, RadVis, and Support Vector Machine.

5.1 Case Study 1: Glass Processing with CPC, APC and SPC

This case study uses 9-D *glass processing data* from UCI Machine Learning Repository (Lichman 2013).

Figures 5.1, 5.2 and 5.3 show two classes of these data (floated processed and non-floated processed) in Collocated, Anchored, and Shifted Paired Coordinates. The Anchored Paired Coordinates (Fig. 5.2) allowed isolating these classes better than two other methods. Such differences show the importance of having multiple lossless n-D visualizations to be able to discover a better one for the given data.

While the visual patterns P_1 and P_2 of classes 1 and 2 are distinct in Fig. 5.2 a simple mathematical and natural language descriptions of them are not obvious. Discovering these patterns P_1 and P_2 analytically requires the mathematical description of the class of the patterns $\{P_i\}$ where a learning algorithm can search for them. Visual representation allows easing or even avoiding this difficult formalization process by direct classification of the new objects in the visual representation.

B. Kovalerchuk, *Visual Knowledge Discovery and Machine Learning*,
Intelligent Systems Reference Library 144,
https://doi.org/10.1007/978-3-319-73040-0_5

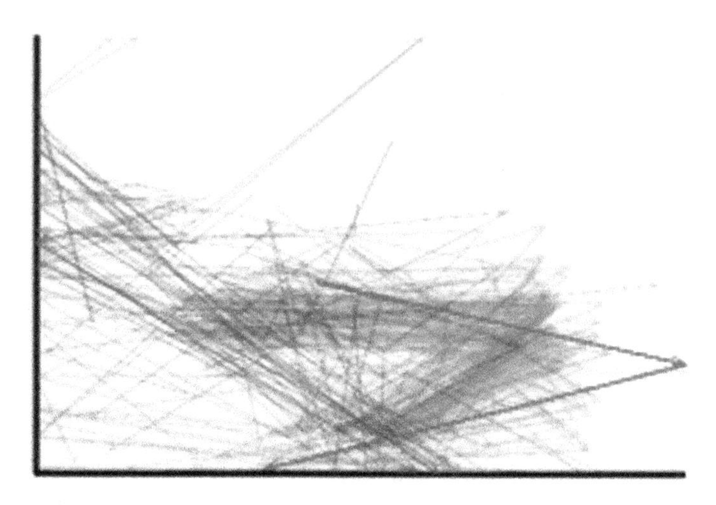

Fig. 5.1 Two classes in Collocated Paired Coordinates

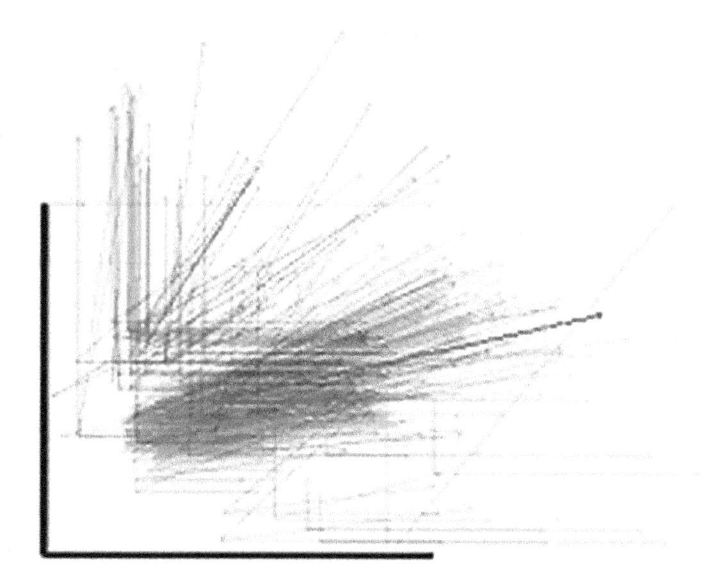

Fig. 5.2 Two classes in Anchored Paired Coordinates (best result)

The use of the natural language to communicate patterns P_1 and P_2 to other experts is also problematic. We do not have ready expressions for them. The description like "elongated almost horizontal pattern" for the red pattern is quite uncertain, fuzzy. It does not tell how much elongated, how close to the horizontal line and how it is related to the blue pattern. For the more complex blue pattern, even such uncertain description is difficult to form. Thus, the use of pictures, like shown in Fig. 5.2, is a better way to communicate these patterns to experts. It is

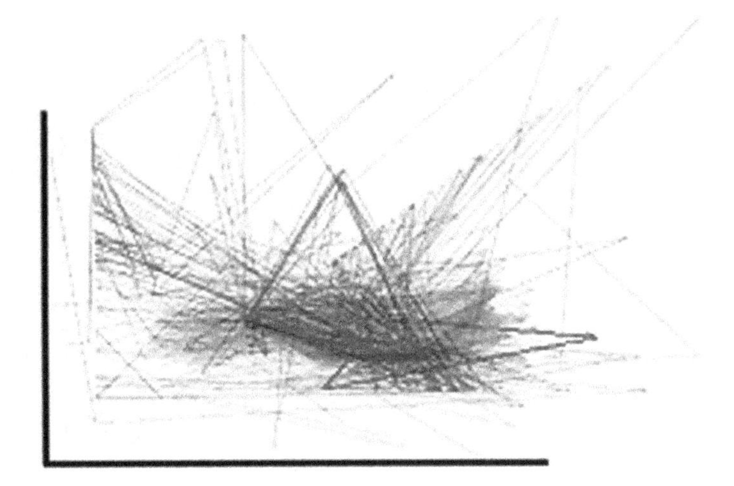

Fig. 5.3 Two classes in Shifted Paired Coordinates

consistent with the Asian proverb: "Better to see something once than to hear about it a thousand times."

5.2 Case Study 2: Simulated Data with PC and CPC

This study uses simulated dataset **A** that is designed as described below. Figure 5.5 shows advantages of visualization of these data in CPC relative to visualization of these data in Parallel Coordinates shown in Fig. 5.4. CPC shows structure of 23-D data simpler than the Parallel Coordinates.

The method for generation of lines is as follows. First an n-D point **a** is generated with a single random number a, $\mathbf{a} = (a, a, \ldots, a)$ for each of 23 dimensions. Then for generation of all other data points $\mathbf{a}_k$ we add a random shift s and a small amount of random noise **e** to each dimension:

$$a_{ki} = a + s + \text{RAND}()^{*}5 - 2.5.$$

This experiment was conducted for $n = 23$. Parallel coordinates indicated linear correlation, as all lines were roughly horizontal (see Fig. 5.4).

CPC show those PC lines as small "knots" located on a straight diagonal line in Fig. 5.5. It resembles a common correlation plot for 2-D data, which makes it a quite familiar and intuitive metaphor. With less resolution those blobs are visible as single points strengthening this analogy.

Next, these blobs occupy much less space than PC lines. Thus, it is scalable to a larger number of n-D points while PC representation will completely cover PC area where the visual pattern of correlated lines will disappear.

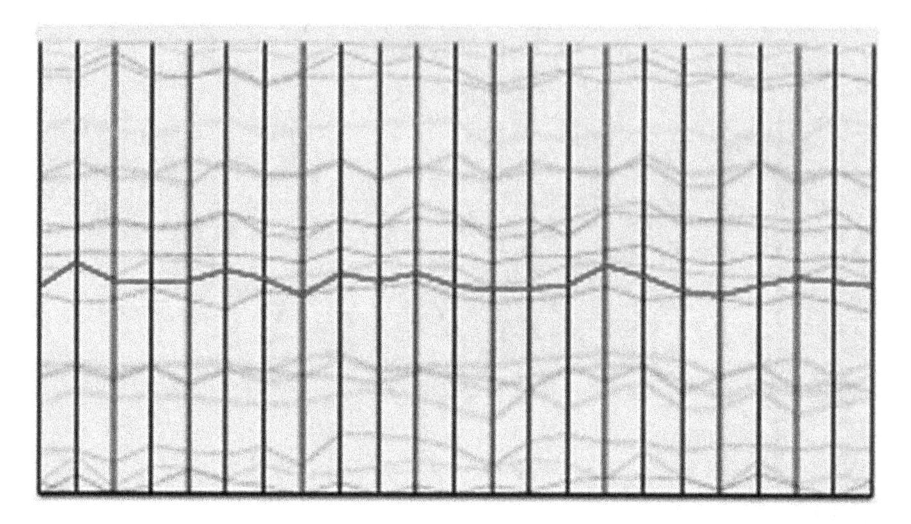

Fig. 5.4 23-D dataset **A** in Parallel Coordinates

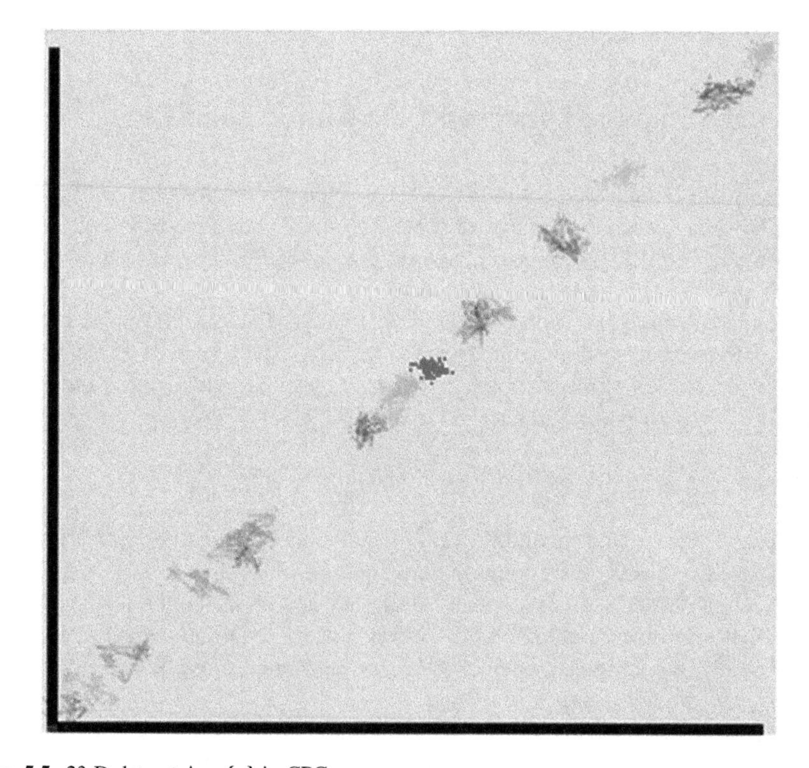

Fig. 5.5 23-D dataset **A** = {**a**} in CPC

5.3 Case Study 3: World Hunger Data

Figure 5.6 provides an example of visual representation in Collocated Paired Coordinates of World Hunger data from the International Food Policy Institute (Global Hunger index 2012). For comparison, Fig. 5.7 shows the same data visualized as traditional time series.

Note that this representation is the same as Parallel Coordinates for such time series. The International Food Policy Institute classifies the hunger, between 10 and 20% as serious and less than 10% as low or moderate. The Global Hunger Index (GHI) for each country measures as,

$$\mathrm{GHI} = (\mathrm{UNN} + \mathrm{UW5} + \mathrm{MR5})/3,$$

where UNN is the proportion of the population that is Undernourished (in %), UW5 is the prevalence of Underweight in children under age of five (in %), and MR5 is the Mortality rate of Children under age five (in %).

This institute considers GHI between 30 and 40% as extremely alarming, and between 20 and 30% as alarming.

CPC representation in Fig. 5.6 complements Tableau visualizations available at the website of this Institute by showing integrated picture in its dynamics for countries with the hunger problem for four time intervals 1990–92; 1995–1997, 2000–2002; 2006–2008.

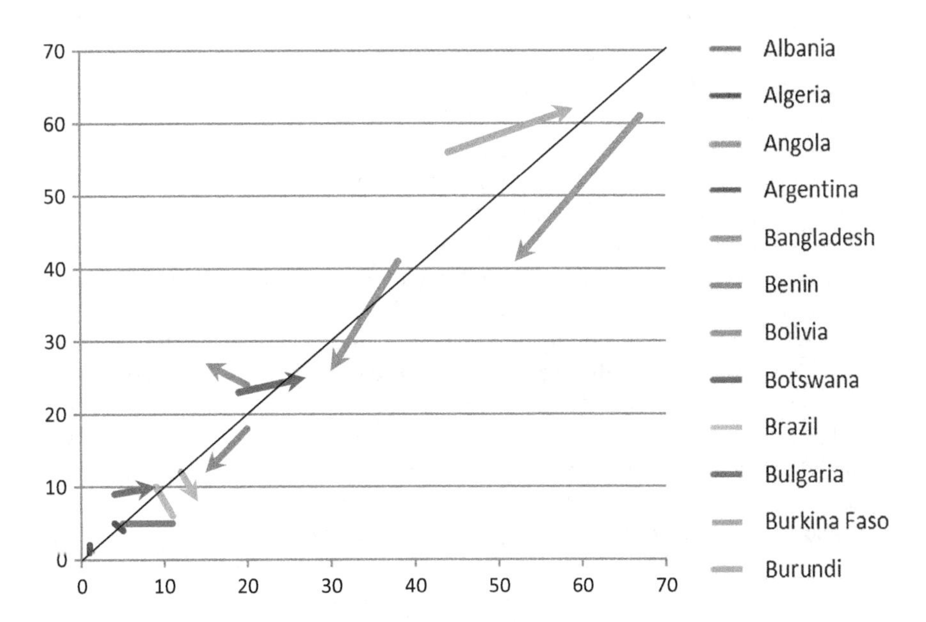

Fig. 5.6 4-D data: representation of prevalence of undernourished in the population (%) in Collocate Paired Coordinates

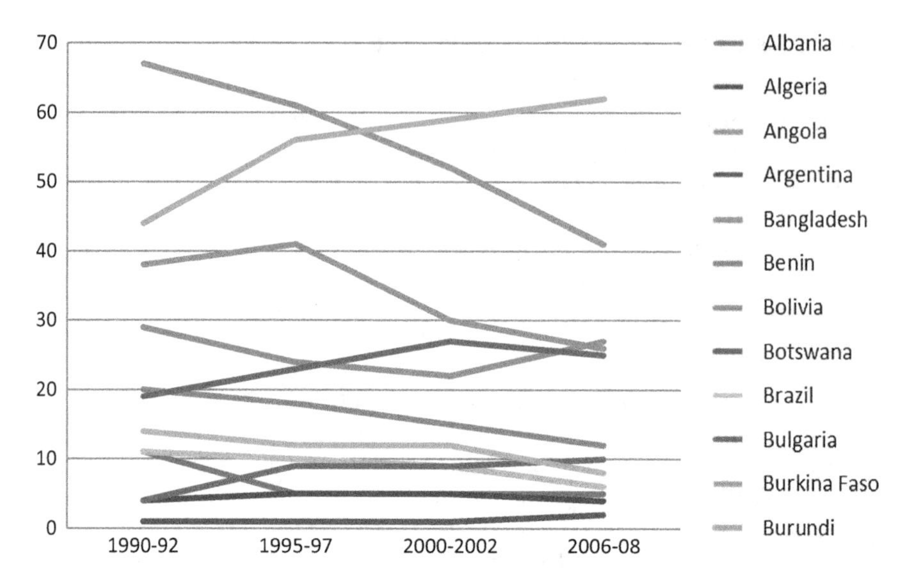

Fig. 5.7 4-D data: representation of prevalence of undernourished in the population (%) in traditional time series (equivalent to Parallel Coordinates for time series)

Figure 5.7 shows these data as the traditional time series. We denote these four time intervals as t_1, t_2, t_3, and t_4. Each arrow in Fig. 5.6 represents GHI at time t_1–t_4 for each country, where the arrow start in the pairs (GHI(t_1), GHI(t_2)), and the arrow end is (GHI(t_3), GHI(t_4)).

Figures 5.6 and 5.7 show that the traditional time series and Parallel Coordinates require *4 points* and *3 lines* for each 4-D vector. In contrast, the CPC required only *one line* (*arrow*). In Fig. 5.6, these lines *do not overlap* and have *no any occlusion* in contrast with the traditional time series in Fig. 5.7. The arrows that go up (e.g., Benin, Botswana) indicate hunger growth and the arrows that go down (e.g., Angola, Bangladesh) in Fig. 5.6 indicate hunger decline from the first pair of years (1990–92; 1995–1997) to the second pair of years (2000–2002; 2006–2008), respectively. Note that Angola consistently improves the situation year after year. It is visible in the traditional time series and in CPC.

In CPC, it is expressed differently from the former—by the fact that Angola's arrow is located below the diagonal. In contrast, the arrow for Bangladesh crosses the diagonal which indicates that the hunger situation was worsened from 1990–92 to 1995–1997, but improved from 1995–1997 to 2000–2002 and 2006–2008. Thus CPC provides a new metaphor and a user needs to learn it to use CPC more efficiently as it is common with any new visual representation.

Several features of CPC metaphor are quite intuitive without special learning. For instance, the arrows at the top in Fig. 5.6 indicate countries with highest level of hunger (above extremely alarming) and arrow at the bottom indicates the countries with the low or moderate level of hunger. A user can project a respective arrow point to the axis of the year of interest to get an estimate of the hunger index for the country interest. As we see, for two countries, GHI is way above 40% (extreme alarming level).

Up or down directions of arrow are also self-explanatory for hunger grows or decline, respectively. Next arrows of similar directions and that are close to each other indicate countries with similar hunger situations.

Figure 5.8a shows Angola in n-Gon (triangular) coordinates that constitute three GHI components (UNN, UW5, MR5) shown within max values of each of them. Figure 5.8b shows the same Angola in n-Gon (triangular) coordinates with filled area.

5.4 Case Study 4: Challenger USA Space Shuttle Disaster with PC and CPC

Challenger USA Space Shuttle O-Ring Dataset on Challenger disaster from the UCI Machine Learning Repository (Draper 1993, 1995; Lichman 2013). The Challenger O-rings data include parameters such as (1) temporal order of flight, (2) number of O-rings at risk on a given flight, (3) number of O-rings experiencing thermal distress, (5) launch temperature (degrees F), and (5) leak-check pressure (psi).

These data have been normalized to be in the [0,1] interval before visualizing them. We considered two different normalizations of the number of O-rings at risk. This number is 6 in all flights. It is mapped to 0 in the first visualization and to 1 in the second visualization.

The used data (Draper 1993) differ from data analyzed in Tufte and Robins (1997, p. 44). The data used by Tufte and Robins include three erosion incidents at temperature 53F, which makes the link between low temperature and large incidents much more transparent. Draper's data are more difficult for revealing this pattern. Figures 5.9 and 5.10 show a visualization of Draper's data in traditional plots. Figures 5.11 and 5.12 present the same data in CPC requiring a single line per record with low overlap and without occlusion issues in contrast with the traditional plots.

Figures 5.11 and 5.12 show three distinct flights #2, #14, and #2 with orientations that differ from other flights. These flights had the maximum value of O-rings at risk. Thus, CPC visually show a distinct pattern in flights that has a meaningful interpretation. The well-known case is #14, which is the lowest temperature (53F) from the previous 23 Space Shuttle launches. It stands out in the Collocated Paired

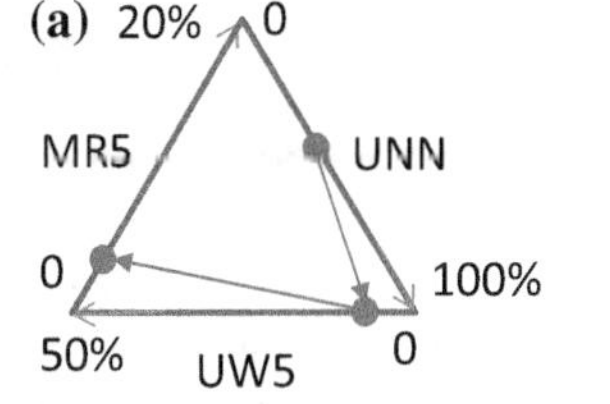

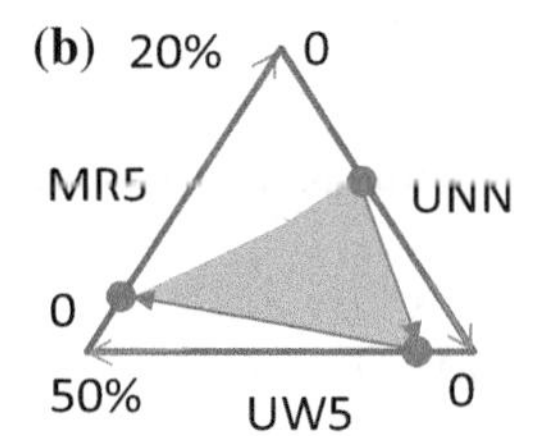

Fig. 5.8 **a** Angola in n-Gon (triangular) coordinates and **b** in n-Gon (triangular) coordinates with filled area

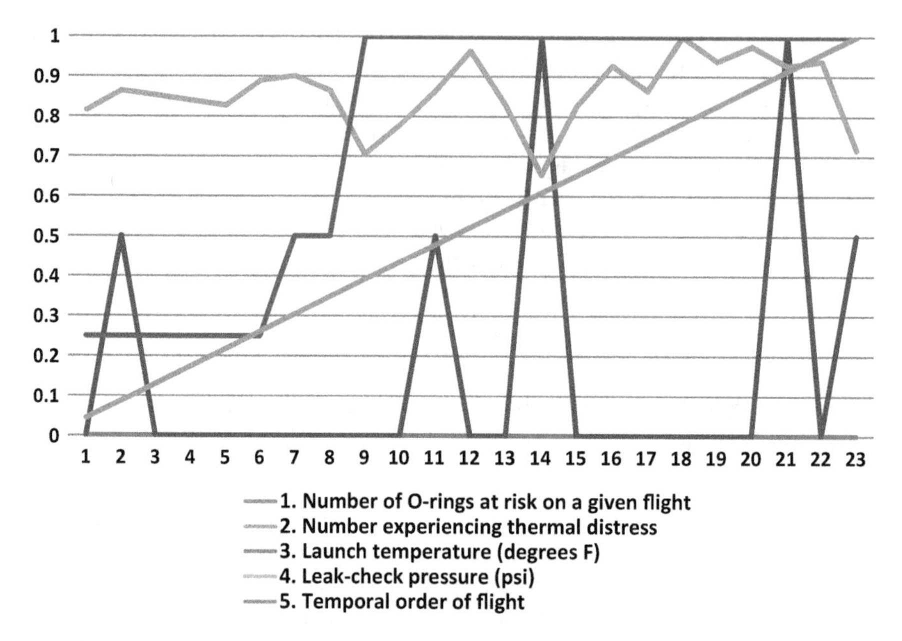

Fig. 5.9 Challenger USA Space Shuttle normalized and mapped O-Ring dataset in a traditional line chart. X coordinate is a temporal order of flights. Each line represents an attribute

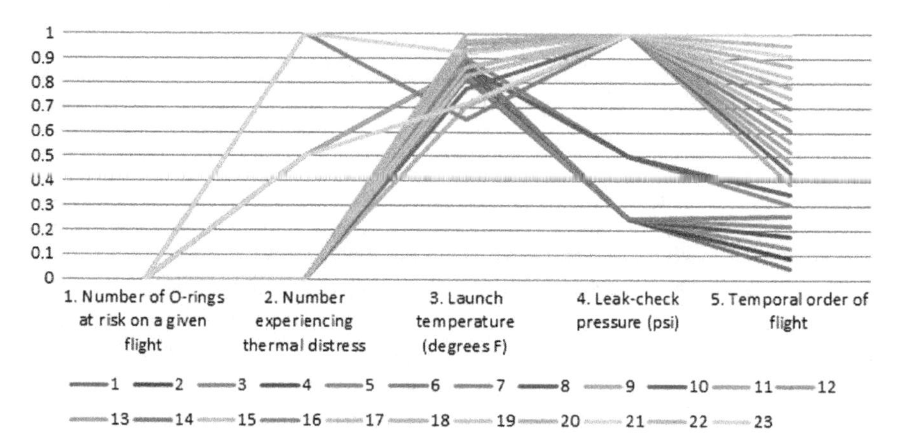

Fig. 5.10 Challenger USA Space Shuttle normalized O-Ring dataset in Parallel Coordinates. Each line represents a flight

Coordinates. In the test at high leak-check pressure (200 psi) it had 2 O-rings that experienced a thermal distress. The case #2 also experienced thermal distress for one O-ring at much higher temperature of 70 F and lower leak-check pressure (50 psi). This is even more outstanding from others with the vector directed down. The case #14 is directed horizontally. All other cases excluding case #21 are directed

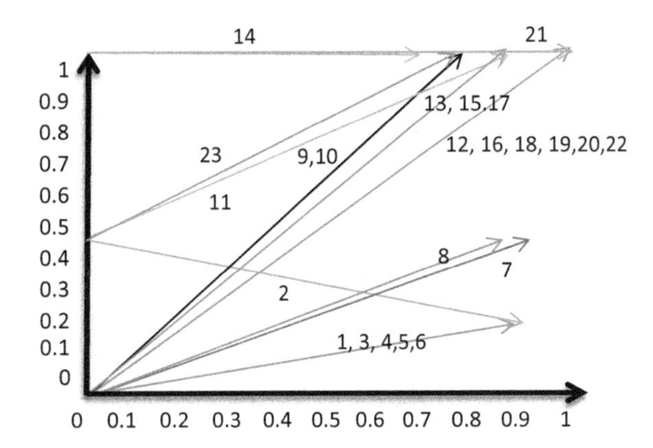

Fig. 5.11 Challenger USA Space Shuttle normalized O-Ring dataset in the Collocated Paired Coordinates. X, Y coordinates are normalized values of attributes. Six O-rings at risk are mapped to 0. Flight numbers are attached to each arrow

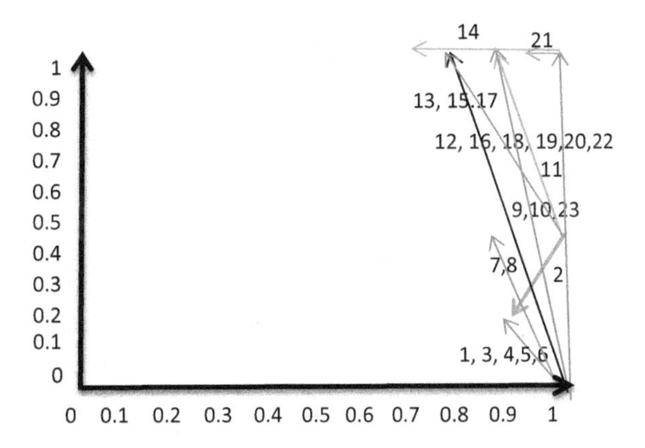

Fig. 5.12 Challenger USA Space Shuttle normalized O-Ring dataset in the Collocated Paired Coordinates. X, Y coordinates are normalized values of attributes. Six O-rings at risk are mapped to 1

up. The case #21 stands out because it had O-rings that experienced a thermal distress at high leak-check pressure (200 psi) and high temperature (75F).

5.5 Case Study 5: Visual n-D Feature Extraction from Blood Transfusion Data with PSPC

While we have situations, where the visualization in General Line Coordinates systems allow separate classes directly in one step, in some other situations more stages are needed. Figure 5.13 shows 748 cases of 4-D data points of two classes from the *Blood Transfusion Service Center data* (Lichman 2013) visualized in Parametrized SPC (PSPC) with the 4-D anchor point in the average point of the red class. Each 4-D point is represented losslessly as an arrow. In this PSPC space two classes and their convex hulls heavily overlap.

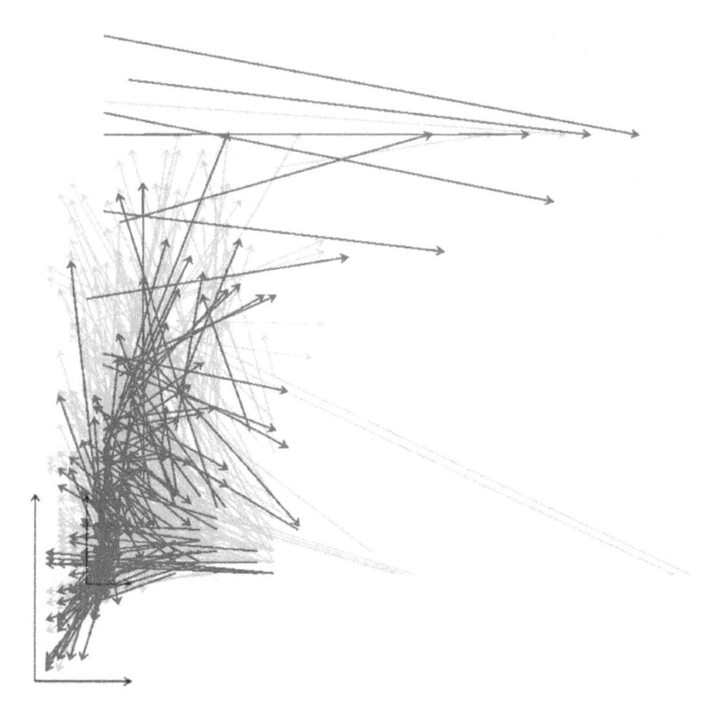

Fig. 5.13 Original 4-D data of two classes in PSPC system

While this PSPC representation does not separate classes, it gives multiple *visual insights* for extraction features that can be more discriminative for these classes. An observer can discover the differences in orientation and length of some arrows of two classes. For instance, west and southwest orientation is common for a large number of red arrows in the bottom that are often shorter than blue arrows with the same direction. The length of the majority of the red arrows is also smaller than the length of the blue arrows with a few exclusions on the top. Therefore, the next step is extracting two new features from PSPC: orientation, y_1, and length, y_2. The formulas for these new features are given below in terms of two variables,

$$v_1 = x_3 + (a_1 - a_3) - x_{1,} v_2 = x_4 + (a_2 - a_4) - x_2,$$

where x_1, x_2, x_3 and x_4 are values of coordinates of a given arrow $\mathbf{x}$ in Fig. 5.13, and (a_1, a_2, a_3, a_4) are coordinates of the 4-D anchor point A which is the average point of the red class.

In these terms, new features y_1 and y_2 are defined as follows:

$$y_1 : \text{if} \quad v_2 \geq \text{ then } \quad y_1 = \quad \arccos(v_1/v_3) \text{ else } \quad y_1 = 2\pi - y_1$$

$$y_2 : y_2 = \quad \left(v_1^2 + v_2^2\right)^{1/2}_{.}$$

Fig. 5.14 Two classes in features y_1 and y_2 extracted using visual insight

Then both y_1 and y_2 are normalized to [0,1] interval for the visualization. The result is shown in Fig. 5.14 as a standard scatter plot of these two attributes. It shows that the majority of the blue cases are in the middle and the upper section of this middle part contains mostly the blue cases. The most overlapping area is the section of the lower section of the middle part. This informal visual exploration gives an insight on the area, where the next stage of feature extraction must be concentrated. This is the area with most of the overlap in Fig. 5.14. The cases from that area can be visualized in the original PSPC system to attempt to extract additional features that can separate blue and read cases in this smaller dataset.

What are the chances to extract features y_1 and y_2 pure analytically using analytical machine learning techniques without any visual insight? From our viewpoint it is not realistic. It would require: (1) to identify a class of functions $\{f\}$ that will include both functions y_1 and y_2 without knowing that these functions can be potentially useful, (2) to develop a formal criterion K that will evaluate all functions from $\{f\}$ to be a good feature, (3) to use significant computing resources to run K on $\{f\}$. To ensue (1) we would need to make the class of functions $\{f\}$ very large if not infinite. This makes (2) and especially (3) extremely unrealistic. This example illustrates the general advantages of visualization approach that in large part substitutes cognition for perception (Munzner 2014).

5.6 Case Study 6: Health Monitoring with PC and CPC

Figure 5.15 illustrates the opportunities of using Parametrized Shifted Paired Coordinates for health monitoring of an individual in comparison with Parallel Coordinates. In this example four health characteristics are monitored: systolic blood pressure, x_1; diastolic blood pressure, x_2; pulse, x_3; and total cholesterol, x_4.

At the initial moment the individual health status is presented by four values of these characteristics (100, 150, 95, 250). A desired health status for these characteristics is identified as (70, 120, 60, 190). The goal is to monitor the progress the individual is making toward this goal with a complex of medical treatments, diet, exercises, and so on.

In Fig. 5.15a in PSPC the goal is presented as a single preattentive point that is simple metaphor to learn because targets quite often are represented as points. This single point is a result of the PSPC design. In contrast, the Parallel Coordinates show the goal as a zig-zag line with 4 points that is not preattentive. Next the current status in Fig. 5.15a for PSPC consists just of a single line (arrow). In Parallel Coordinates it is again a zig-zag line with three segments. PSPC uses a standard Cartesian representation for pairs (X_1, X_2) and (X_3, X_4) that is familiar to everybody with high school background. In contrast, the Parallel Coordinates need to be learned.

The only novelty that a user needs to learn in Fig. 5.15a is a shift of coordinates (X_3, X_4) relative to (X_1, X_2). The coordinates are labeled, thus, it is quite intuitive, and dotted lines help to trace values in (X_3, X_4) coordinates that interactive software can provide if needed.

Figure 5.15a exploits the PSPC property that n-D points with values of coordinates that are similar to the values of coordinates of the anchor n-D point are visualized as smaller graphs (see mathematical statements in Chap. 3 and Figs. 5.18

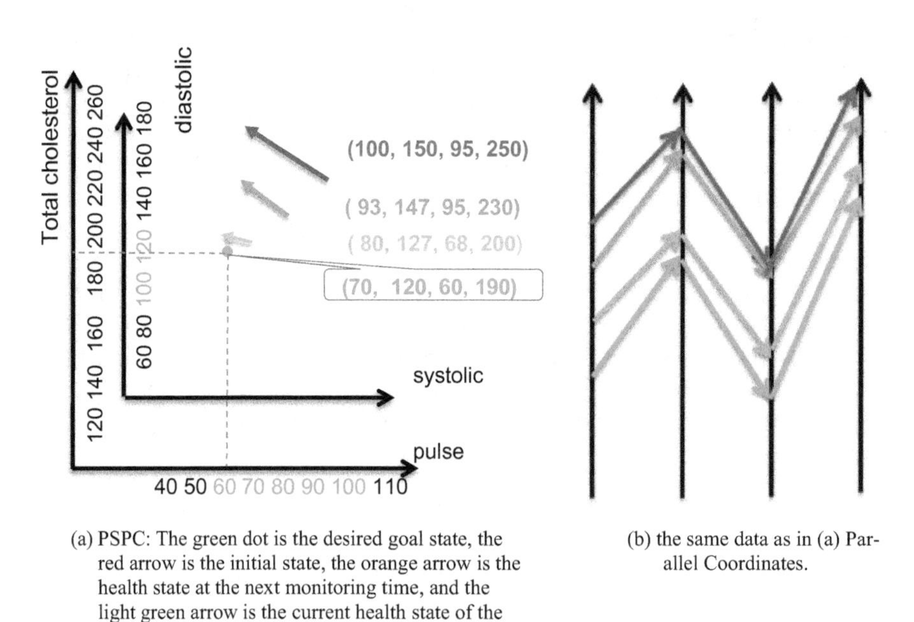

(a) PSPC: The green dot is the desired goal state, the red arrow is the initial state, the orange arrow is the health state at the next monitoring time, and the light green arrow is the current health state of the person.

(b) the same data as in (a) Parallel Coordinates.

Fig. 5.15 4-D Health monitoring visualization in PSPC **a** and Parallel Coordinates **b** with parameters: systolic blood pressure, diastolic blood pressure, pulse, and total cholesterol at four time moments

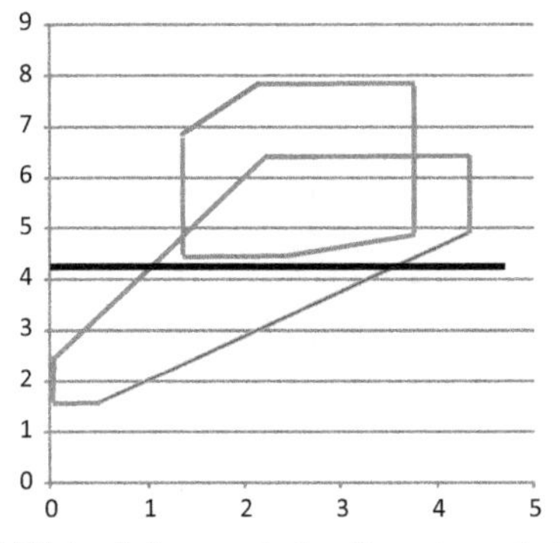

(a) Extended convex hulls of two classes in 4-D. The blue convex hull represnts class 1 and red convex hull represents class 2.

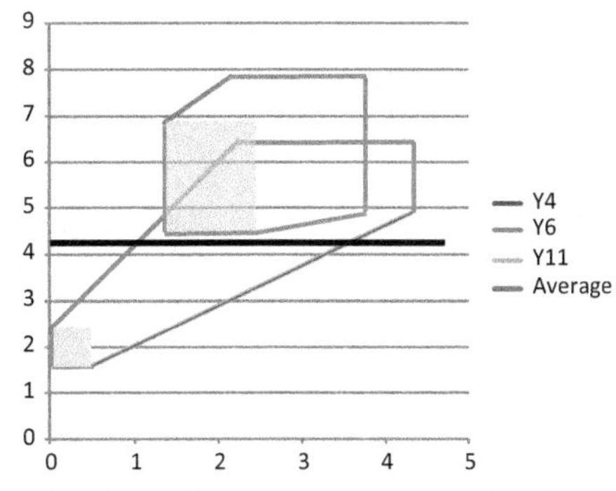

(b) Class unique areas: blue area for class 1 and yellow area for class 2.

Fig. 5.16 Convex hulls for Iris data

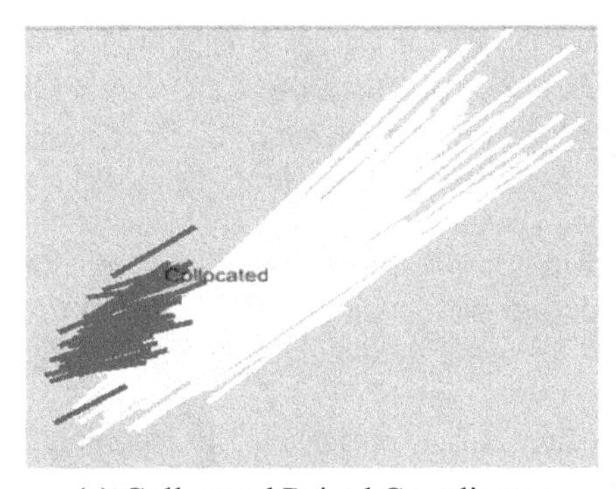

(a) Collocated Paired Coordinates

(b) Anchored Paired Coordinates.

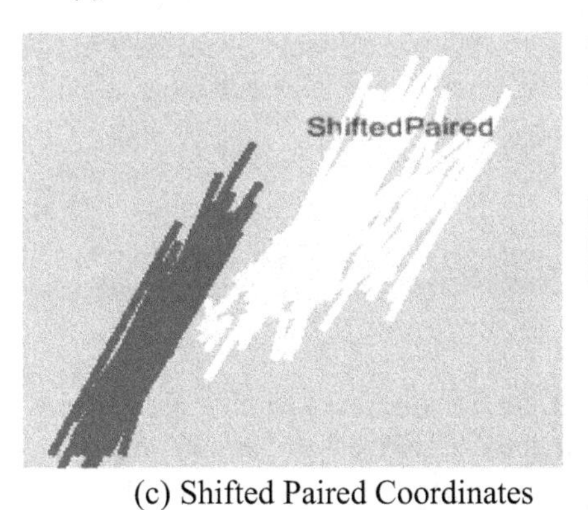

(c) Shifted Paired Coordinates

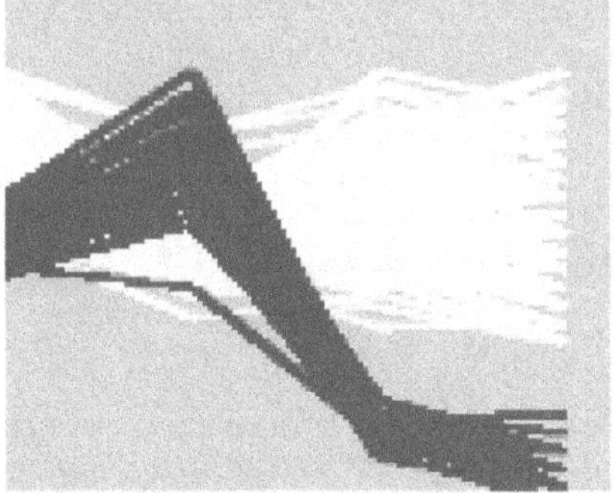

(d) Parallel Coordinates

Fig. 5.17 Iris data. Red is Iris-setosa class

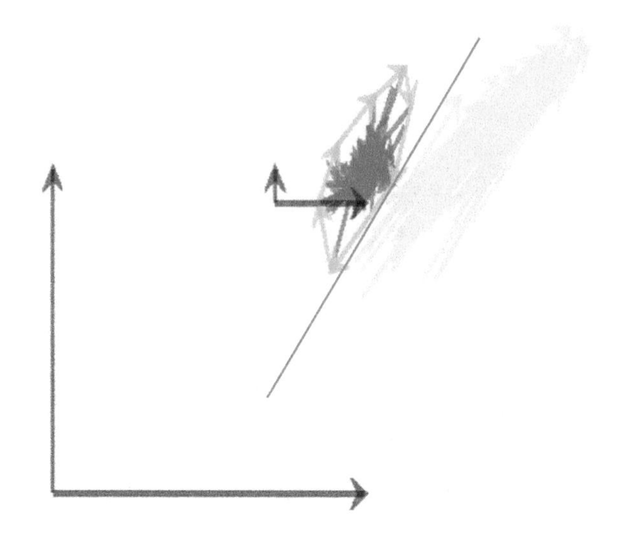

Fig. 5.18 Iris data in Parametrized Shifted Paired Coordinates—PSPC anchored in class 1. The class 1 convex hull is in green

and 5.19 below as other examples). In 4-D case it means smaller arrow as Fig. 5.15a shows. This is also a quite intuitive metaphor that cases that are closer to the goal are more similar to the goal in its visual representations.

Next, Fig. 5.15 uses color to indicate the progress in reaching the goal. The initial health status is shown as a red arrow. Then the arrows that are closer to the goal are shown in yellow and light green with the goal shown as a dark green dot. A few informal experiments that we conducted with participants had shown that people very quickly grasp how to use PSPC for such health monitoring. Studies that are more formal will be conducted later.

While this example used 4 health indicators it can be expanded to incorporate more such indicators. For instance, adding two more health indicators will just add another pair of shifted Cartesian Coordinates. The goal still will be a single dark green 2-D dot with each graph that represents the status at time t consisting of two connected arrows. These graphs became smaller when they approach the goal point. We will illustrate this situation in some cases studies below.

5.7 Case Study 7: Iris Data Classification in Two-Layer Visual Representation

Below we present a visual process to construct the Machine Learning classification model using GLCs with *Iris data* from the UCI Machine Leaning repository (Lichman 2013) that are commonly used to demonstrate new Machine Learning methods. These data represent three classes of Iris. Each record is described by four attributes: SL: sepal length, SW: sepal width, PL: petal length, and PW: petal width.

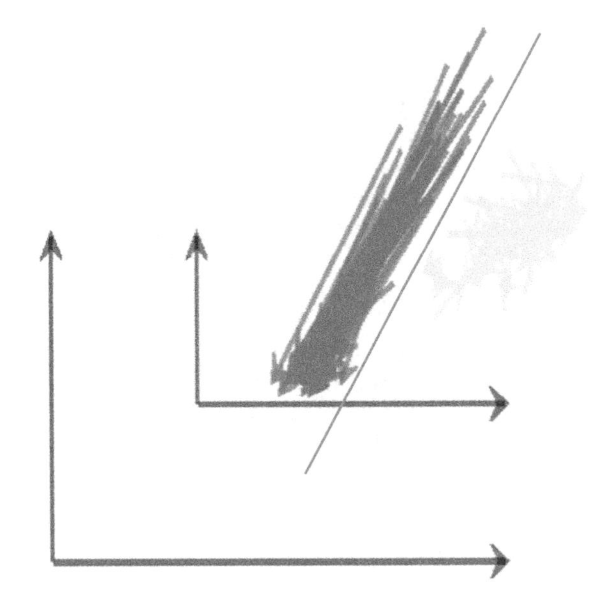

Fig. 5.19 Change number. Iris data in Parametrized Shifted Paired Coordinates (PSPC) anchored in class 2

5.7.1 Extended Convex Hulls for Iris Data in CPC

In this section, we explore whether we can separate classes visually by comparing hyper-rectangles built around each class. We form a 4-D hyper-box (hyper-rectangle) with the center at average points of the respective class. Each 4-D hyper-box is fully described by its 16 corners. To ensure that each hyper-box includes all points of the respective class the corners are defined as 4-D points by all 16 combinations of min and max values of 4 coordinates (PW, PL, SW, SL) of all points from the respective class. We will call these hyper-boxes *extended convex hulls* for the classes because the actual convex hulls of these classes are contained in them.

Figure 5.16a shows extended convex hulls of the first two classes of Iris data in Collocated Paired Coordinates. These 2-D convex hulls overlap as Fig. 5.16a shows. However, it does not imply that 4-D data are also overlap. It is similar to the case of usual projections: overlap of projection to any coordinate does not imply the overlap of n-D data. There is also an important difference between CPC and usual projections.

In general, a single 2-D projection of n-D point does not allow restoring this n-D point, but CPC allow this, because each n-D point is encoded as a graph in 2-D with n/2 edges not as a single 2-D point. Below we explain the way to discover that classes do not overlap in 4-D.

Consider the black horizontal line in Fig. 5.16a. At first glance it cannot discriminate classes with 100% accuracy due to overlap. This could be the case if 2-D graphs of some 4-D points are *fully located in the overlap area*. Otherwise classes can be fully discriminated by the black horizontal line in the example.

If there are graphs from two classes that are partially or fully in the overlap area, then we search visually for subareas $\{E_{1i},\}$ and $\{E_{2i}\}$ in CPC coordinates, where only nodes of graphs of one class present and nodes of graphs of another class are absent, respectively.

A simple version of this situation is when there is an area E_m where nodes $\{Q_{i1}, Q_{i2},\ldots, Q_{ik}\}$ of *all training data of only one class* C_m are present. For this situation the discrimination rule R has a simple form for n-D point $\mathbf{x} = \{x_1, x_2,\ldots, x_n\}$:

$$R : \text{If}\{Q_{i1}(\mathbf{x}), Q_{i2}(\mathbf{x}), \ldots, Q_{ik}(\mathbf{x})\} \subseteq E_m \text{ then } \mathbf{x} \in C_m.$$

This is the case in the example for the data from Fig. 5.16, see the blue area (E_1) and light orange area (E_2) in Fig. 5.16b. The first nodes of all 4-D points of class 1 are in the blue rectangle and the first nodes of all 4-D points of class 2 are in the light orange rectangle.

5.7.2 *First Layer Representation*

Below we explore the separation of Iris class 1 from classes 2 and 3 in Parallel. Collocated Paired, Anchored Paired, and Shifted Paired Coordinates defined in Chap. 2. Figure 5.17 shows the results of the comparison of PC, CPC, APC, and SPC for these data that contain 150 4-D iris records.

In contrast with the Parallel Coordinates (Fig. 5.17d), in new Collocated Paired visualizations (Fig. 5.17a, b, c) class 1 (red) and classes 2 and 3 (white) almost do not overlap. The Iris-setosa class (class 1) is clearly separated from the other two classes in these new visualizations. In CPC, classes slightly overlap and in SPCs, they touch each other, but do not overlap. Here in APC and SPC, the anchor and shifts are fixed and selected in advance without using this dataset to assign them. Later we will show parametrized SPC visualization with shifts adjusted for the given Iris data.

Note that CPC, APC, and SPC paired visualizations need only one 2-D segment to represent a 4-D data record. In contrast, the Parallel Coordinates require three segments per 4-D record. The larger number of segments leads to more overlaps among the lines in Parallel Coordinates.

Figure 5.18 shows the results of representation of Iris classes 1 and 2 in PSPC with the anchor point as the middle point of cases of class 1 and Fig. 5.19 shows the same two classes with the anchor point as the middle point of cases of class 2.

The middle 4-D point is computed as $(\min(x_i) + \max(x_i))/2$ for each attribute in the respective class. In both pictures, classes 1 and 2 are clearly visually separated while the separation is better in (b). The blue lines in both figures are separation lines discovered visually. Those lines can be formalized for analytical linear discrimination of these classes.

In Figs. 5.18 and 5.19 Iris class 1 is shown in red and Iris class 2 is shown in yellow. Figures 5.18 and 5.19 also show a convex hull (in green) for Iris class 1. In addition to the black lines, a user can easily construct other discrimination lines visually at the different levels of generalization, i.e., how far from the given points each class is extended. These extensions can be non-convex hulls, then convex-hulls, extended convex hulls, and straight (linear) discrimination lines shown in Figs. 5.18 and 5.19. Comparison of classes in Figs. 5.18 and 5.19 show that the class that is used as a source of the anchor point in PSPC is visually represented as a smaller and more compact blob. It is in full accordance with PSPC concept and methodology. In Fig. 5.18 it is a small red blob and in Fig. 5.19 it is a small yellow blob. The anchor point A is represented as a single 2-D point and 4-D points that are close to it are represented as small graphs around it.

Next, we show separation of classes 2 and 3. Figure 5.20 shows graphs of the n-D points of classes 2 and 3 in PSPC with the anchor point as the average point of class 2. The graphs of n-D points of class 2 are in yellow. In Fig. 5.20 the graphs of class 3 that are fully within the convex hull of class 2 are in orange. At first glance, here classes 2 and 3 heavily overlap. In fact, this is not the case.

Most of the end points (x_3, x_4) of the arrows that represent 4-D points of class 3 are on the right of one red line and above another red line in Fig. 5.20. Respectively there is a simple rule that separates class 2 from class 3 based on these two red lines in the coordinates (X_3, X_4). The first line is a vertical line $x_3 = e$, where e is some constant extracted from the Fig. 5.20. The second line has an equation $d_3x_3 + d_2x_4 + d_{12} = 0$, where coefficients also extractable from Fig. 5.20. Thus, this rule is as follows for a 4-d **x** point:

$$\text{If } (x_3 > e) \text{ or } (d_3x_3 + d_2x_4 + d_{12} > 0) \text{ then } \mathbf{x} \text{ belongs to class 3} \quad (5.1)$$

An alternative and more conservative (less generalized) way to separate the cases of class 3 that are only partially within the convex hull of class 2 is building a rule that directly uses the convex hulls:

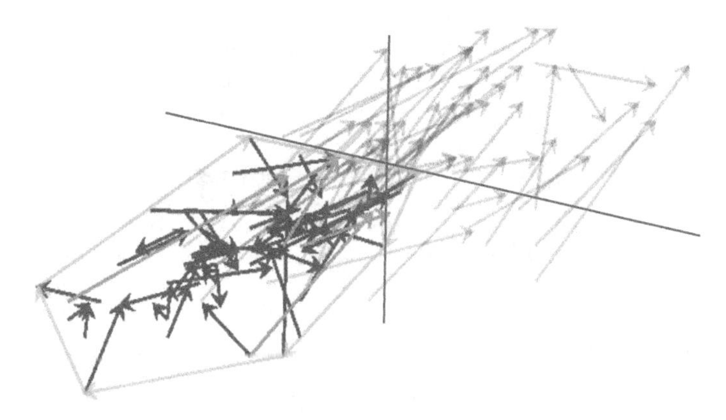

Fig. 5.20 Graphs of n-D points of classes 2 and 3 in PSPC. Graphs of 4-D points of class 2 are dark blue, of class 3 are light blue, and graphs of class 3 that are fully within the convex hull of class 2 are orange

$$\text{if}(x_1, x_2) \notin H_1 \;\&\; (x_1, x_2) \notin H_2 \text{ then } \mathbf{x} \text{ is in class 3, else } \mathbf{x} \text{ is in class 2} \quad (5.2)$$

Here (x_1, x_2) is a part of the 4-D point $\mathbf{x} = (x_1, x_2, x_3, x_4)$, H_1 and H_2 are convex hulls of classes 1 and 2, respectively, i.e., if pair (x_1, x_2) of coordinates of the 4-D point **x** is not in the convex hulls of classes 1 and 2 then **x** is in the class 3. The accuracy of both rules for classes 2 and 3 is the same (50 + 45)/(50 + 50) = 0.95 with 50 cases in class 2 and 50 cases in class 3, due to the fact that 5 cases of class 3 are fully within the convex hull of class 2 (see orange cases in Fig. 5.20).

5.7.3 Second Layer Representation for Classes 2 and 3

The next step is an attempt to improve this accuracy by using *the second layer of visual discovery* by:

- Extracting visually the features from graphs of class 2 and from misclassified graphs of class 3 in Fig. 5.20 that can potentially discriminate these cases, and then
- Discovering new classification rules visually based on those new secondary features.

While below we demonstrate this approach for Iris data the concept of the two-layer representation is a part of a general concept of *multilayer visual knowledge discovery*. When classes are not fully separated in the visualization in original features, this first layer visualization serves as a source of information for extracting features of the second layer to be used to fully separate classes. Similarly, the features of the second layer serve as a source for the third layer and so on if needed.

The visual analysis of Fig. 5.20 shows that orange arrows are closer to the anchor point **a** (in the middle of the dark blue arrows) than other arrows from class 3. This leads to the extraction of the distance between **x** and **a** and the distance between the ends of **x** and the anchor point **a** in coordinates (X_1, X_2). Another subtle feature is associated with the length of orange lines relative to arrows of class 2. Several of them are longer than arrows of class 2. This leads to extracting lengths of arrows. To simplify computations we computed horizontal and vertical projections of length of arrow. The results of this visual feature extraction written for 4-D point **x** and the 4-D anchor point $\mathbf{a} = (a_1, a_2, a_3, a_4)$ are:

$y_2 = ((x_3 - a_3)^2 + (x_4 - a_4)^2)^{1/2}$—the distance between points **x** and **a**.
$y_1 = ((x_1 - a_1)^2 + (x_2 - a_2)^2)^{1/2}$—the distance between the ends of **x** and the anchor point **a**;
$y_3 = x_1 - x_3$—horizontal coordinate difference;
$y_4 = x_2 - x_4$—vertical coordinate difference.

Figure 5.21 shows the results of the representation of 50 cases of class 2 and 5 misclassified cases of class 3 in y_1, y_2, y_3, y_4 coordinates in PSPC, with the anchor in the average case of class 2 in y_1, y_2, y_3, y_4 coordinates.

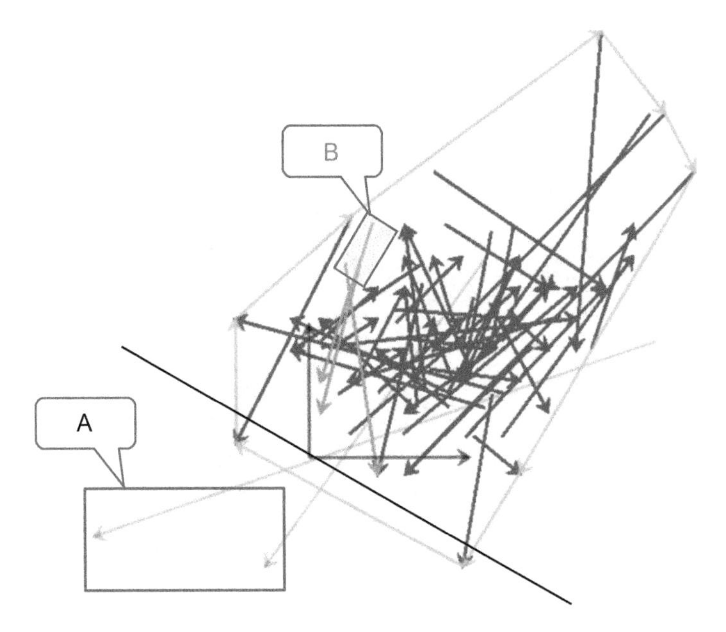

Fig. 5.21 Second layer visual representation: dark blue—50 cases of class 2, orange and light blue—5 cases of class 3 in y_1, y_2, y_3, y_4 coordinates in PSPC

Figure 5.21 allows visual finding two areas A and B where only points of class 3 are present. This leads to a rule:

$$\text{If}(y_3, y_4) \in \text{A or } (y_1, y_2) \in \text{B} \quad \text{then class 3} \tag{5.3}$$

It classifies all 5 cases of class 3 correctly with total 100% accuracy. Instead of (y_3, y_4) ∈ A we can use in (5.3) another more general and robust condition

$$(y_3, y_4) \notin H_2 \,\&\, \mathbf{y} \notin H_1,$$

where H_1, H_2 are the convex hulls of classes 1 and 2. The next robust, but less accurate rule for Fig. 5.21, is rule (5.4) with 3 cases of class 3 in area B misclassified based on the black line L that separates area A from the green convex hull of class 2,

$$\text{If } L(y_3, y_4) > 0 \text{ then class 2 else class 3} \tag{5.4}$$

5.7.4 Comparison with Parallel Coordinates, Radvis and SVM

Figures 5.22, 5.23 and 5.24 show all three Iris classes in Parallel Coordinates (Dzemyda et al. 2012; Gristein et al. 2002) and Radvis (Rubio-Sánchez 2015). Both

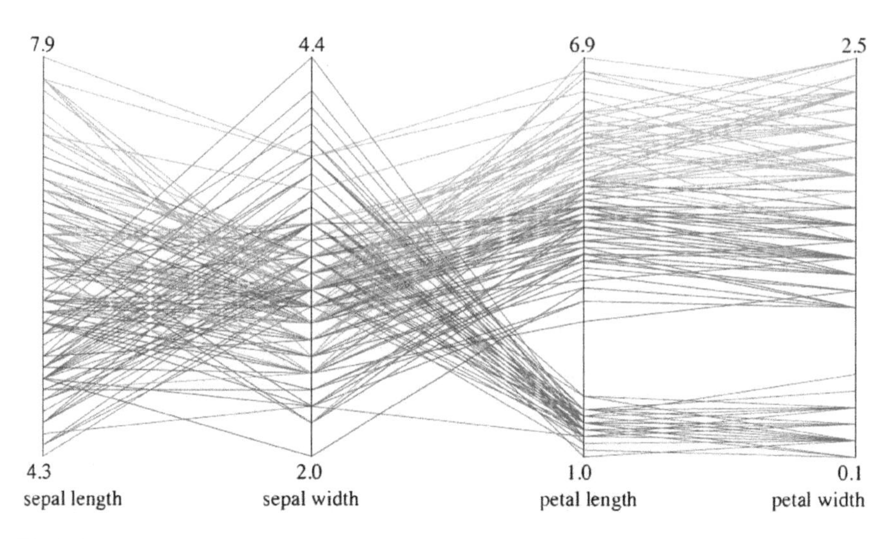

Fig. 5.22 Iris data in Parallel Coordinates representation (Dzemyda et al. 2012)

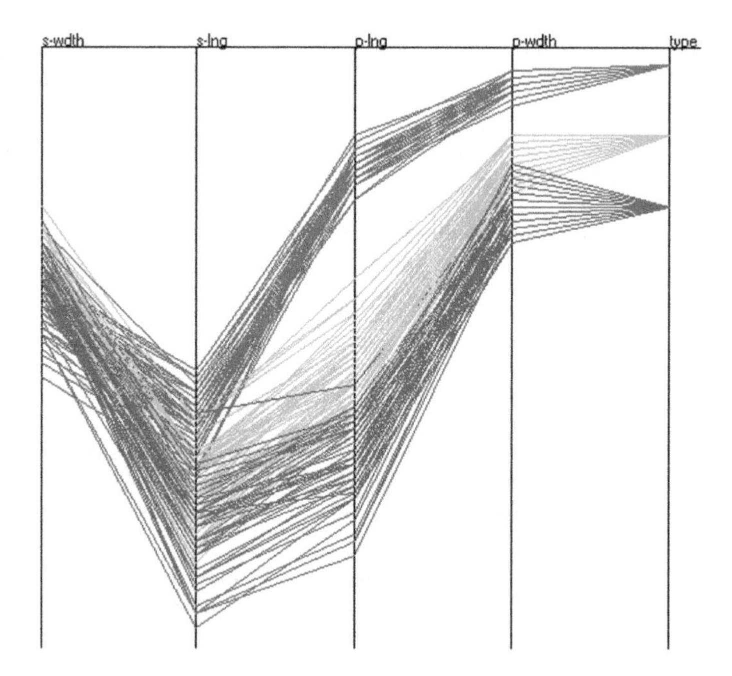

Fig. 5.23 Iris flowers in parallel coordinates (Grinstein et al. 2002)

clearly show the abilities to separate only class 1 from classes 2 and 3, but do not separate well the classes 2 and 3. Also, note that while Figs. 5.22 and 5.23 show the same data the same visualization method—Parallel Coordinates, the classification pattern is visible better in Fig. 5.23 due to showing original (not normalized) data.

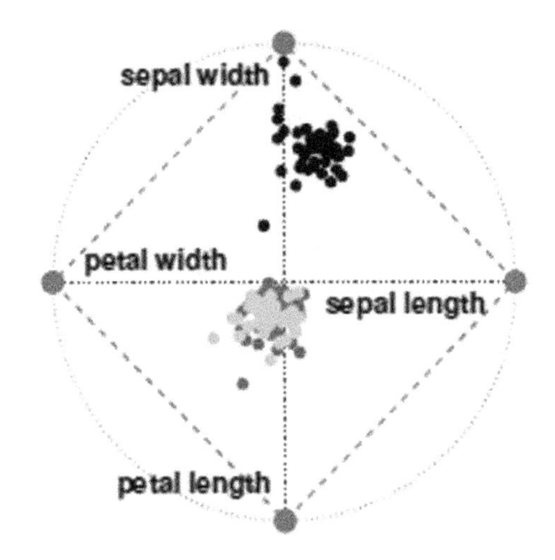

Fig. 5.24 Iris data in RadVis representation (Rubio-Sanchez et al. 2016)

Table 5.1 shows results of SVM algorithm on the same Iris data (Kaur 2016; Mafrur 2015). In both SVM models from (Kaur 2016) the total number of errors is 5, which are in classes 2 and 3.

This is exactly the same number of errors that we obtained in Fig. 5.20 before the second layer of visual discovery. Results from (Mafrur 2015) with 4 errors are slightly better than our result and from (Kaur 2016), but it uses 59 support vectors (over 1/3 of all cases) that may indicate overfitting. For class 2 it has 23 support vectors. Our solution uses only 9 4-D border points of the convex hull. Our alternative solution with two 2-D linear discrimination lines in Fig. 5.20 only needs 6 scalar coefficients with very similar accuracy as in (Mafrur 2015).

Note that while our second layer brings 100% accuracy it may be overfitting for the area B in Fig. 5.21 that is responsible for eliminating three errors. In contrast, a linear discrimination line (a black line in Fig. 5.21) that needs just three scalar coefficients can generalize the area A. The accuracy of this solution is 98% (147/150). The advantage of visual analysis is that a user can see areas A and B and judge how artificial and complex they are to decide which one to ignore to avoid overfitting.

Table 5.1 SVM confusion matrixes for Iris data

Radial kernel (Mafrur 2015)				Radial kernel (Kaur 2016)				Polynomial kernel (Kaur 2016)			
Real class	Predicted class			Real class	Predicted class			Real class	Predicted class		
	1	2	3		1	2	3		1	2	3
1	50	0	0	1	50	0	0	1	50	0	0
2	0	48	2	2	0	47	3	2	0	46	4
3	0	2	48	3	0	2	48	3	0	1	49

While this analysis does not involve cross-validation for testing models, the presented GLC visualizations are informative to the possible success or failure of cross-validation. It is visible in Figs. 5.20 and 5.21 that the common leave-one-out cross-validation will not significantly change the accuracy produced by the GLC methods.

Next, SVM will classify every 4-D point as belonging to one of three classes being trained on them, while Figs. 5.20, 5.21 and 5.22 show significant areas without any 4-D point from the given 150 4-D points.

Therefore, the refusal to classify 4-D points in such areas can be justified. In this case, the points outside of the convex hulls in Figs. 5.20 and 5.21 will not be classified to these 3 classes. When new genetically modified Iris will be introduced, the 4-D points in these areas can get their classification, but to another class, not to one of these 3 classes.

5.8 Case Study 8: Iris Data with PWC

This section demonstrates the use of Paired Crown Coordinates defined in Sect. 2.2.6 in Chap. 2. In this section, the PWC step that orders odd coordinates is omitted. It allows visualizing multiple n-D points with different orders in a single plot. This case study also uses 4-D Iris dataset from UCI Machine Learning repository (Lichman 2013). It includes 150 4-D records of three classes with 50 records per class. PWC are lossless which means that using it we are preserving and showing every part of Iris data without skipping anything.

The first phase includes the selection of a shape of the closed figure ("crown"). In this study, the crown is a circle with radius r. The normalized values of odd coordinates X_1 and X_3 of each 4-D point $\mathbf{x} = (x_1, x_2, x_3, x_4)$ are plotted on the perimeter of the circle. The formulas for mapping x_i with odd i, (i = 1, 3) of 4-D point $\mathbf{x}$ to the 2-D location (p_1, p_2) on the circle are:

$$p_{i1} = c_1 + r \times cos(a(x_i))$$
$$p_{i2} = c_2 + r \times sin(a(x_i))$$

where (c_1, c_2) is the center of the circle, r is the radius of the circle, and $a(x_i)$ is the angle. The angle $a(x_i)$ is proportional to the distance $D_i = ||x_i - x_{min}|$ on the circle from x_{min} to x_i relative to the *total distance*

$$D = ||x_{max} - x_{min}|| + d_g$$

where d_g is a gap between location of x_{max} and x_{min} on the circle.

Similarly, the formulas for mapping x_i with even i, (i = 2, 4) of 4-D point $\mathbf{x}$ to the 2-D location (q_1, q_2) are:

$$q_{i1} = c_1 + (r + x_i) \times \cos(a(x_i))$$
$$q_{i2} = c_2 + (r + x_i) \times \sin(a(x_i)).$$

Thus, two 2-D points $Q_1 = (q_{11}, q_{12})$ and $Q_2 = (q_{21}, q_{22})$ for pairs (x_1, x_2) and (x_3, x_4) from each 4-D point **x** are generated.

The final phase is connecting point Q_1 to point Q_2 by the directed edge (arrow). The Iris dataset with three classes were passed through all phases of the PWC process. Figures 5.25, 5.26 and 5.27 show these classes separately in PWC. Each 4-D point is visualized in PWC as a single dotted pink edge between yellow nodes Q_1 and Q_2.

The projections of Q_1 and Q_2 to the circle are blue dots on the circle. The difference in graphs of three classes can be easily determined in these figures. First, the *distances* from the crown (circle) to graphs in all of three classes are different. Class 1 practically has no distance from the circle in this figure, class 2 has some distance, and class 3 has comparatively larger distance. This difference allow immediately construct a classification rule R1 to separate class 1 from classes 2 and 3:

$$\text{R1}: \text{If } Q_1 \text{ and } Q_2 \text{ are in the circle with radius } r + e \text{ then } (Q_1, Q_2) \text{ belongs to class } 1,$$

where e is some positive constant that can be derived from Figs. 5.25, 5.26 and 5.27. Next, in all three classes, Q_2 is closer to the circle than Q_1 with edged go down towards the circle with some subtle differences in angles for different classes. Such properties also potentially can be used for designing another classification rule.

Also for the comparison, plots with pairs of classes were generated. Figure 5.28 shows the comparison plot of classes 2 and 3. As we have seen from individual plots, classes 2 and 3 differ more from class 1 than from each other. Figure 5.28 shows classes 2 and 3 together with coloring class 2 in blue and class 3 in yellow.

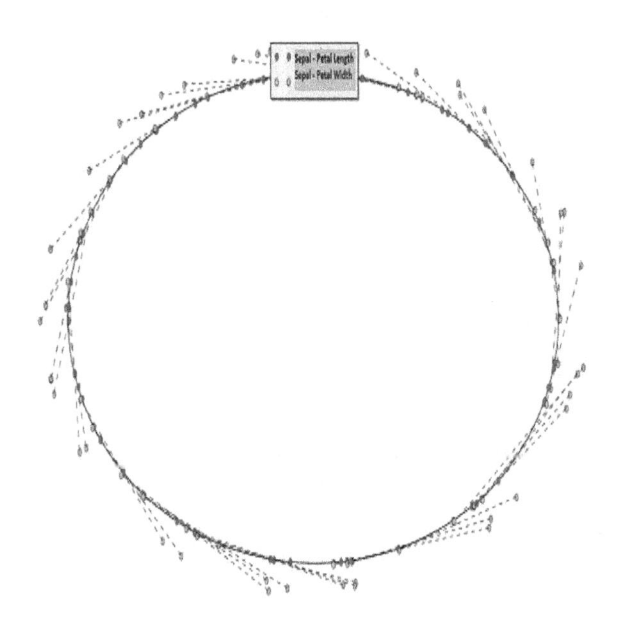

Fig. 5.25 Class 1 of Iris dataset in PWC

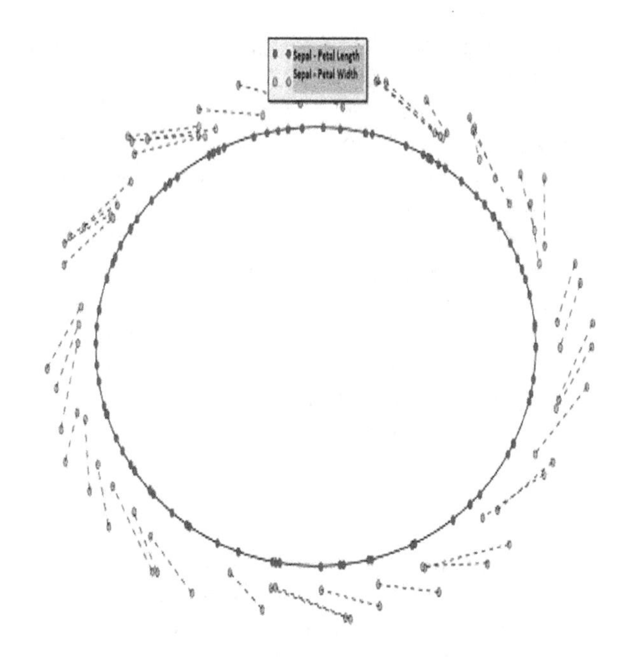

Fig. 5.26 Class 2 of Iris dataset in PWC

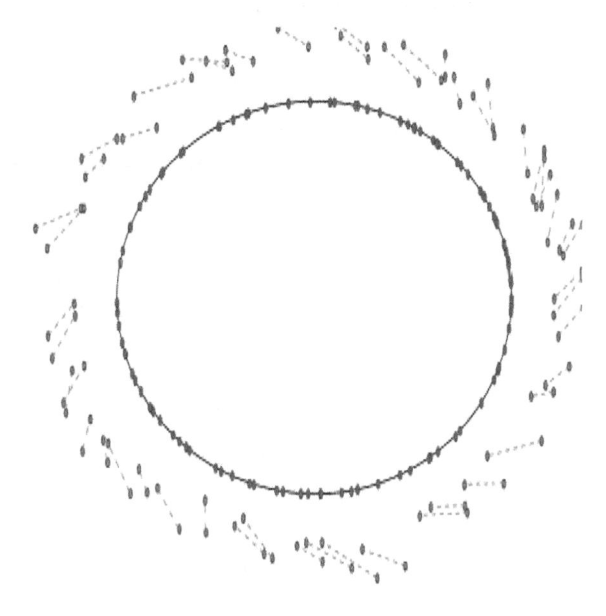

Fig. 5.27 Class 3 of Iris dataset in PWC

At the first glance, we do not have a visible pattern that separates classes 2 and 3 in Fig. 5.28. However, adding another circle (black circle in Fig. 5.29) the pattern became visible better.

Fig. 5.28 4-D points of Iris class 2 (blue) and class 3 (yellow)

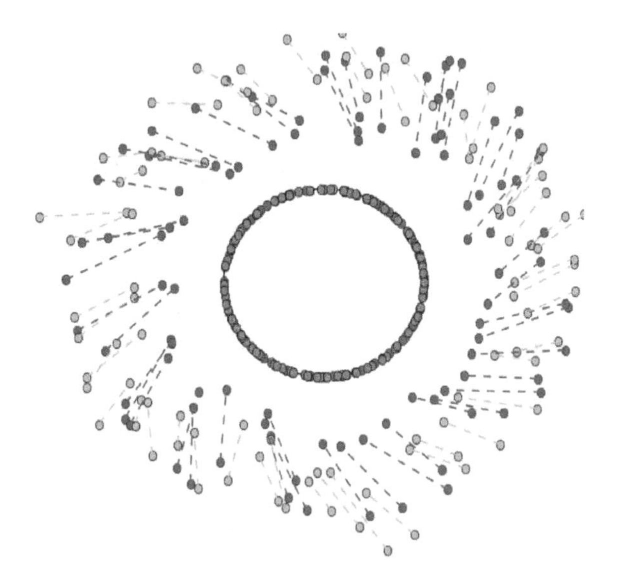

The data points inside the black circle that we denote as B show some unique properties. Most of those data points are blue, i.e., belong to class 2. This allows designing a classification rule R2,

R2 : If (Q_1, Q_2) is in circle B and (Q_1, Q_2) does not belong to class 1 then (Q_1, Q_2) belongs to class 2.

Figure 5.30 shows a comparison plot of classes 1 and 2, and Fig. 5.31 shows a comparison plot of classes 1 and 3 with another black circle E that discriminates class 1 from classes 2 and 3.

These plots are showing a clear distinction between classes where data points lying inside or outside the circle belongs to class 1 and data points lying outside the circle belongs to classes 2 and 3. Thus, we can specify rule R1 by setting a radius using Figs. 5.30 and 5.31.

An alternative PWC visualization of the same data with an *ellipse* as a crown is presented in Fig. 5.32. It shows the same result regarding accuracy of classification of the dataset. Data points that lie inside the black ellipse belong to class 1 and those lie outside the ellipse belong to classes 2 or 3.

The motivations for using an elliptic shape instead of a circle could be (1) giving more space for graphs in the area where graphs are densely located, (2) giving more space for areas where more important data are located data, or (3) shrinking space in areas with less number of graphs or less important data. It can be done expanding or shrinking the circle to the ellipse in respective directions. In essence, it works as local zooming similar to fish eye zooming.

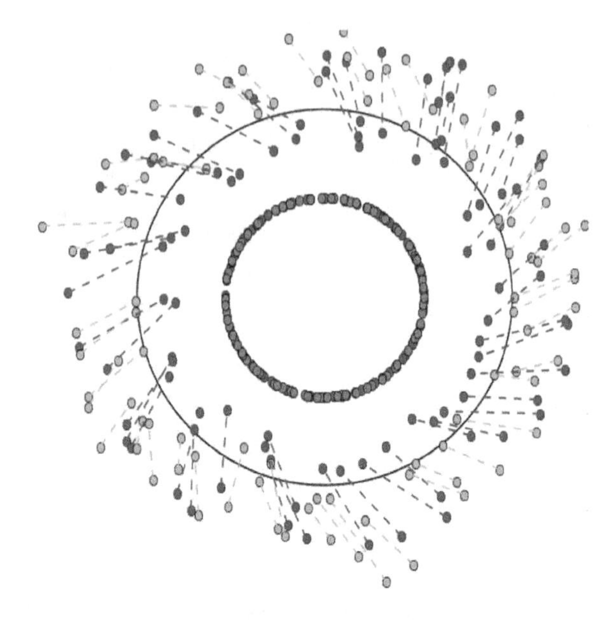

Fig. 5.29 4-D points of Iris class 2 (blue) and class 3 (yellow) with discrimination circle (thin black)

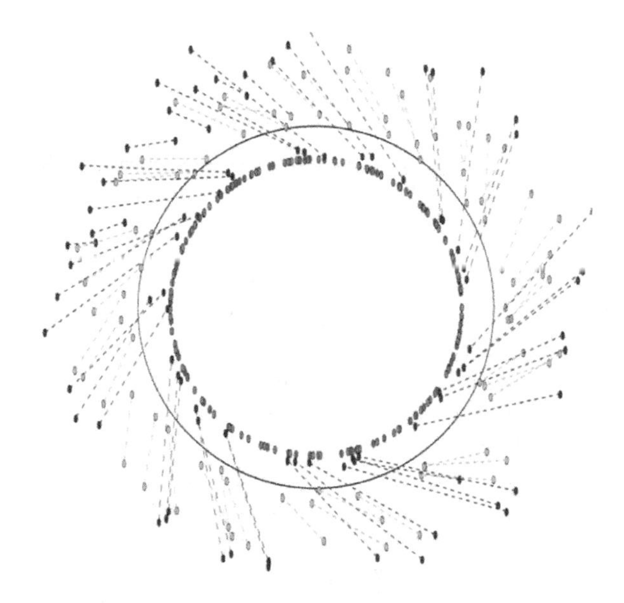

Fig. 5.30 4-D points of class 1 (magenta) and class 2 (yellow)

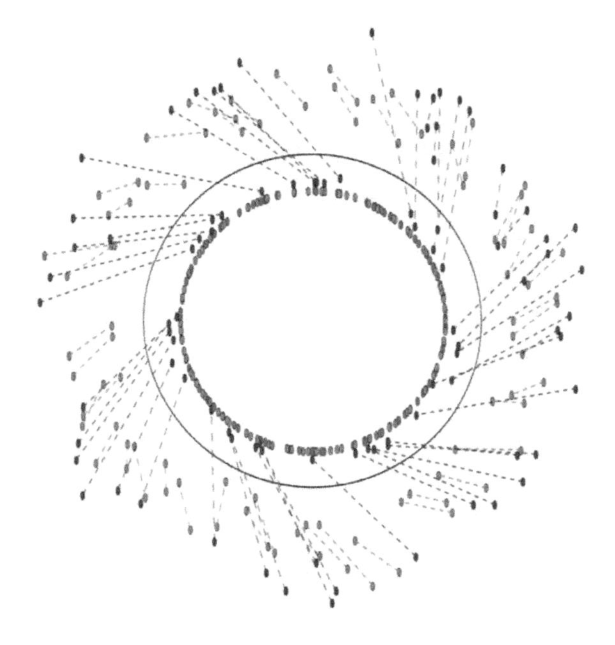

Fig. 5.31 Iris 4-D points of class 1 (magenta) and class 3 (green)

5.9 Case Study 9: Car Evaluation Data with PWC

This case study used 6-D *Car evaluation data* from UCI Machine Leaning repository (Lichman 2013) with four classes. Figures 5.33 shows 6-D points each class individually. Car evaluation data have six dimensions, which leads in PWC to three 2-D points for each 6-D point and two arrows connecting three points on a 2-D plane in its visualization.

Figure 5.33 allows one to see some patterns with respect to arrows direction. All arrows are coming *towards* the circle for third data point and all the arrows are going *away* from the circle for second data point, which shows that first and third data points lie close to the circle whereas second data points lie away from circle. The same "triangular" pattern is present in all four classes of data.

Features that are more interesting are revealed in this dataset if we plot each node of the graphs in different color (see Fig. 5.34). In the class 2 (see Fig. 5.34b), all the starting nodes that are shown in red are making a circle and all the nodes that are shown in blue come under that circle.

Therefore, based on this observation, rule R3 was generated to be tested for each $\mathbf{x} = (x_1, x_2, x_3, x_4, x_5, x_6)$

$$\text{R3 : If } x_2 > x_6 \text{ then } \mathbf{x} \text{ belongs to Class 2.}$$

Similarly, by looking into data visualization of classes 1 and 3 rule R4 was generated to be tested:

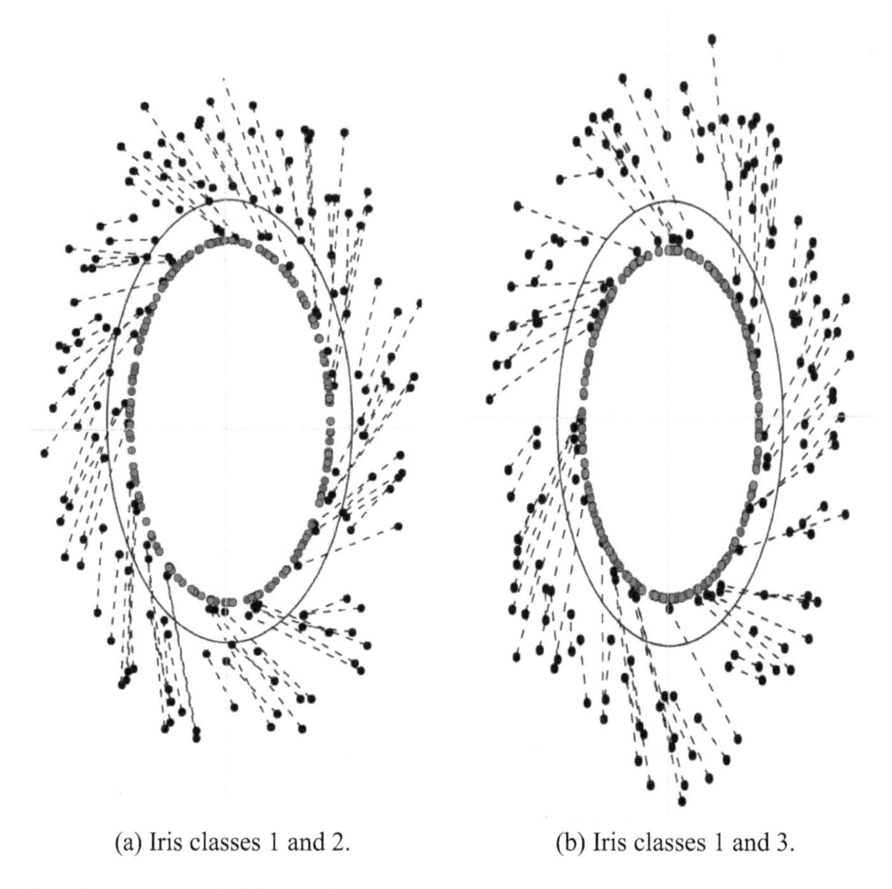

(a) Iris classes 1 and 2. (b) Iris classes 1 and 3.

Fig. 5.32 Iris data in PWC with elliptic crown

$$\text{R4 : If } x_6 \geq x_2 \text{ AND } x_6 < x_4 \text{ then } \mathbf{x} \text{ belongs to Class 1 OR Class 3.}$$

By looking into class 4 another distinct rule R5 was generated by considering the *difference* Δ between positions of x_2 and x_6 in visualization ($\Delta x_2 x_6$), which is the same constant C for all nodes of class 4:

$$\text{R5 : If } x_6 \geq x_2 \text{ AND } x_6 < x_4 \text{ AND } \Delta x_2 x_6 = C \quad \text{then } \mathbf{x} \text{ belongs to Class 4.}$$

By the design of this visualization the difference $\Delta x_2 x_6$ is equal to x_6 and the rule is simplified:

$$\text{R5 : If } x_6 \geq x_2 \text{ AND } x_6 < x_4 \text{ AND } x_6 = C \quad \text{then } \mathbf{x} \text{ belongs to Class 4.}$$

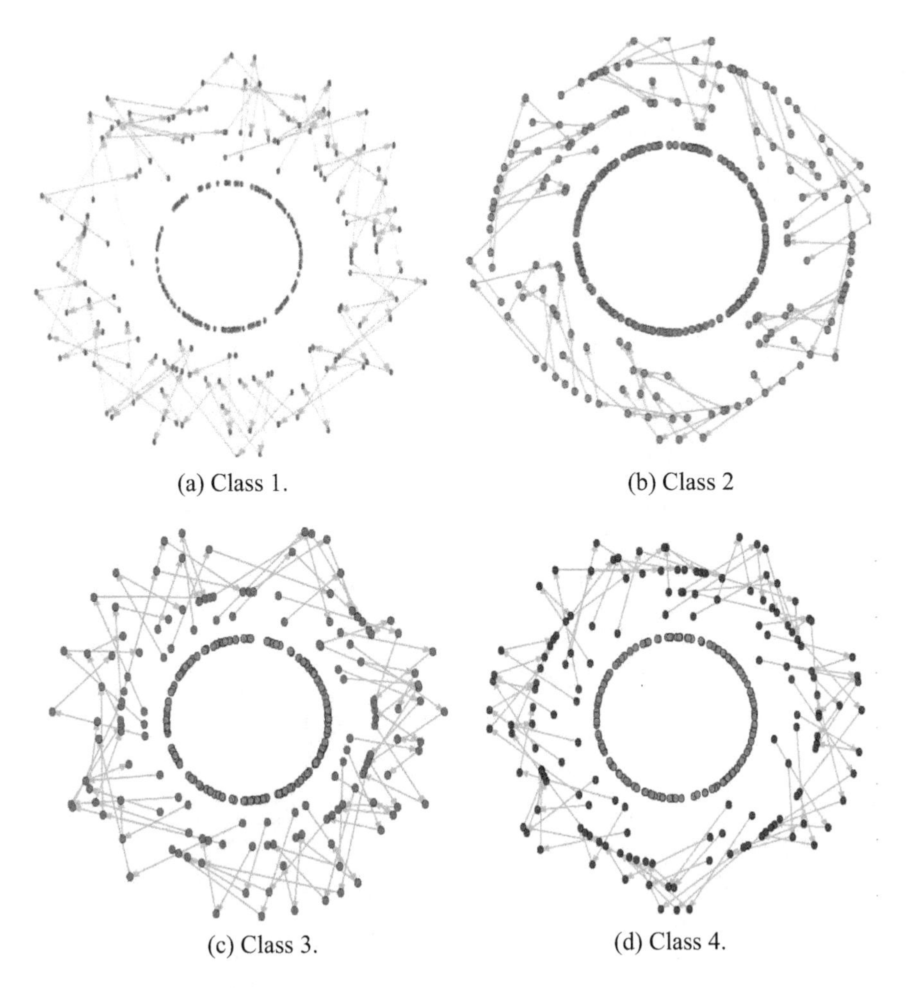

Fig. 5.33 Car data in PWC

Here x_6 is one of two values (x_5, x_6) mapped to the third node (blue in Fig. 5.34), x_4 is one of two values (x_3, x_4) mapped to the second node (green in Fig. 5.34), and x_2 is one of two values (x_1, x_2) mapped to the first node (red in Figs. 5.34),

The combination of rules R4 and R5 allows deriving the rule for class 1:

$$\text{R6 : If R4}(\mathbf{x}) \text{ and not R5}(\mathbf{x}) \text{ then } \mathbf{x} \text{ belongs to class 1,}$$

i.e., if rule R4 is true for **x**, but rule R5 is false for **x** than **x** is in class 1. This rule R6 was false on 4 cases other rules were correct on tested cases.

To derive these illustrative rules only about 11% of the data have been used. This shows the opportunities to build and validate more complex rules with more data used in the same visual representation of data in PWC.

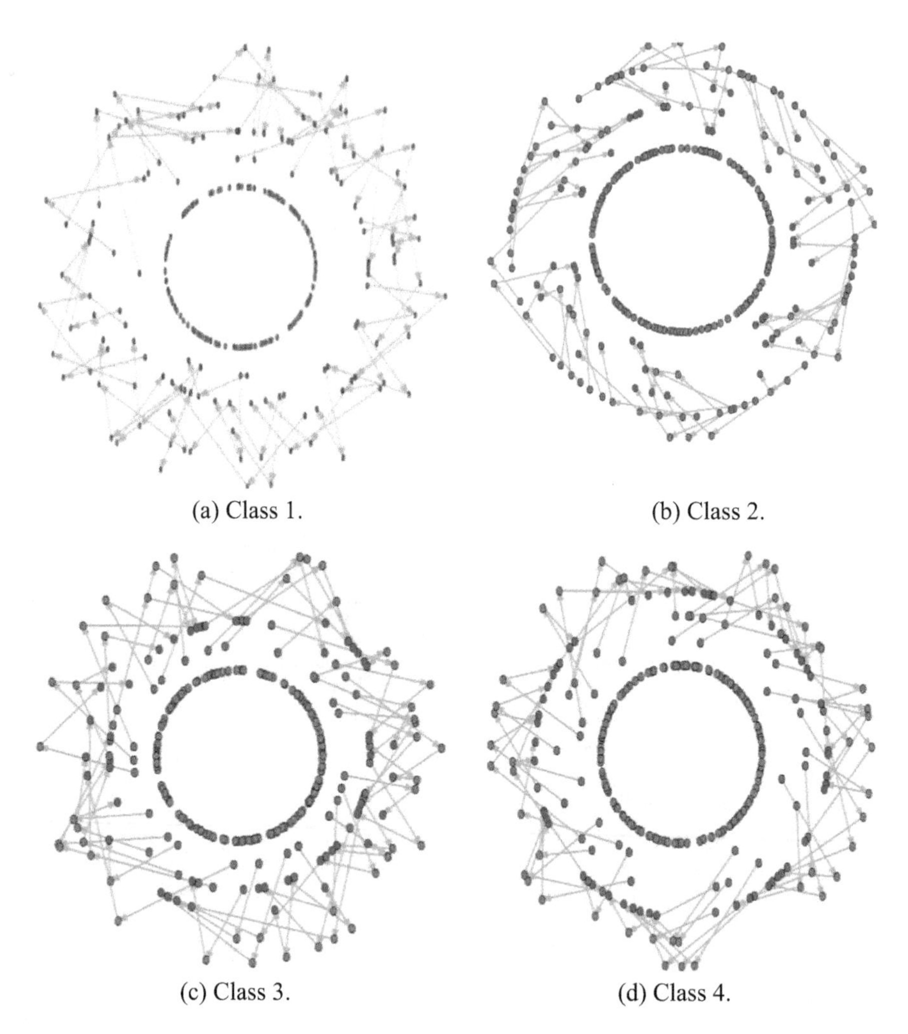

Fig. 5.34 Car data in PWC with differently colored nodes for each 6-D point

5.10 Case Study 10: Car Data with CPC, APC, SPC, and PC

In this section the same car data as in Case study 9 are visualized with CPC, APC, SPC, and PC. Visualizing the data separately for each class makes a clear distinction between classes when compared altogether.

As mentioned above, classes 2 and 4 have unique features in PWC with unique rules. These classes also exhibit unique patterns in some of GLCs used in this case study. Classes 1 and 3 share some patterns in all four visualization methods similarly to observed in PWC visualization before.

In Fig. 5.35 in CPC, each class exhibits a distinct pattern with class 2 as the most distinct from other classes. While these patterns are visible and describable, they are not an immediate guide for classification of individual cases. How can we use these distinct patterns? We can check whether the same unique visual pattern is present in training, validation, and testing data. If this is the case, then training data are correct data for the given validation and testing data. This justifies applying analytical machine learning algorithms for discovery discrimination rules on these training data for expected validation and testing data. We discuss this fundamental opportunity and demonstrate its application in Chap. 7.

In Fig. 5.36 in APC, also each class exhibits a distinct pattern with class 2 as the most distinct from other classes. Similarly to Fig. 5.35, we can

- check whether the same unique visual pattern is present in training, validation, and testing data, and
- justify the use of these training data, or
- reject these training data and construct new training data.

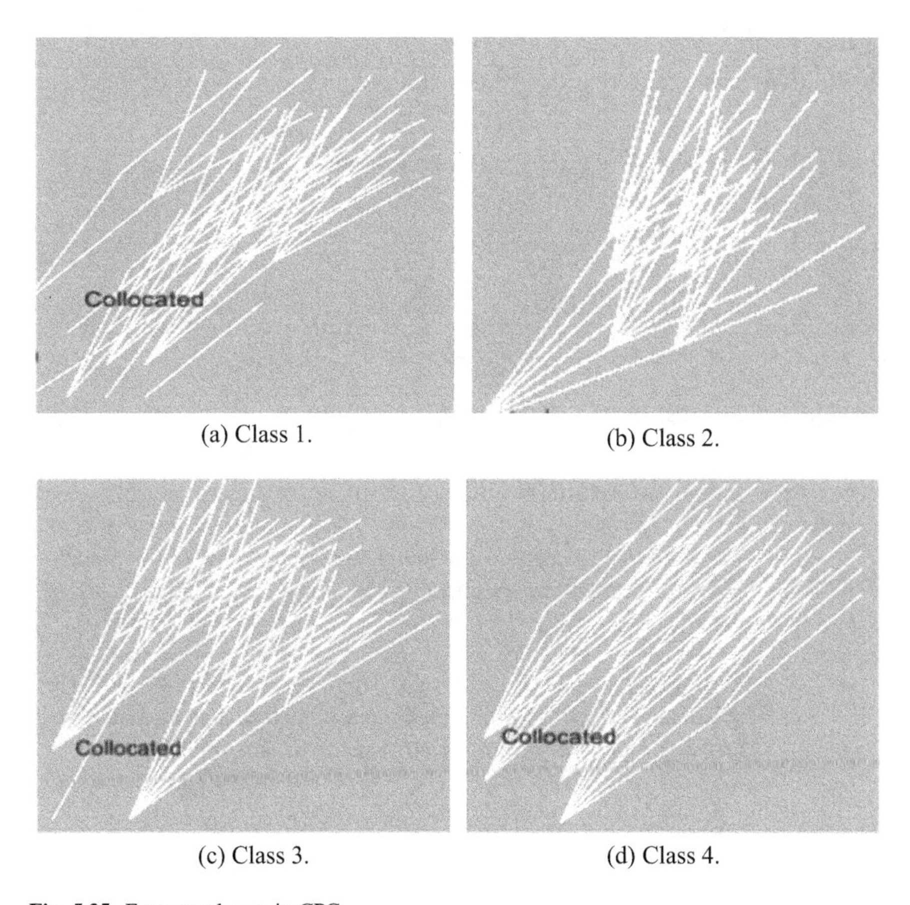

(a) Class 1.

(b) Class 2.

(c) Class 3.

(d) Class 4.

Fig. 5.35 Four car classes in CPC

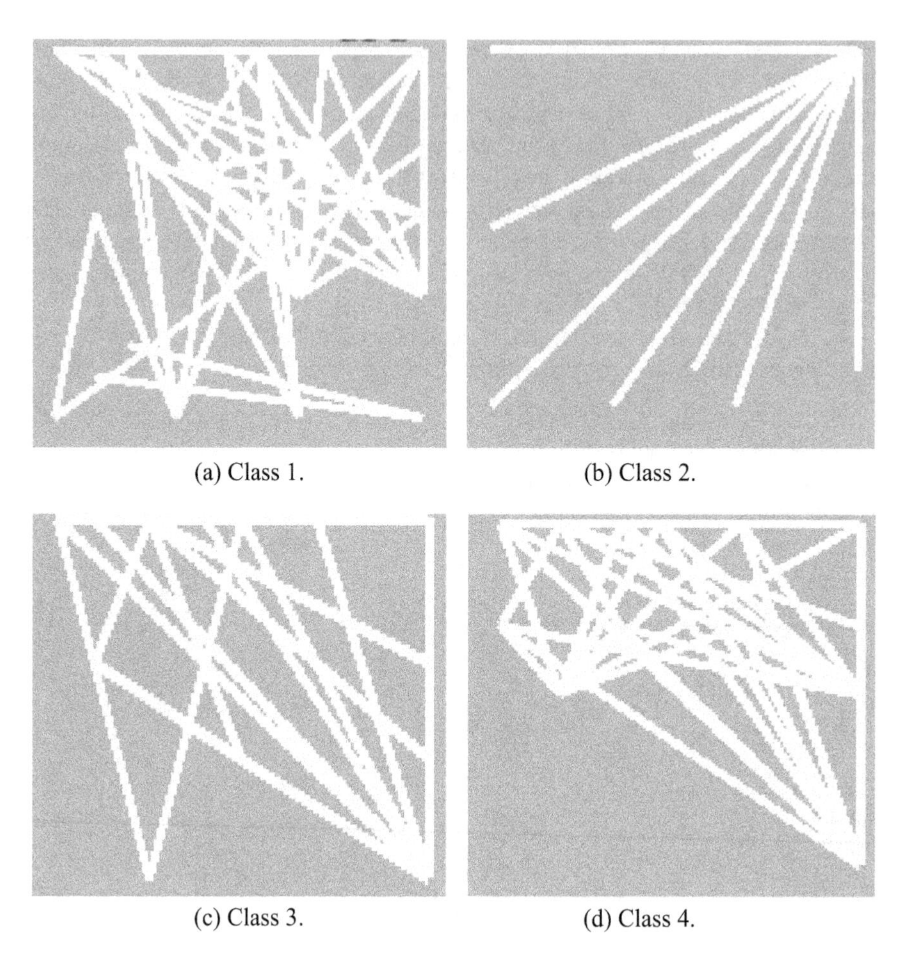

Fig. 5.36 Four car classes in APC

Note, that in contrast with CPC, in Fig. 5.35, here the pattern in APC can be used directly to build and test a classification rule using unique directions of edges.

In Fig. 5.37 in SPC, the distinction of visual patterns is less obvious, while class 3 and 4 are most distinct from classes 1 and 2. The classification rule for classes 3 and 4 can be constructed from Fig. 5.37c using unique directions of edges from the bottom.

In Fig. 5.38 in PC, the distinction of visual patterns is also less obvious, while class 2 is most distinct. The classification rule for class 2 can be constructed directly from Fig. 5.38c using unique directions of edges from the bottom.

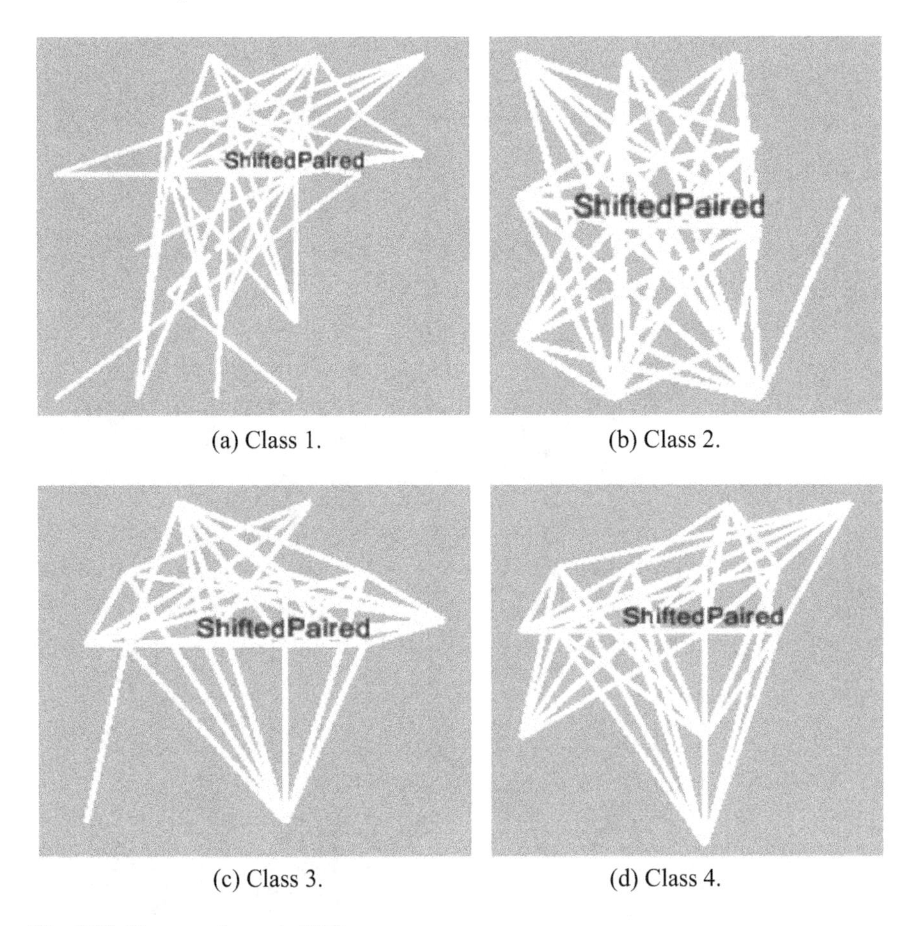

(a) Class 1.
(b) Class 2.
(c) Class 3.
(d) Class 4.

Fig. 5.37 Four car classes in SPC

5.11 Case Study 11: Glass Identification Data with Bush Coordinates and Parallel Coordinates

This case study illustrates Bush Coordinates (defined in section in 2.1.6 in Chap. 2) using *Glass identification data* from UCI Machine Learning repository (Lichman 2013). Bush Coordinates are defined in Chap. 2. The full Glass dataset consists of 214 instances represented by ten attributes and the type of glass. Figure 5.39 shows a hundred of 10-D records of three types glass (in red, green and blue) from this dataset in Parallel and Bush Coordinates.

The comparison of visualizations in Parallel and Bush Coordinates shows that in Bush Coordinates:

- blue lines for all coordinates are more clustered, and
- red and green lines that connect coordinates X_1 and X_2 are more clustered.

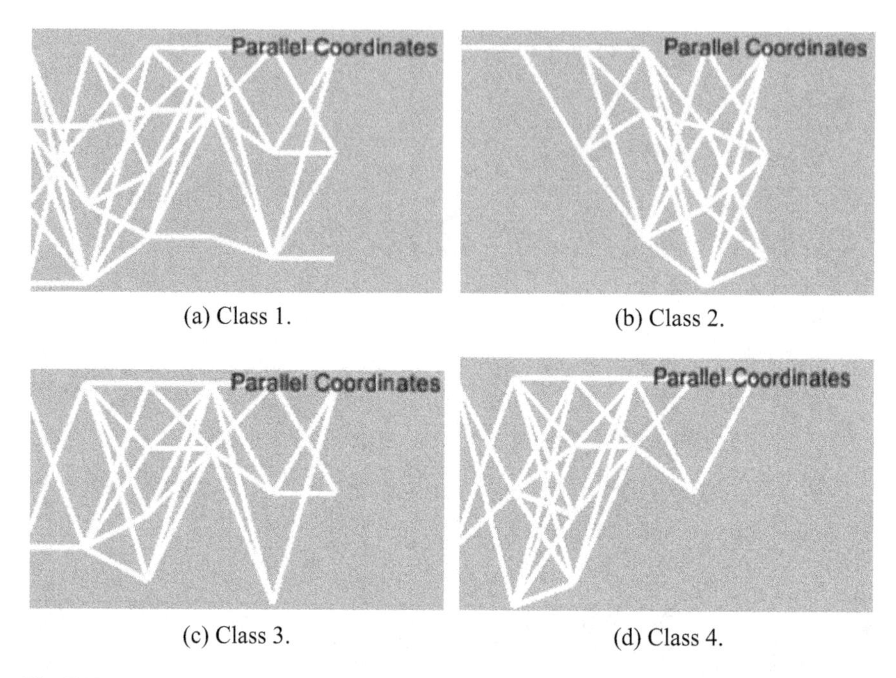

(a) Class 1. (b) Class 2.

(c) Class 3. (d) Class 4.

Fig. 5.38 Four car classes in PC

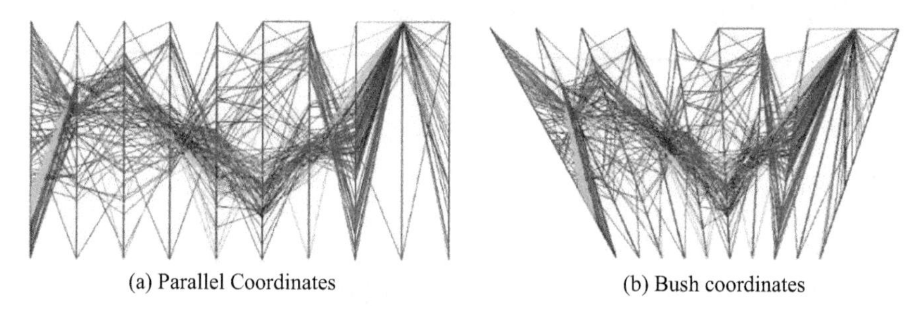

(a) Parallel Coordinates (b) Bush coordinates

Fig. 5.39 Three classes of Glass visualized in Parallel and Bush Coordinates

As a result, on Bush Coordinates we can see faster and better

- clusters of red and blue lines,
- negative correlation of red and green lines connecting X_1 and X_2, and
- two opposing correlation in blue lines connecting X_7 and X_8 coordinates.

This example demonstrates that tilt of coordinates allows improving the perception of visual patterns.

5.12 Case Study 12: Seeds Dataset with In-Line Coordinates and Shifted Parallel Coordinates

This section explores *seed data* from UCI Machine Learning Repository (Lichman 2013) with Parallel Coordinates and In-Line Coordinates.

Shifted Collapsing Parallel Coordinates. These coordinates are defined in Chap. 2 in the section on Bush and Shifted Parallel Coordinates. While commonly, Parallel Coordinates are drawn vertically, in Fig. 5.40 they are rotated and made horizontal for better comparison with In-Line Coordinates later in this section. In this figure, data of class 1 (seeds of diameter 1 mm) are on the bottom (red) and data of the class 2 (seeds of diameter 2 mm) are on the top (green). Each record of each class is 7-dimensional. For this case study, 100 randomly selected records out of total 210 records are used.

In this figure, the green data are grouped more tightly together. The pattern of red data is somewhat similar except for the few curves that are further down the line showing that some of the seeds of this class have more varying attributes. Some differences in patterns of two classes are visible. However, it is not clear how to build a discrimination function from them. Therefore, other visualizations are explored below.

In-Line Coordinates. Below we use in-line coordinates with triangles and Bezier curves defined in Chap. 2 in the section on In-Line Coordinates. In Fig. 5.41

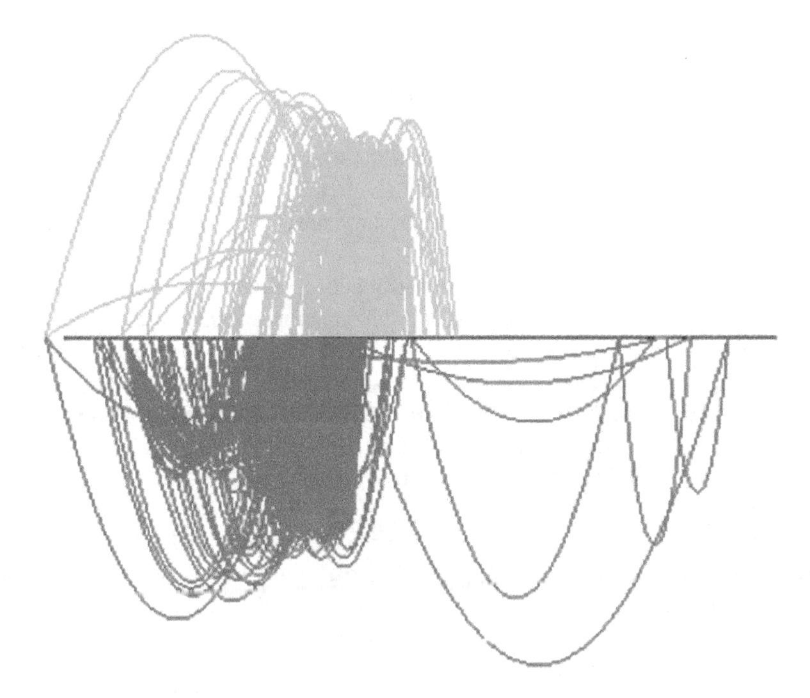

Fig. 5.40 7-D seed data in Shifted Collapsing Parallel Coordinates with rotated axis

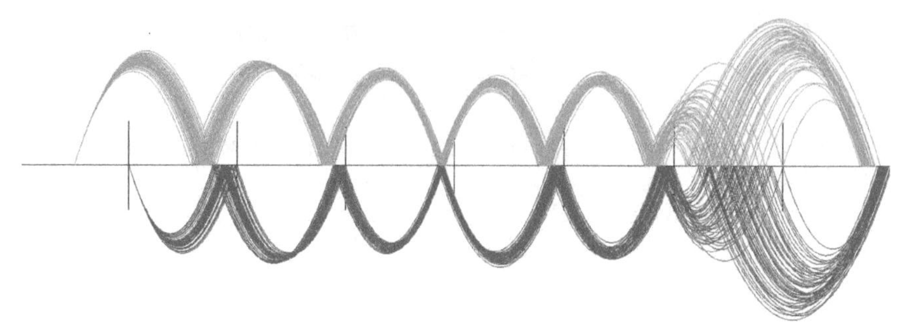

(a) Seeds data in In-Line Coordinates with Bezier curves.

(b) Seeds data in In-Line Coordinates with triangles.

Fig. 5.41 Seeds data in in-line Coordinates with Bezier curves and triangles. The orange curves are the seeds of class 1 and the blue curves are the seeds of class 2. The coordinate X_3 with smallest variation is put first. The last coordinate X_8 is class

the size and shape of the Bezier curves and the triangles are specified by properties L, P, C and O of ILC defined in Chap. 2. Bezier curves and triangles are ordered by increasing variance of coordinates in the given dataset with the coordinate X_3 with smallest variance put first.

Figure 5.41 shows that the seeds are grouped up tightly for both the blue and orange curves. The grouping stops for both blue and orange curves at the end where the attribute with the highest variance is located. The orange line starts sooner than the blue line, which shows a significant difference in values of the respective attribute between classes. It is also visible very well, that majority of attributes have significant difference in values of attributes between classes. This gives a good visual guidance to build a classification rule with using most attributes that are most separated visually.

Figure 5.42 shows the zoomed overlap area from Fig. 5.41. It illustrates the *advantages* of ILC with using both sides of its base line to *draw class 1 on one side and class 2 on the other side*. It is visible on the right in Fig. 5.42 that, while there is a heavy overlap in values, the patters for blue and orange classes differ. It is more difficult to see such difference in Parallel Coordinates where both classes are *overlaid* by drawing one on the top of another. Traditional PCs lack ILC capability

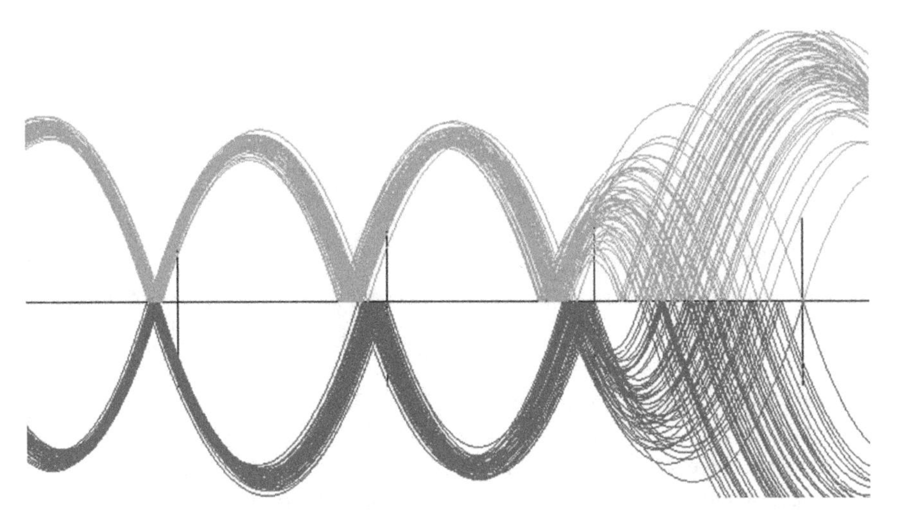

Fig. 5.42 Zoomed overlap area from Fig. 5.41

of using two sides of the base line. Note, that in Fig. 5.40 we used the expanded capabilities of Shifted Parallel Coordinates by drawing classes on two sides of the Parallel Coordinates (which are collapsed to a single middle coordinate) and by using Bezier curves to draw lines. However, it is less informative than ILC visualizations in Figs. 5.41 and 5.42.

5.13 Case Study 13: Letter Recognition Dataset with SPC

This case study is based on the *Letter Recognition Data Set* from the UCI Machine Learning Repository (Lichman 2013). It contains 20,000 records that represent extracted 16 features of letters. This case study uses all 16-D records of two classes (letters T and I) to classify them. These 1550 records were split with 1085 records to training data (70%) and 560 records (30% to validation data). Then centers $\mathbf{c}_1$ and $\mathbf{c}_2$ of two classes on the training data are computed in two steps: (1) computing average 16-D points $\mathbf{e}_1$ and $\mathbf{e}_2$ of each class, and (2) finding two centers $\mathbf{c}_i$ as 16-D points with the smallest Euclidian distance to the respective $\mathbf{e}_i$.

The method used in this case is described in detail in Sect. 4.5 in Chap. 4. Figure 5.43 shows colored SPC graphs of these two points $\mathbf{c}_i$ and the colored nodes of all graphs from the vicinity sets (hypercubes) $V(\mathbf{c}_i, T_i)$. The edges of graphs are omitted to decrease clutter. Here a half of the length of the side of the hyper-cubes for letter T is $T_1 = 0.3$ and for letter I it is $T_2 = 0.2$. As this figure shows 16-D points in each $V(\mathbf{c}_i, T_i)$ belong only to training data of the respective class without any point from the other class. In other words, the 16-D hyper-cubes defined by pairs $(\mathbf{c}_i, T_i)$ separates these training data with 100% accuracy. Moreover, this separation allows constructing simple and accurate discrimination functions in SPC

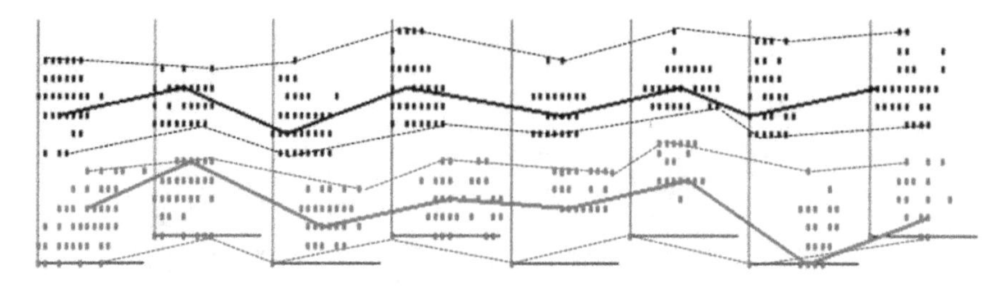

Fig. 5.43 Phase 1: 16-D training data subset of the letter data in SPC. Centers of classes are colored lines. Dots are nodes of graphs

as lines between dotted black lines in Fig. 5.43. These dotted lines connect most distant nodes of graphs in each vicinity (hypercube) $V(\mathbf{c}_i, T_i)$. This part of the training data covers 282 cases (18.19%) of all training cases. Figure 5.44 shows centers $\mathbf{c}_1$, $\mathbf{c}_2$ and all validation 16-D points that are within the hypercubes found on training data from Fig. 5.43. These points are shown by nodes of their graphs in the colors of their classes. As Fig. 5.44 shows, all these points in each hypercube belong only to a single class, indicating 100% accuracy of classification.

Each next phase repeats the same process for unclassified data that are remained from the previous phases. This includes computing new centers $\mathbf{c}_1$, $\mathbf{c}_2$ and new vicinity hypercubes $V(\mathbf{c}_i, T_i)$. Total it required 5 phases with 5 letters T and 7 letters I remained unclassified (Fig. 5.47). Table 5.2 shows results of all phases numerically. Figures 5.45 and 5.46 illustrate phase 4. As Table 4.2 shows, all phases produced classification of both training and validation data with 100% accuracy.

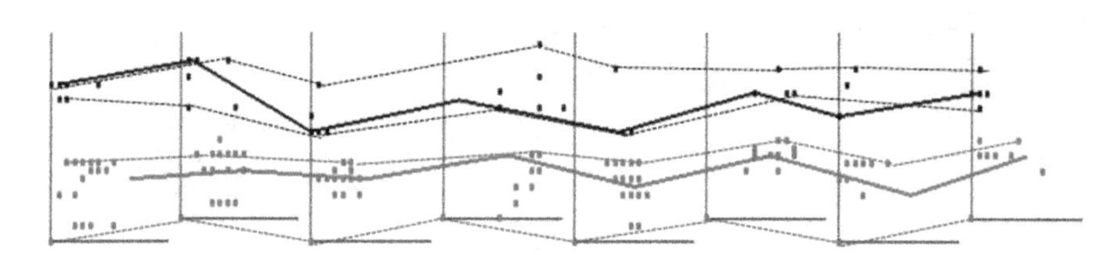

Fig. 5.44 Phase 1: 16-D validation data subset of the letter data in SPC. Centers of classes (colored lines) are from training data. Dots are nodes of graphs

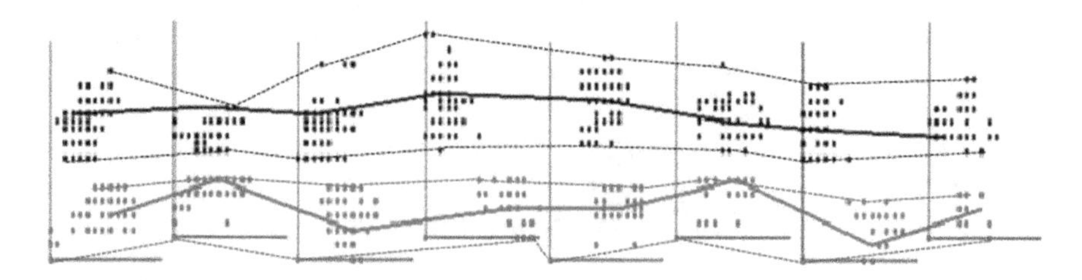

Fig. 5.45 Phase 4: 16-D training data subset of letter dataset in SPC. Centers of classes are colored lines. Dots are nodes of graphs

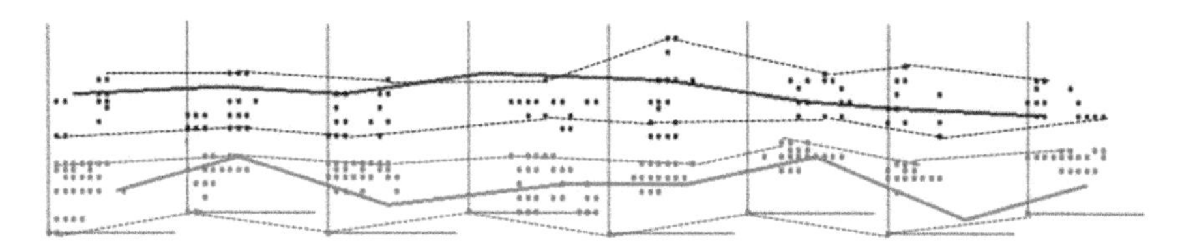

Fig. 5.46 Phase 4: 16-D validation data subset of letter dataset in SPC. Centers of classes (colored lines) are from training data. Dots are nodes of graphs

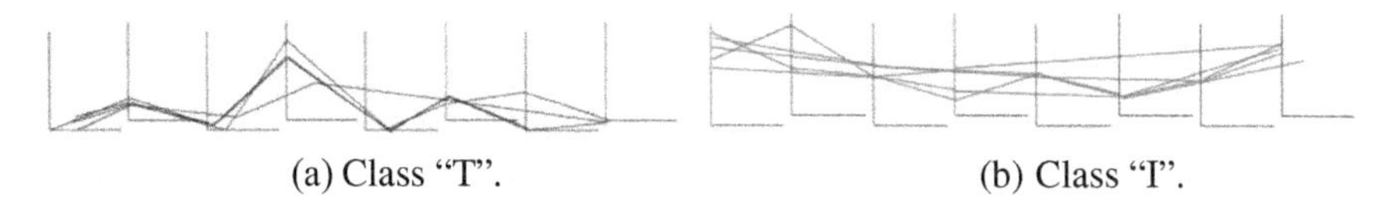

(a) Class "T". (b) Class "I".

Fig. 5.47 Remaining 16-D training and validation subsets of letter dataset

Table 5.2 All phases of visual classification of letters T and I in SPC

	Letter T	Letter I	Total	Data %	Accuracy %
Total training	560	525	1085	70.00	100
Total Validation	235	230	465	30.00	100
Phase 1 training	178	104	282	18.19	100
Phase 1 validation	31	17	48	3.10	100
Phase 2 training	102	154	256	16.52	100
Phase 2 validation	70	53	123	7.94	100
Phase 3 training	91	127	218	14.06	100
Phase 3 validation	81	65	146	9.42	100
Phase 4 training	159	99	258	16.65	100
Phase 4 validation	44	72	116	7.48	100
Phase 5 training	26	35	61	3.94	100
Phase 5 validation	9	21	30	1.94	100
Remaining training data	5	7	12	0.77	
Remaining validation data	0	0	0	0	

Thus, the classification rule R_V: for this dataset is

$$R_V : \text{If } \mathbf{x} \in V_1(\mathbf{c}_i, T_i) \cup V_2(\mathbf{c}_i, T_i) \cup V_2(\mathbf{c}_i, T_i) \cup V_4(\mathbf{c}_i, T_i) \cup V_5(\mathbf{c}_i, T_i) \text{ then } \mathbf{x} \in \text{class } i.$$

where $V_j(\mathbf{c}_i, T_i)$ for $j = 1{:}5$ are hypercubes learned in phases 1–5, respectively. Remaining 5 "T" denoted as $\mathbf{a}_k$, $k = 1{:}5$, and 7 "I" denoted as $\mathbf{b}_t$, $t = 1{:}7$ are outside of found hyper-cubes. These cases can be classified by memorizing them without making the rule significantly overfitted, because they are only 0.77% of the dataset, with a new rule R_{ab} to augment R_v:

$$R_{ab} : \text{If } \exists i\ \mathbf{x} = \mathbf{a}_i \text{ then } \mathbf{x} \in \mathit{class}\ ''\mathrm{T}'', \text{ else if } \exists t\ \mathbf{x} = \mathbf{b}_t \text{ then } \mathbf{x} \in \mathit{class}\ ''\mathrm{I}''$$

5.14 Conclusion

This chapter provided examples of applying multiple GLCs to a variety of datasets. It shows the advantages of having multiple options to visualize the same data that GLCs offer. GLCs allowed extracting different kind of features present in datasets. The challenge for the future is expanding ways to extract features present in a dataset. The patterns that discovered in these studies range from circular/closed shape patterns, to relations between locations of different nodes of the same graph and between different graphs, and clouds of other graphs around a given graph. Multiple graphs share similar patterns, which allowed discovering and validating classification rules. Visual representation with GLCs also helps to see and resolve conflicts among the rules to maximize accuracy of classification. The results produced in several case studies are quite comparable with results produced by analytical algorithms such as SVM. Hence, GLC expands the important visual component in machine learning.

References

Draper, D.: Challenger USA Space Shuttle O-Ring Data Set. UC Irvine Machine Learning Repository. http://archive.ics.uci.edu/ml/machine-learning-databases/space-shuttle/ (1993)

Draper, D.: Assessment and propagation of model uncertainty (with Discussion). J. Roy. Stat. Soc. B. **57**, 45–97 (1995)

Dzemyda, G., Kurasova, O., Žilinskas, J.: Multidimensional Data Visualization: Methods and Applications. Springer Science & Business Media (2012)

Global Hunger index: https://www.water-energy-food.org/news/2012-10-16-global-hunger-index-2012-ensuring-sustainable-food-security-under-land-water-and-energy-stresses/ (2012)

Grinstein, G.G., Hoffman, P., Pickett, R.M., Laskowski, S.J.: Benchmark development for the evaluation of visualization for data mining. Information visualization in data mining and knowledge discovery, pp. 129–176 (2002)

Kaur, P.: Support Vector Machine (SVM) with Iris and Mushroom Dataset. http://image.slidesharecdn.com/svm-131205085208-phpapp01/95/support-vector-machinesvm-with-iris-and-mushroom-dataset-15-638.jpg?cb=1471176596 (2016)

Lichman, M.: UCI Machine Learning Repository (http://archive.ics.uci.edu/ml). Irvine, CA: University of California, School of Information and Computer Science, 2013

Mafrur, R.: SVM example with Iris Data in R. http://rischanlab.github.io/SVM.html (2015)

Munzner, T.: Visualization analysis and design. CRC Press (2014)

Rubio-Sánchez., M., Raya, L., Díaz, F., Sanchez, A.: A comparative study between RadViz and Star Coordinates. Visualization and Computer Graphics, IEEE Transactions on **22**(1), 619–628 (2015)

Tufte, E.R., Robins, D.: Visual explanations. Graphics Press (1997)

Chapter 6
Discovering Visual Features and Shape Perception Capabilities in GLC

All our knowledge has its origins in our perceptions.
Leonardo da Vinci.

6.1 Discovering Visual Features for Prediction

Features. Analysis of data visualized with different GLCs in previous chapters show that multiple visual features could be estimated for each *individual graph*. These *features* include, but are not limited by:

- types of angles (e.g., sharp angle),
- orientation and direction of straight lines and angles,
- length of the straight lines,
- color of straight lines,
- width of the lines (as representation of the number of edges with similar values),
- width, length and color of the curves (e.g., Bezier curves),
- number of crossings of edges of a graph,
- directions of crossed edges,
- shape of an envelope that contains the graph,
- a "type" of the graph (dominant direction or absence of it: knot, L-shape, horizontal, vertical, Northwest, etc.).

The analysis of these features can be split between different *agents* in the *collaborative visualization* process to speed up processing and use skills of analysts most efficiently.

Relations. Many *relations between graphs* also can be estimated visually by individual or collaborative agents and can be split between agents to find relations such as:

- parallel graphs,
- graphs rotated, shifted or affine transformed relative to each other,
- percentage of overlap of graphs,

B. Kovalerchuk, *Visual Knowledge Discovery and Machine Learning*,
Intelligent Systems Reference Library 144,
https://doi.org/10.1007/978-3-319-73040-0_6

- the size and shape of the area of the overlap of envelopes of two graphs,
- the distance between two graphs.

Some of these relations are illustrated in Fig. 6.1 in Parallel Coordinates and CPC for six 8-D data points $\mathbf{a}_1, \mathbf{a}_2, \ldots, \mathbf{a}_6$, where for each $\mathbf{a}_i$ values of all a_{ij} are equal to a constant c_i, $a_{ij} = c_i$, $j = 1, 2, \ldots, n$.

In Parallel Coordinates (on the top of Fig. 6.1) these points are shown as quite intuitive and preattentive parallel horizontal *lines*. In CPC (on the bottom of Fig. 6.1) these points are shown as preattentive single 2-D *points*. These points are simpler than PC lines in Fig. 6.1.

However, the metaphor used to know that values are equal is a new one and needs to be learned along with CPC visualization. In spite of these differences, it is easy to see in PC that the structures of 8-D points are the same—equal distant horizontal lines. In CPC it is shown by equal distant points on the same line.

Figure 6.2 provides examples in Parallel Coordinates and CPC for other six 8-D data points $\mathbf{b}_1, \mathbf{b}_2, \ldots, \mathbf{b}_6$ with unequal values and more complex structure. The common structure is less evident in Parallel Coordinates on the top of Fig. 6.2 than in CPC on the bottom of Fig. 6.2.

This example shows there is no "silver bullet" visualization that is the best in revealing the data structures for all datasets. The best one for the given data must be discovered by exploring alternative visual representations.

Feature discovery. Now assume that 8-D points in Fig. 6.1 belong to class 1 and in Fig. 6.2 belong to class 2 and we need to find not any features of these datasets, but discriminating ones. In PC, a discriminating feature is whether the

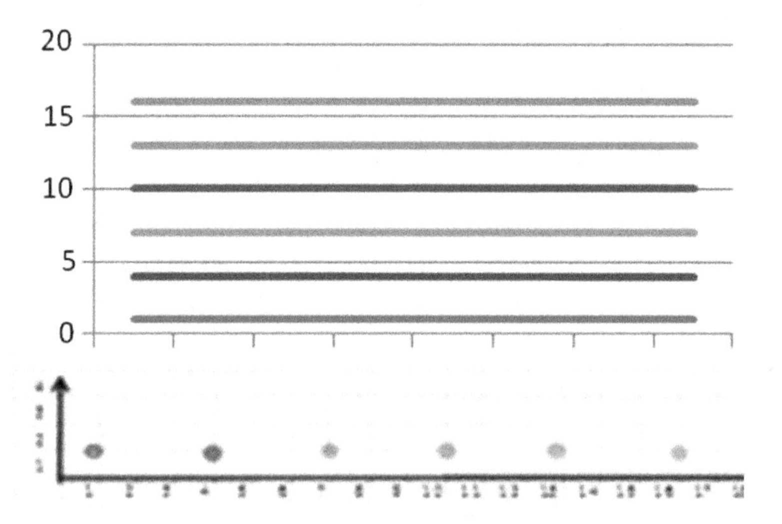

Fig. 6.1 Six 8-D points $\{\mathbf{b}_i\}$ in Parallel Coordinates (top) and CPC (bottom)

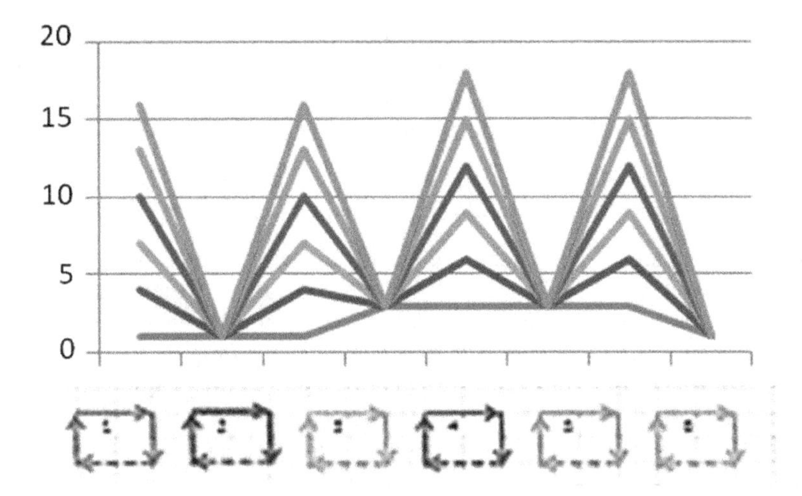

Fig. 6.2 Six 8-D points {$\mathbf{b}_i$} in Parallel Coordinates (top) and CPC (bottom)

shape is a horizontal line or not. Another one is whether the shape has peaks or not. In CPC a discriminating feature is whether shape is a rectangle or not, and another one is whether the shape is a single point or not.

These examples illustrate the difference between generic *feature observing* and *discovery* in contrast with discovering features that help *discriminating cases* from different classes. This book, including this chapter, focuses on discovering underlying features of data structure to be used for supervised data *classification* and data *class prediction* with the use of visual means.

The visual means potentially include both the *human visual system* and *computer vision systems*. In this sense our goal here is similar to *feature extraction* in image processing and computer vision.

Goal and approach. This chapter is to explore and evaluate how efficiently the human visual system can discover features in Closed Contour Paired Coordinates (traditional Stars/Radial Coordinates, and CPC Stars) in comparison with Parallel Coordinates. The approach includes:

- Random generation of a few base classes of n-D data (hyper-tubes, hyper-spheres, hyper-planes, hypercubes);
- Experimental evaluation of average time that subjects use to discover complex features of shapes of these classes;
- Interpretation of these features as properties of data structures;
- Generalization to wider classes of n-D data by involving affine, projective and other transformation of base data classes.

The essence of the study in this chapter is visual shape recognition by humans. It is a drastically flexible and a very complicated process. Many years of psychology development produced very general *qualitative* Gestalt laws, but only very limited *quantitative* estimates of perception thresholds for simple shape features such as line length, box size or brightness.

The reasons of this limited progress are:

- *mutual influence* of many features in recognition of complex form, and
- *deficiencies of modeling theories* for such complex processes.

As a result, building of quantitative applied models of vision is extremely difficult and time consuming. Therefore, this study focuses on experimental *qualitative ranking* test on finding data displays that are essentially better than others are. Our experiment on a modeled data structure allows extending its results only *qualitatively* onto data with similar structures. The actual numbers from such tests are very dependent on many factors (objective and subjective).

This dependence and multiple external assumptions of the statistical theory limit applicability of the statistical theory and criteria to such type of data (Trafimow and Marks 2015; Trafimow and Rice 2009; Valentine et al. 2015).

The goals of the experiments are to:

1. Test *effectiveness* of some GLCs for visual discovery of n-D data features and data structures for different data dimensions in classification tasks;
2. Identify *advantages* of Radial Coordinates (traditional stars) and Star Collocated Paired Coordinates (Star CPC);
3. Further expose the advantages of *modeled data approach* of visual analytics, as allowing results generalization vs. getting results applicable only to very specific analyzed data.

Modeled data approach. First, we outline the modeled data approach to data generation for the experiments. Testing new visualization methods is possible in two fundamentally different ways. The first one is generating data with *given mathematical properties*. When the method is successful in experiments on these modelled data this method can be successfully applied to other data with the same mathematical properties. The second way is to experiment with data without known mathematical properties, which is common in uncontrolled real world data. In the last case, the success of visual representation of such data does not help to know how successful this method can be on another data, because the properties of data were not formulated.

Thus, the judgment about the method effectiveness is quite limited under such testing. Only if the solution for these specific data has its own value, beyond the illustration of the method success, then such tests are beneficial by themselves, but not for other datasets. However, not all real data are such “self-beneficial” data Therefore a wide use of modeled data will be beneficial for testing many other new methods not only GLC.

6.2 Experiment 1: CPC Stars Versus Traditional Stars for 192-D Data

Hyper-tubes. This section quantitatively evaluates the hypothesis denoted as H_1 that CPC stars have advantages over traditional stars in easiness of discovering visual patterns by humans. It is shown that these advantages grow with data dimension n. To test H_1 we consider the task of discovering n-D points with identical features in a given set of n-D points using their representations as CPC stars and traditional stars. Subjects are asked to find these identical features in both visual representations. The success is measured by comparing accuracy of pattern discovery in CPC stars and traditional stars.

Five data classes have been generated as points $\{\mathbf{x}\}$ in the n-D data space R^n. with $n = 192$. Points of each class C_k are located in a separate hyper-tube T_k around its randomly generated direction A_k from the space origin, where $k = 1, \ldots, 5$ is a class label. We generated from 5 to 15 192-D points randomly within each hyper-tube.

Then these points have been additionally randomized (either by Gaussian multiplicative or additive n-D "noise" with given standard distributions). The distance of each point from tube central line (generatrix) is one of the factors that specify the variation of shapes of stars. Stars are similar in shapes and differ in sizes for narrow hyper-tubes of this type.

The design of this experiment is representative for other n-D data structures because many classes of n-D data can be represented as combinations of tubes. For instance, a curved tube can be approximated by a sequence of linear tubes. A hyper-sphere ("ball") is a "tube" around a single point. Another important aspect is that partial affine invariance of human shape perception allows detecting the shapes that are rotated, shifted and resized. Thus, it opens wide possibilities for visual recognition and interpretation of complex nonlinear n-D structures.

Experiment setting and results. Each subject was asked to find a few features separating five tubes (with 6 n-D points in each tube) from others tubes by using CPC stars randomly placed on a sheet of paper.

In addition, the subject solved the same task for traditional stars that represent the same n-D data points with the same placement on a paper sheet. The experiment was repeated for different levels of "noise", i.e. tube width (10, 20, 30% of maximum possible value of each coordinate of an n-D point).

All subjects were volunteers and the time of each test was limited to 20-30 min. It essentially restricted the number of features that can be analyzed. Time and errors of solutions are shown in Tables 6.1 and 6.2. Due to obvious qualitative visual advantages of CPC Stars versus traditional stars especially for $n = 192$, the first quantitative tests involved only two subjects to roughly estimate these advantages. The tests results clearly show that for $n = 192$ traditional stars cannot compete with CPC.

Subject #1, with some previous experience in similar tests, did these tests first for CPC stars with 10% noise, then for traditional stars with 10% noise and then for

Table 6.1 Subject 1: Time (mean/standard deviation) and errors of one feature detection for 5 tubes (sec)

		Subject #1	
Noise		Stars	CPC stars
10%	Time	124/68	92/52
	Features	3	4
	Errors	2	2
20%	Time	n/a	119/74
	Features	n/a	4
	Errors	n/a	3

Table 6.2 Subject 2: Time (mean/standard deviation) and errors of one feature detection for 5 tubes (sec)

		Subject #2	
Noise		Stars	CPC stars
10%	Time	107/48	60/33
	Features	3	5
	Errors	3	0
20%	Time	159/71	84/42
	Features	3	5
	Errors	4	1
30%	Time	n/a	197/105
	Features	n/a	5
	Errors	n/a	3

CPC stars with 20% noise. During 30 min time limit, the respondent detected four discriminative features of each of five classes for CPC stars and three such features for traditional stars.

However, in the last test only *wide* and *large features* were detected while comparison of 80–90% of features that consist of the sets of narrow peaks was drastically more difficult. Therefore, subject #1 denied test for stars with 20 and 30% of noise.

This experiment had shown that for n = 192 traditional stars display data *perceptually lossy*, while these stars actually preserved and display all information contained in n-D points that the stars represent.

This is an example of *mathematical lossless*, but *perceptually lossy* traditional stars when the perception time was limited and stars were small. In contrast, placing one star on a full A4 sheet of paper makes everything visible without losses. Thus, magnification can improve performance of the common stars, but with fewer stars in a vision field.

Recognition of these features with 30% noise is so time consuming that even very experienced Subject #2 could not separate more than two tubes for 30 min and only by using large local and integral features.

In contrast, **CPC Star** tests showed acceptable time (1–5 min) for all figures and noise up to 30% (see Table 6.3). Subject #2 with advanced skills performed two times better with CPC stars versus traditional stars, especially for wide tubes with 20–30% of noise.

Table 6.3 Time and errors of grouping pictures by their whole shapes

sd/m %	0/0%		5/10%		10/15%	
Subject	Sec	Errors	Sec	Errors	Sec	Errors
Stars						
#1	280	0	585	4	780	5
#4	173	0	312	3	539	4
Parallel Coordinates						
#1	985	0	1020	7	2185	11
#4	742	0	823	4	1407	8

The success of CPC stars versus traditional stars is due to: (1) two times less dense placement of the points, and (2) a specific mapping of pairs x_i, x_{i+1} into a contour.

6.3 Experiment 2: Stars Versus PC for 48-D, 72-D and 96-D Data

6.3.1 Hyper-Tubes Recognition

This section quantitatively evaluates the hypothesis denoted as H_2 that Stars have advantages versus PC lines. Consider an n-D Euclidian data space E^n with n-D points $\{\mathbf{x} = (x_1, x_2, \ldots, x_n)\}$. A *linear hyper-tube* (*hyper-cylinder*) is a set of n-D points in E^n defined by its radius. The axis A of a linear hyper-tube is given by an n-D orientation vector $\mathbf{v}$ and a start n-D point $\mathbf{x}_s$). In the hyper-tube the distance d from each point $\mathbf{x}$ of the hyper-tube to its axis is no greater than its radius R, $d(\mathbf{x}, \text{A}) \leq R$.

For this experiment, five data classes were generated as n-D points in five linear hyper-tubes with the randomly generated orientation vectors $\mathbf{v}_\text{k}$ of these hyper-tubes. Axis of all hyper-tubes go through the origin point (0, 0, …, 0). Initially all given n-D points of each hyper-tube lay on its axis before “noise” was added, i.e., any two of these points related linearly $\mathbf{y} = a\mathbf{x}$, where coefficient a represents the proportion of their distances to the origin. This leads to the similarity of shapes of stars for $\mathbf{x}$ and $\mathbf{y}$, which visually represent their sizes proportional to the distance from the origin. Shapes of PC lines are also similar for the same reason.

Three independent datasets were generated with total 90 n-D points. Each dataset includes 30 n-D points (6 points from each of 5 hyper-tubes):

- In dataset 1 points of each hyper-tube T lay equidistantly on the axis A of T, $\mathbf{x}_k = k\mathbf{x}_1$ (k = 1, 2, … 6).
- In dataset 2 a “normal noise” was added to each of 6 equidistant points with mean m = 10% and standard deviation, sd = 5% of the norm $|\mathbf{x}|$ of each n-D point.

- In dataset 3 the "normal noise" was added in the same way with m = 15% and sd = 10%.

Such random deviations of points from axis mask similarity of shapes. See Fig. 6.3. In addition, these "central" tubes are representative for other data structures such as arbitrary tubes with the axes that cross at any point of the data space. These tubes are transformed to central tubes by shifting the origin to the crossing point.

Curved hyper-tubes (curved tubes for short) are defined by substituting a constant orientation vector **v** to a vector that is defined as a vector **v**(t) that depends on the parameter t, where **v**(0) and **v**(1) are the orientation vectors at the beginning and end of the curved tube. A curved tube can be approximated in a piece-wise fashion by linear tubes. Humans can detect compact clusters of curved tubes due partial affine invariance of human shape perception in 2-D.

This setting opens wide possibilities for complex nonlinear structure detection and their interpretation as a relation between different sets of coordinates. Figure 6.3 clearly shows the advantages of stars versus PC lines, especially for n = 96. To estimate roughly these advantages, again only two subjects were tested. Subjects have been given 30 stars of five hyper-tubes points on an A4 paper sheet. They were asked to group them into five groups by shape similarity. Similarly, they have been given 30 PC graphs of n-D points from five hyper-tubes on an A4 paper sheet and have been asked to answer the same question as for stars.

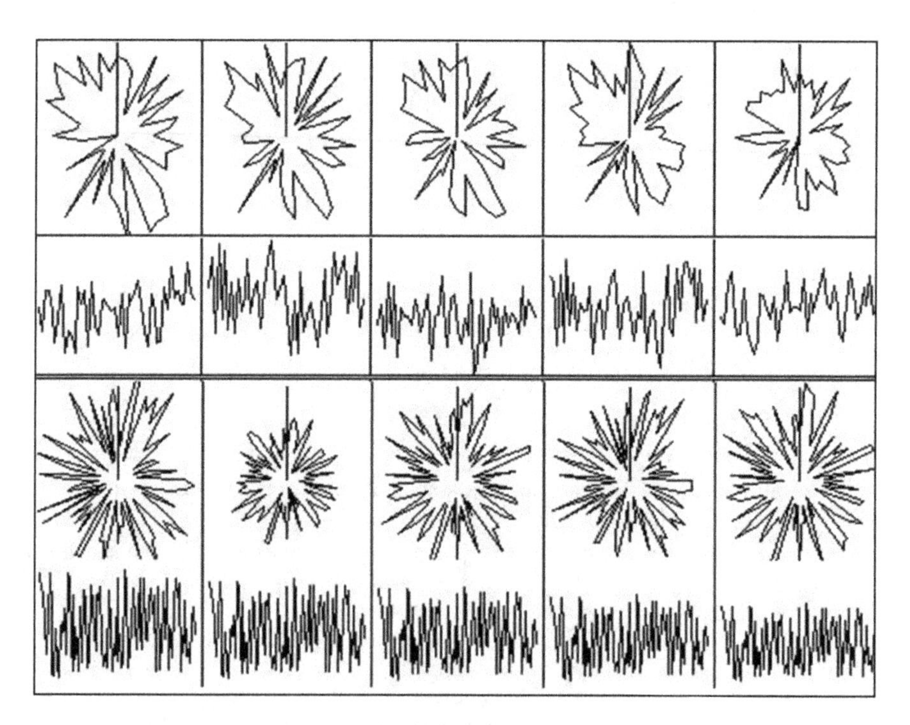

Fig. 6.3 Examples of corresponding figures: stars (row 1) and PCs lines (row 2) for five 48-D points from two tubes with m = 5%. Row 3 and 4 are the same for dimension n = 96

Table 6.3 shows the task performance time (in seconds) and errors amount in each test for stars and PCs with different levels of noise. The first two columns show results *without noise*. Columns 4 and 5 show results with *moderate* noise ($sd = 5$, mean of noise = 10%) and the last two columns show results with a *higher* level of noise ($sd = 10$, mean of noise = 15%).

With moderate noise, performance, time and errors are almost doubled relative to the tests without noise. Similarly, with higher noise, performance time and errors have grown relative to the tests with moderate noise. Also Table 6.3 shows that performance of the stars 2–3 times better than performance of PC lines with and without noise.

6.3.2 Feature Selection

Pattern recognition is based on discovering features of shapes that are common for majority of the pictures of one class and are not typical for pictures of others classes. Therefore, in this experiment stars and PC lines are compared by a set of local features that are common for certain classes of figures. To get such figures, n-D data points were generated as follows:

(1) Select m ($5 \leq m \leq 9$) consecutive coordinates $X_a, X_{a+1}, \ldots, X_b$ (that we will call *feature location*), $b = a + m - 1$, and assign some values to these coordinates. Thus, each feature location is identified by a sequence of coordinates, $X_a, X_{a+1}, \ldots, X_b$ that is given by the index interval $[a,b]$.
(2) Repeat this process k times for $3 \leq k \leq 5$, i.e., assign some values to consecutive coordinates in k other feature locations of the n-D point for different intervals $[a,b]$. In this way, several coordinates of the point are assigned and form a deterministic pattern.
(3) Repeat (1)–(2) for different m and k with assigning different values to selected coordinates. This creates several different deterministic patterns, where each pattern is represented by a single incomplete n-D point with only coordinates that are in the assigned feature locations are identified. The number of coordinated defined in this way varies from 20 to 63.
(4) Generate several complete n-D points in each class by giving *random* values to the undefined coordinates of incomplete n-D points (see Fig. 6.4). The random fragments (different for each figure) were placed between feature locations identical for all figures of a given class. These fragments add noise to the shape that is *not an additive noise*.

Consider a point $\mathbf{x} = (x_1, x_2, \ldots, x_n)$ of the data space S^n as shown on Fig. 6.4. It can be described by the set of Elementary Conjunctions (EC) of coordinates. If all points of given class have the same values of coordinates from i-th to $(i + m - 1)$-th, then all stars that represent these points have identical shape fragment E with identical location on these stars. Appearance of such fragments can be considered as the true value of elementary conjunction (EC),

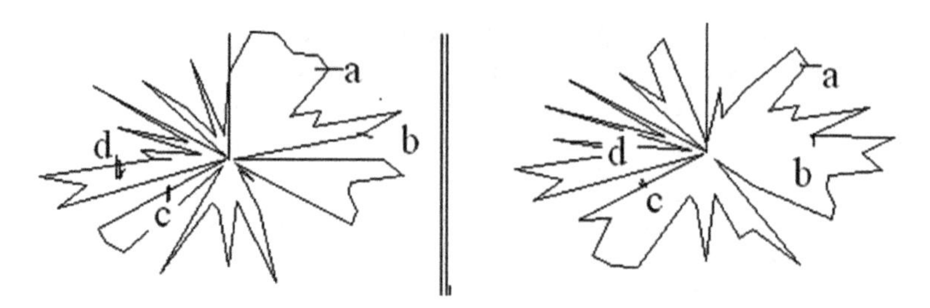

Fig. 6.4 Two stars with identical shape fragments on intervals [a,b] and [d,c] of coordinates

$$x_i = c_i \ \& \ x_{i+1} = c_{i+1} \ \& \ \ldots \ \& \ x_{i+m-1} = c_{i+m-1},$$

where c_i, c_{i+1} and other constants are value of respective indexed coordinates, and m is the length of this fragment.

Consider a class of stars with a few common fragments, E_1, $E_2 \ldots$, E_n. These fragments identify structures in the data space that can be encoded as EC. These fragments can be recognizable by humans if they are large enough to be above a visual perception threshold for shape discrimination.

Several such fragments have been generated separately (see Figs. 6.5 and 6.6) for different dimensions n and different sample sets of n-D points A, …, F from two data classes with 10 n-D points in each class. These fragments were placed at different feature locations in the stars.

Subjects knew that the stars or PC graphs of the first class are placed in the first two rows of the screen and second one in the last two rows. They were instructed to find all first class fragments not existing in second class and vice versa. Tests were done separately for PC lines and stars displays of the same classes. Subjects were required to find complete set of coordinates that form each common fragment.

Table 6.4 shows performance times for different dimensions, classes and displays of figures for each subject in the same format as before. In this table, the

Fig. 6.5 Samples of some class features on stars for n = 48

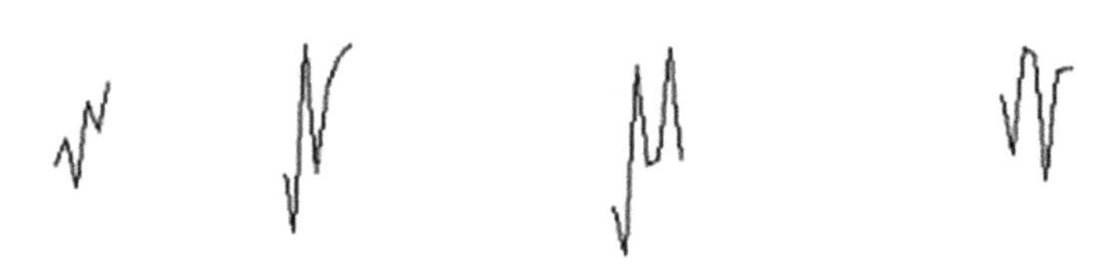

Fig. 6.6 Samples of some class features on PCs for n = 48

Table 6.4 Feature selection time (mean/standard deviation in seconds)

Subjects	Figures	Dimensions				
		48			72	96
		Samples			Sample	Sample
		A	D	F	A	A
1	Stars	32/14	36/12	46/27	104/26	59/18
	PC lines	79/38	93/31	117/79	133/36	91/27
2	Stars			96/12	116/60	164/99
	PC lines			274/113	374/240	241/110
3	Stars			51/21	78/26	99/47
	PC lines			138/53	119/33	142/56
4	Stars	30/13	33/11	44/23		56/15
	PC lines	64/21	89/24	108/49		114/30
5	Stars		54/25	61/30		93/27
	PC lines		122/37	83/35		167/49
Average subject	Stars	31/14	38/16	60/21	99/19	94/44
	PC lines	72/30	101/15	144/75	209/14	151/58

average feature selection time is the first number and its standard deviation is the second number.

For dimension n = 48, different shapes for samples A, D, F have been generated to verify the impact of fragment shapes on performance time. As expected, subjects used more time to perform tests for larger dimensions. However, time increase was less than expected due learning by subjects during the tests. In addition, studies had shown that other factors also influenced performance. These factors include: subject's individuality, a number of tests with him/her, and others. Despite these differences, all subjects had shown **at least 2 times faster** selection of informative features with stars than with PC lines.

6.3.3 Unsupervised Learning Features for Classification

The above mentioned data samples of 72 and 96 dimensions with two classes by 10 figures in each of them were visually represented as stars and lines in PCs on separate sheets. These figures were placed randomly to roughly estimate the feature selection time for self-learning by subjects. Besides, Gaussian noise was added, as in the previous experiment. Thus, the subjects did not know the class of any picture and sorted the plots into two sets based on visible similarity.

In this case, subjects spent a significant time to find the first feature for separating 10 figures in two classes. Then subjects searched for other separating features with a similar performance time in the same way as in another experiment. For example, searching for the first feature in sample with dimension n = 72 required

112 s for stars and 274 s for PC lines. For both classes the average time of feature detection without noise was:

- 54 s (mean)/17 s (standard deviation) for stars and 128/22 s for PC lines for $n = 72$;
- 44/23 s for stars and 92/42 s for PC lines for n = 96.

Again, despite dimension increase, performance was enhanced due to learning of a subject on preceding tests. When each data point of sample with $n = 96$ has been distorted by 10% additive noise, the participants spent 4 min 44 s to find the first valid feature. Then they spent on average 3 min 16 s to find the next feature with stars, and refused to continue after 27 min work with PC lines, when only 7 figures were classified. This provides further evidence confirming the advantages of stars versus PC displays for these types of tasks. These results can convince analysts to use CPC stars and regular stars more actively, not only using Parallel Coordinates.

6.3.4 Collaborative N-D Visualization and Feature Selection in Data Exploration

Our experiments above have shown that agents recognize the same features with different speed. The experiment below reveals that the teamwork of agents leads to significant time saving to solve the task.

Collaborative feature selection. In this experiment, the figures are compared by selecting sets of local features, which are common for classes of figures. Figures of each class include 4–6 identical shape features formed by 5–9 consecutive coordinates. Different random fragments were placed between these informative features in each figure forming "noise". Subjects knew that the stars or Parallel Coordinates graphs of the first class are in the first two rows and second one in the last two rows. Subjects were instructed to find all first class fragments not existing in the second class and vice versa. Subjects were required to find all coordinates that create each common fragment.

Subjected collaborated by *discussing feature discovery, guiding each other* when looked at the *same data source* and then transitioning together to discuss the next data source. Table 6.5 shows performance time for 96 and 48 dimensions. It shows the average time of feature selection (first number) and the standard deviation of this value (second number).

Table 6.5 Individual and collaborative feature selection time (mean/standard deviation in sec.)

	96 dimensions	48 dimensions		
	C	A	D	F
Individual	94/44	31/14	38/16	60/21
Collaborative	33/12	13/11	14/13	15/12

To verify the impact of shapes of fragments on the performance time we used different shapes for samples A, D, F in the dimension n = 48. Table 6.5 shows that collaborative feature selection is **2–4 times faster** than in case of individual work for all these sets of features and both dimensions (96 and 48).

6.4 Experiment 3: Stars and CPC Stars Versus PC for 160-D Data

This section experimentally explores human abilities to recognize 160-D linear patters with Gaussian noise. Those patterns are represented by n-D points within linear hyper-tubes (hyper-cylinders) in 160-D where each hyper-tube represents a data class. The axis of the hyper-tubes are given as $A_k + t\mathbf{u}_k$, where A_k is a starting n-D point of the axis, $t \in [0,1]$, and n-D vector $\mathbf{u}_k$ sets up a direction of the hyper-tube. The noise level defines the width of each linear hyper-tube.

6.4.1 Experiment Goal and Setting

Data Modeling steps implemented in this experiment include:

Step 1. Randomly generating the linear hyper-tubes (hyper-cylinders) that cross the origin of the n-D data space of dimension n = 160. All hyper-cylinders are normalized to length 1 with the axis $A + t\mathbf{u}$, where A is its starting, $t \in [0,1]$ and n-D vector $\mathbf{u}$ is a hyper-tube direction. Both A and $\mathbf{u}$ are randomly generated.

Step 2. Computing randomly equidistant points on axis of these hyper-tubes in the range from $t = 0.3$ to $t = 1.0$, i.e. forming a set of vectors $\{\mathbf{v}_i\} = \{(v_{i1}, v_{i2}, \ldots, v_{in})\}$ from the origin to these points.

Step 3. Computing n-D vector $\mathbf{k}_G = (k_{G1}, k_{G2}, \ldots, k_{Gn})$ of Gaussian noise with standard deviation $\sigma \in [0.1,0.3]$ for each $\mathbf{v}_i$ separately.

Step 4. Computing n-D points $\mathbf{w}_i = (k_{G1}v_{i1}, k_{G2}v_{i2}, \ldots, k_{Gn}v_{in})$, i.e., vectors $\mathbf{v}_i$ with multiplicative noise.

Data Visualization steps implemented in this experiment include:

Step 1. Selecting visualization method M_k: Regular Stars, CPC Stars, and Parallel Coordinates.

Step 2. Displaying each generated n-D point $\mathbf{w}_i$ using M_k in a separate window.

Step 3. Tiling these windows in the random order (see Figs. 6.7, 6.8 and 6.9). Please note that the Figures are renumbered to ensure sequential order of citations. Please check and confirm the change.Thank you. Your refrences are correct. I just changes only one below.

Step 4. Repeating steps 1–3 for other selected visualization methods M_k.

Fig. 6.7 Twenty 160-D points of 2 classes represented in star CPC with noise 10% of max value of normalized coordinates (max = 1) and with standard deviation 20% of each normalized coordinate

Fig. 6.8 Twenty 160-D points of 2 classes represented in Parallel Coordinates with noise 10% of max value of normalized coordinates (max = 1) and with standard deviation 20% of each normalized coordinate

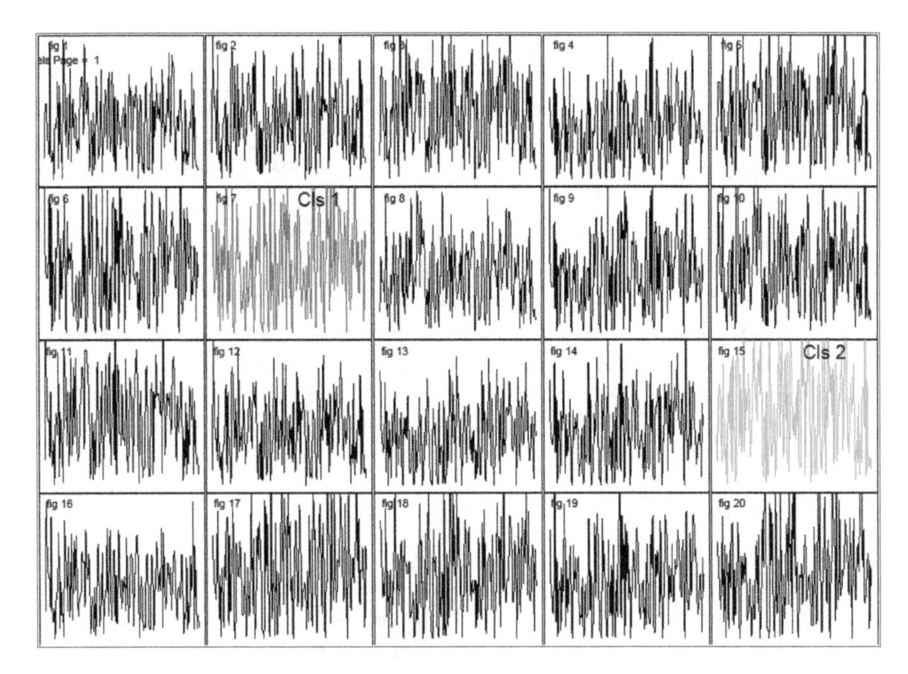

Fig. 6.9 Twenty 160-D points of 2 classes represented in Parallel Coordinates with noise 10% of max value of normalized coordinates (max = 1) and with standard deviation 20% of each normalized coordinate

6.4.2 Task and Solving Hints

In the actual experiment, the participants observed 160-D data. In the introduction session participants observed other data and of a smaller dimension n = 100. The goal of introduction session was to make participants familiar with the experiment set up not data of actual experiment. The actual 160-D training data where provided to participants only during the experiment as two labeled figures on the same sheets of paper where the 18 test cases where present (see Figs. 6.7, 6.8 and 6.9). Each participant received the three sheets of paper in A4 format with 20 figures in each sheet generated by a given method as shown in the Figs. 6.7, 6.8 and 6.9. Radial Coordinates, Star CPC, and Parallel Coordinates methods are used in these sheets to visualize n-D points. These 20 figures split equally between 2 classes. Participants are informed about equal number of figures of two classes, but only one figure of each class is labeled by the class number and distinctly colored. These two figures serve as training data. The locations of figures of the same class in the three sheets are randomized to eliminate the impact of location on results of the experiment.

The goal of the participant in the actual experiment is to classify unlabeled 18 figures using two labeled training figures within 20 min per sheet (one hour total for three sheets). In the pilot study, we found that it is not required more than an hour for the images of this complexity. Each participant is asked to write the class number next to each unlabeled figure. It was also recommended to participants to circle up to 3–5 found local patterns in unlabeled figures that match to training figures. Each participant worked with total 18*3 = 54 stimuli to be recognized. These stimuli are presented in three sheets shown in Figs. 6.7, 6.8 and 6.9.

In the introduction session with 100-D data to help participants to better understand examples of patterns of radial directions, angles, convexities, concavities, different forks, figure symmetries, envelops, orientations of parts, elongations, and so on distorted and not distorted by noise have been provided.

Figure 6.10 shows samples of these data. This was a part of the lecture delivered to participants (all participants are students majoring in Computer Science). The lecture explains to them the design of all three visualization methods with several examples.

6.4.3 Results

The experiment was conducted with two groups of computer science students at two universities. Over 100 sets of forms were distributed in one university and 15 sets in another one. Total 75 sets were returned from the first set and 14 sets were returned from the second set. Not all forms were fully filled. Tables 6.6, 6.7 and 6.8 show the results of this experiment based on these responses.

Total 18 students did not make any errors and six students made one error in all tests. It shows that there is a room to increase both noise level and dimensions. It also show that to reveal differences in difficulties in these three displays (PCs, Stars, CPCs) for these students the noise level and/or dimension has to be increased in further tests.

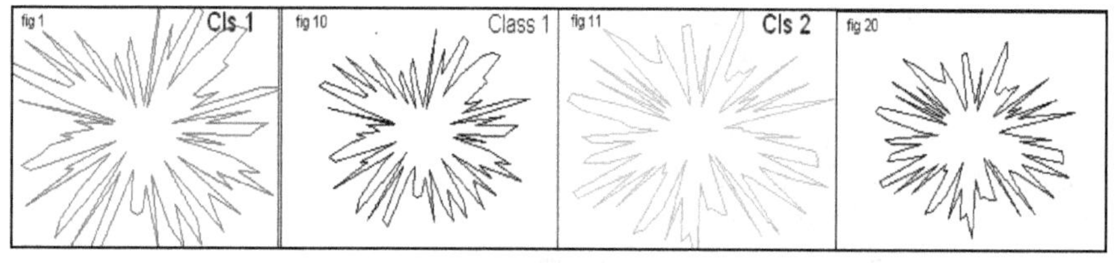

(a) Initial 100-D points without noise for Class (Hyper-tube) #1 and Class (Hyper-tube) #2

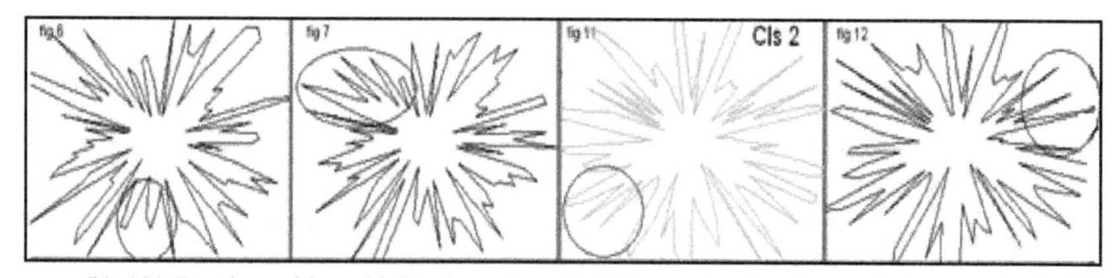

(b) 100-D points with multiplicative noise: circled areas are the same as in upper star.

Fig. 6.10 Samples of 100-D data in Star CPC used to make participants familiar with the task

Table 6.6 PC results of the experiment with 160-D data based on answer by 60 respondents who filled all forms on 160-D data classification

	PC			
	Errors	Refusals	Errors + refusals	Correct
Answers	189	35	224	856
Mean per person	3.15	0.58	3.73	14.27
Mean %	17.50	3.24	20.74	79.26
Stand. Dev.	2.52	1.44	2.42	3.03

Table 6.7 Stars results of the experiment with 160-D data based on answer by 60 respondents who filled all forms on 160-D data classification

	Stars			
	Errors	Refusals	Errors + refusals	Correct
Answers	78	10	88	992
Mean per person	1.30	0.17	1.47	16.53
Mean %	7.22	0.93	8.15	91.85
Stand. Dev.	1.87	0.46	1.57	1.9

Table 6.8 CPC Stars results of the experiment with 160-D data based on answer by 60 respondents who filled all forms on 160-D data classification

	CPC stars			
	Errors	Refusals	Errors + refusals	Correct
Answers	59	9	68	1012
Mean per person	0.98	0.15	1.13	16.87
Mean %	5.46	0.83	6.30	93.70
Stand. Dev.	1.8	1.22	1.77	1.88

On the other side, three students could not solve all three tests at all. One student made Stars and CPC Stars tests without errors, but could not make PCs test and one student could not solve two tests.

The actual reasons are not known while possible reasons can be rushing, lack of motivation, insufficient test time and time to become familiar with test setting, and other personal reasons.

A total 60 students provided *legible answers in all* PCs, Stars, and CPCs displays but some figures may left unanswered. The figures left without answers were interpreted as refusals and incorrect class labeling was interpreted as an error. The results of these 60 students are shown in Tables 6.6, 6.7 and 6.8. Figure 6.11 visualizes some data from Tables 6.6, 6.7 and 6.8. The results in Tables 6.6, 6.7 and 6.8 and Fig. 6.11 show that:

- Respondents are able to find multiple noisy patterns in 160-D data presented in all three visualization methods in a short period of time (within one hour).
- Classification in Parallel Coordinates was three times less accurate that in Radial Coordinates and Star CPC (224 versus 88 and 68 errors and refusals).

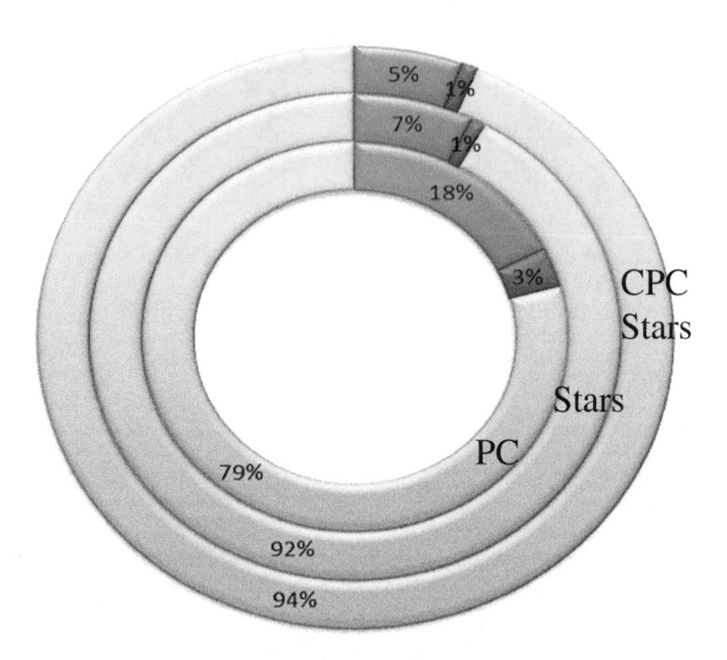

Fig. 6.11 Comparison of results of the experiment on PC, Stars and CPC Stars

- Classification in Radial Coordinates was slightly less accurate (14%) than in Star CPC (total 88 versus 68 errors and refusals), but many students solved CPC Star tests faster than Stars.
- The number of refusals in Parallel Coordinates was 2–3 times greater than in Radial Coordinates and Star CPC (35 versus 10 and 9 refusals).

In an informal interview after the experiment, a number of respondents stated that classifying figures and finding patterns in Star CPC was *easier* than in Radial Coordinates and in both was easier than in Parallel Coordinates. Respondents also stated that they have *more confidence in their decision in Star CPC than in other methods due to less* complexity of the figures. Only the 9 refusals in Star CPC classifications confirm such informal statements.

6.5 Experiment 4: CPC Stars, Stars and PC for Feature Extraction on Real Data in 14-D and 170-D

6.5.1 Closed Contour Lossless Visual Representation

Figures 6.12 and 6.13 show, respectively, traditional and CPC Stars for 5 classes: healthy (black), and 4 diseases (colored) from *the cardiology data* from UCI

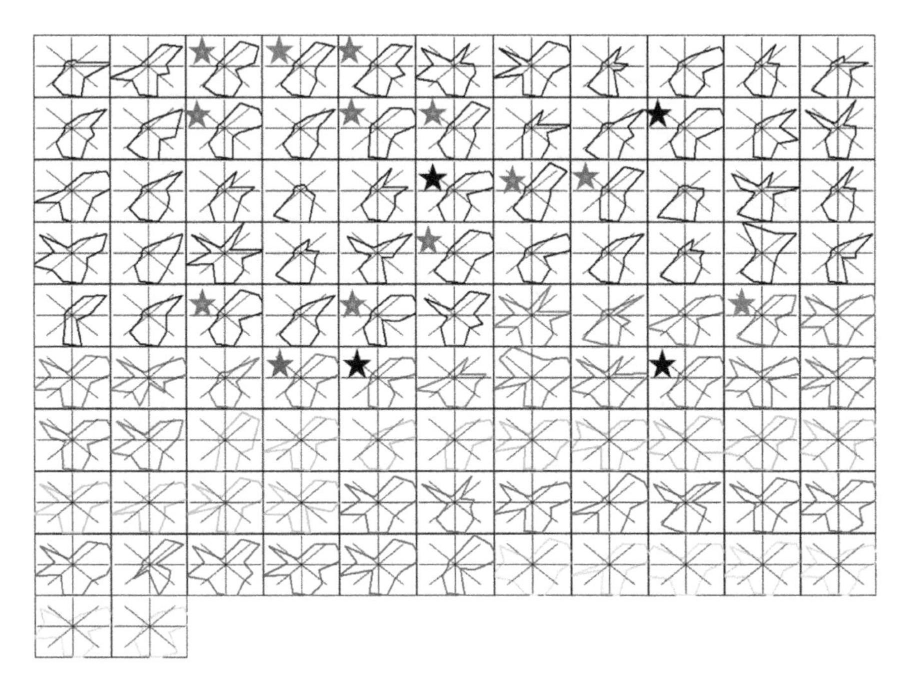

Fig. 6.12 Samples of 14-D data from 5 colored classes represented by closed contours (stars) in traditional Radial Coordinates. 17 stars mark similar forms found in the black and red classes (13 in black class and 4 in red class). The found pattern is dominant in the black class (76.5% accuracy). Red stars mark most similar forms found in these opposite classes

Machine Learning Repository (Lichman 2013). This dataset includes 14 attributes selected by a cardiologist from 47 registered attributes.

In Fig. 6.12, some diseases have visible differences from healthy patients such as more fragments of rectangles and different symmetry axes. This first visual clue is a guide for the next analytical steps. On the next analytical steps the analyst checks the clue on the entire dataset to provide confidence in the discovered pattern.

These figures also show that CPC stars are more compact than traditional stars. It is visible in Fig. 6.13 where all not black cases are more "horizontal" and black cases are mostly vertical with Northwest orientation.

The difference between classes is less evident in traditional stars. CPC stars allow getting better patterns and finding them faster. Figure 6.14a shows a traditional star for an n-D point **p** from the black class. The traditional stars from Fig. 6.12 that are close to **p** were found visually and are presented in Fig. 6.14b–e from each colored class.

Similarly, Fig. 6.15a shows the CPC star for the same point **p** and Fig. 6.15b–e present respective close CPC stars from each colored class. The overlay of stars (a) and (b) from Fig. 6.14 is captured in Fig. 6.16 showing real closeness of these closed forms.

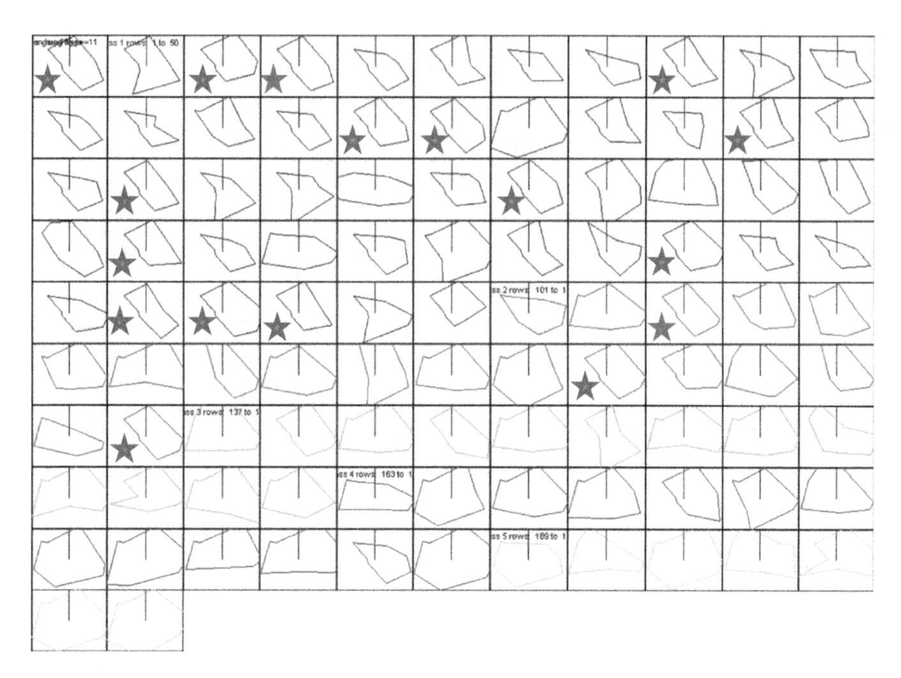

Fig. 6.13 Samples of 14-D data from the 5 colored classes represented by closed contours (CPC stars) in CPC Radial Coordinates. 17 stars mark similar forms found in black and red classes (14 in black class and 3 in red class). The found pattern is dominant in the black class (82.4% accuracy). Red stars mark the most similar forms found in these opposite classes

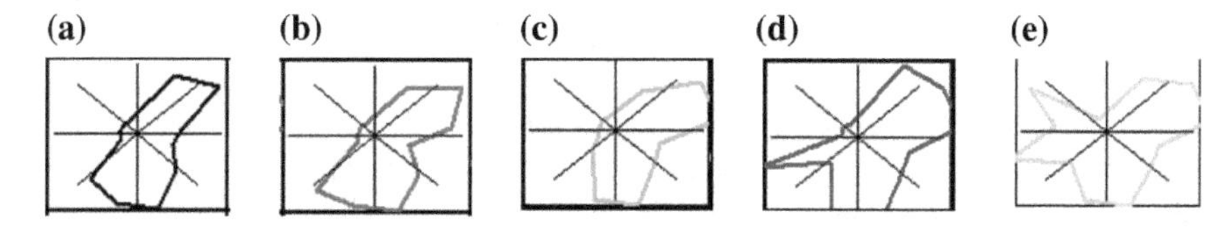

Fig. 6.14 Closest CPC stars from 5 classes from Fig. 6.12

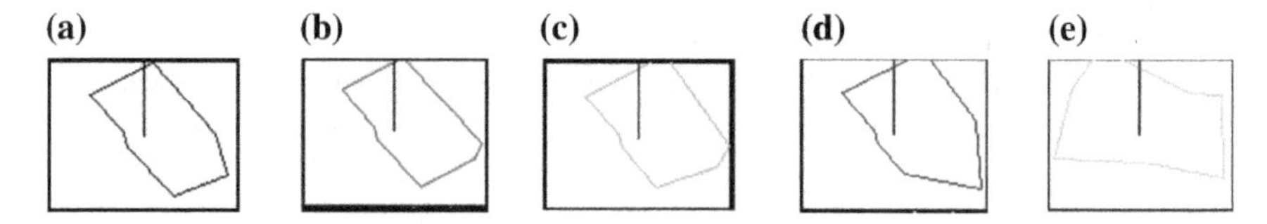

Fig. 6.15 Closest CPC stars (a) and (b) from Fig. 6.13

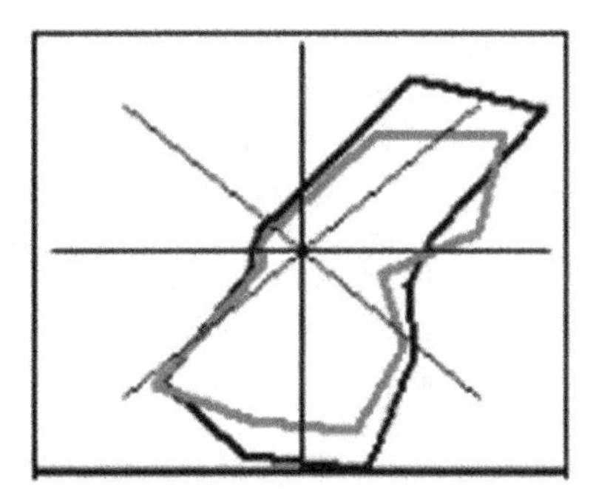

Fig. 6.16 Overlay of stars (a) and (b) from Fig. 6.12

6.5.2 *Feature Extraction Algorithm*

Below we describe the algorithm for extraction of discrimination features using the data explained above. We start from an arbitrarily n-D point $\mathbf{p}_1$ from class C_1 (e.g., black class), and find the n-D point $\mathbf{p}_2$ in class C_2 (e.g., red class), which is most similar to $\mathbf{p}_1$ using a lossless closed contour representation of points in 2-D. See Fig. 6.14, where black Fig. 6.14a represents $\mathbf{p}_1$ and red Fig. 6.14b represents $\mathbf{p}_2$ for Fig. 6.12. Then we search for the n-D points in both classes, which are most similar to $\mathbf{p}_1$ and $\mathbf{p}_2$. These points are marked by stars in Figs. 6.12 and 6.13. Next, we evaluate distribution of these points between C_1 and C_2 classes. In Fig. 6.12, it is 13:4 (76.5% in C_1) and in Fig. 6.13, it is 14:3 (82.4% in C_1).

Respectively the algorithm steps are:

1. Randomly select an arbitrarily n-D point $\mathbf{p}_1$ from class C_1
2. Find all the n-D points in both classes that most similar to $\mathbf{p}_1$ and $\mathbf{p}_2$.
3. Evaluate distribution of these points between C_1 and C_2 classes.
4. Remove these points from the dataset.
5. Select another point in C_1 from the remaining C_1 points and repeat the visual search for this point as we did for $\mathbf{p}_1$ and $\mathbf{p}_2$. This process continues until all points from C_1 and C_2 are processed.
6. Enhance visual patterns to improve separation. For points $\mathbf{p}_1$ and $\mathbf{p}_2$, this is finding features that differentiate them.
7. Formalize found visual patterns to be able computing class of new objects without a human expert who needs to analyze visual patterns.

Below we discuss step 6 in more details. Consider $\mathbf{p}_1$ and $\mathbf{p}_2$ as shown in Fig. 6.14a, b. The upper line in the black case $\mathbf{p}_1$ is going *down*, but in the red case $\mathbf{p}_2$, it is *horizontal*. Next, we test this visually discovered property on its ability to separate better those 17 cases. We have two cases with horizontal line in each class C_1 and C_2 among 17 cases that are similar to p_1 and p_2. Thus, this feature is not a good feature to improve the separation of these 17 cases. Another visual feature must be found.

Having CPC star representation we can try to find separation features in CPC stars. We can see in Fig. 6.15b (red case) a *very short line* on the right, which is almost vertical. This line is present in all three red cases and is not present in any of the 14 black cases that we try to separate. Thus, this is a perfect feature to improve the separation of 17 very visually close cases with 100% accuracy.

Next, we turn to Step 7 to find an analytical form of that visual feature. Denote the start and end points of that line as $\mathbf{w}_s$ and $\mathbf{w}_e$. Their distance $d(\mathbf{w}_s, \mathbf{w}_e)$ serves as a discrimination feature

$$\text{If } d(\mathbf{w}_s, \mathbf{w}_e) > d \text{ then class } C_1 \text{ else class } C_2, \tag{6.1}$$

where d is a distance threshold computed from Fig. 6.15b. Assume for simplicity of notation that we started the graph in Fig. 6.15 from point $\mathbf{w}_s$. Then our start and end points are

$$w_{s1} = f(x_1, x_2), w_{s2} = g(x_1, x_2) \tag{6.2}$$

$$w_{e1} = f(x_3, x_4), w_{e2} = g(x_3, x_4), \tag{6.3}$$

where x_1–x_4 are first four original n-D coordinates of an n-D point that we consider. Here f and g are functions that are used to map x_1–x_4 to CPC star coordinates as we presented in Sect. 6.1. Thus, formula (6.1) will be rewritten as with use of (6.2)–(6.3):

$$((w_{s1} - w_{e1})^2_+ (w_{s2} - w_{e2})^2)^{1/2} > d \text{ then class } C_1 \text{ else class } C_2 \tag{6.4}$$

$$\left((f(x_1, x_2) - f(x_3, x_4))^2 + (g(x_1, x_2) - g(x_3, x_4))^2\right)^{1/2} > d \text{ then Class } C_1 \text{ else Class } C_2 \tag{6.5}$$

Discovering (6.5) demonstrates the power of visual analytics, which combines visual and computational methods in Visual Knowledge Discovery. Discovering (6.5) purely analytically by Machine Learning methods without a visual clue would be extremely difficult. We would need to guess somehow a set of models that includes (6.5). What could be the base for such a guess? It is hard to expect some knowledge for this guess. In these particular data, we did not have such prior knowledge. Next, if the guessed set of models includes (6.5), the machine learning algorithm may not learn it. It may not be the winning model on the given training data for the given ML algorithm.

How general is this algorithm? Why is it not an ad hoc one? Steps 1 and 4 are quite general for any training dataset with the classes of n-D points identified. The success in Steps 2, 3, 5 and 6 depends on 2-D representation of n-Data, perceptual abilities of the viewer, allotted time and amount of data. The step 7 is also quite general and its success depends on success in previous steps and on mathematical skills of the analyst. So far, experiments with CPC Stars show that all these steps are doable successfully for real data providing a consistent framework for visual analytics in Data Mining and Machine Learning. Further research and experiments are needed to specify steps 1–7 more and data types where this algorithm will be efficient. It includes training data scientists in visual features search.

6.5.3 Comparison with Parallel Coordinates

Figure 6.17 shows the same data in Parallel Coordinates as in Fig. 6.13 and Fig. 6.18 shows a subset of similar Parallel Coordinates curves from Fig. 6.17. We do not see a separation pattern between the classes in these figures, but a separation pattern is visible in CPC stars in Fig. 6.13.

Please check and confirm if the inserted citation of Fig. 6.18 is correct. If not, please suggest an alternate citation. Please note that figures and tables should be cited sequentially in the text.Done

The difference between Traditional Stars, CPC stars, and Parallel Coordinates is even more visible in Figs. 6.19 and 6.20 in the higher dimension (n = 170). Figure 6.19 shows traditional 170-D stars in the first two rows: musk chemicals (first row), and non-musk chemicals (second row) from *Musk data* from UCI Machine Learning repository (Lichman 2013). Respectively, the third and fourth rows in Fig. 6.19 show CPC 170-D stars from the same dataset: musk chemicals (third row) and non-musk chemicals (fourth row).

A specific pattern on the right of each star is visible on rows 2 and 4, which represent non-musk chemicals. Multiple other distinct features can be extracted from Fig. 6.19, which can assist in separating the two classes. In contrast, in Parallel Coordinates in Fig. 6.20, it is very difficult to identify and separate features of two classes with four points from the black class, and five points from the red class.

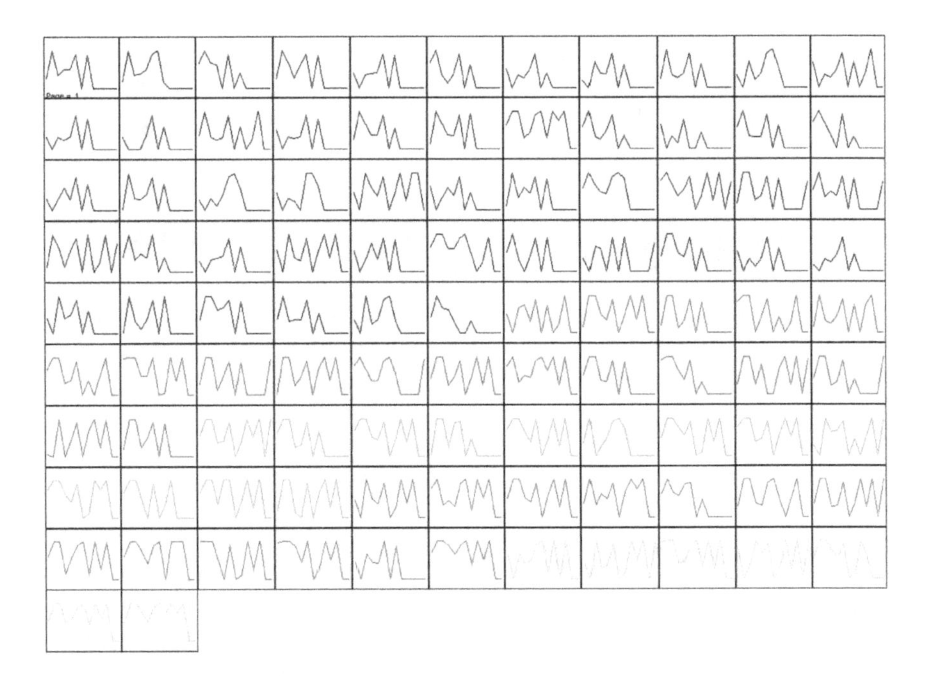

Fig. 6.17 Samples of 14-D data from 5 colored classes in Parallel Coordinates

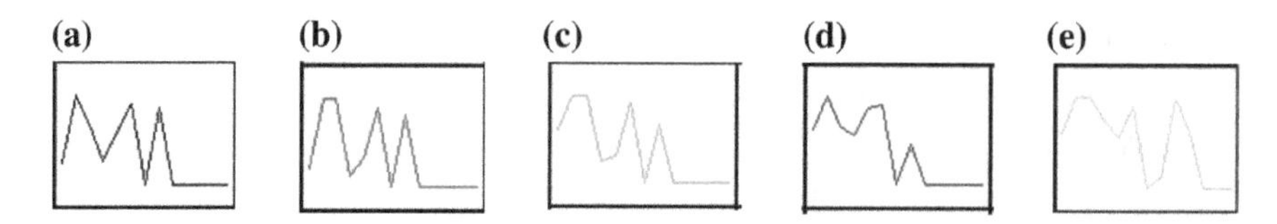

Fig. 6.18 Similar Parallel Coordinates curves from Fig. 6.17

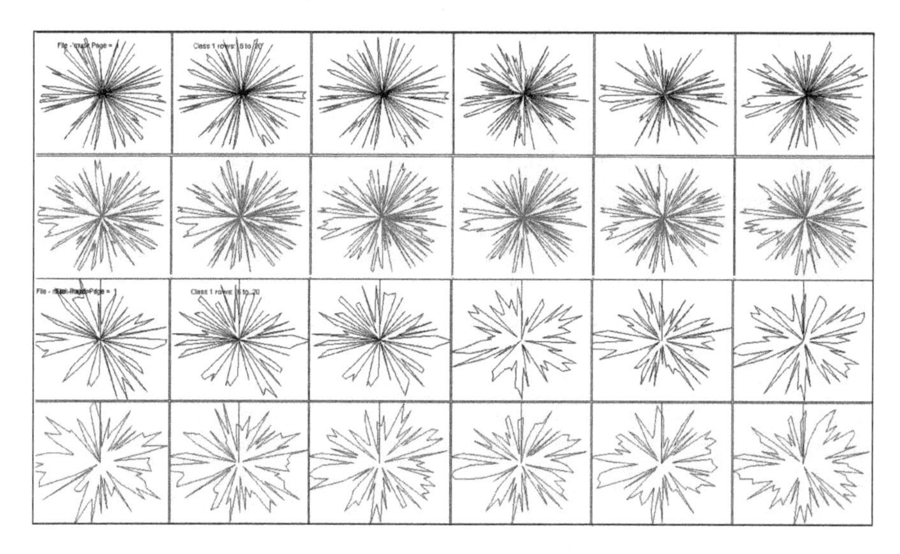

Fig. 6.19 Traditional 170-D stars: class "musk" (first row) and class "non-musk chemicals" (second row). CPC 170-D stars from the same dataset: class "musk" (third row) and class "non-musk chemicals" (fourth row)

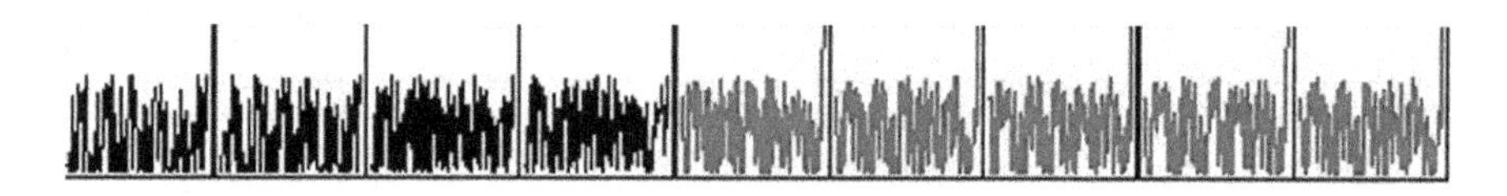

Fig. 6.20 Nine 170-dimensional points of two classes in Parallel Coordinates

6.6 Discussion

6.6.1 Comparison of Experiments 1 and 3

Below we compare the results of experiments 3 with experiment 1 that has less number of respondents, but more difficult data of dimension n = 192.

The experiment 1 was repeated for different levels of "noise", i.e., hyper-tube width (up to 10, 20, 30% of maximum possible value of each coordinate of an n-D point). Each respondent had 20 min for each GLC representation type.

Experiment 1 had shown that with increased noise level from standard deviation 10–30% that respondents not only produced more misclassifications, but one of the respondents completely refused answering using Radial Coordinates for data with 20 and 30% of noise and another one with advanced skills refused answering using Radial Coordinates for data with 30% of noise. In contrast, CPC Star tests showed acceptable time (1–5 min) for all figures with noise up to 30%. The success of CPC stars versus Radial Coordinates was due to two times less dense placement of the points as a result of a specific mapping of pairs (x_i, x_{i+1}) into a closed contour in CPC stars.

The difference in experiments 1 and 3 is not only in the dimensions and the number of classes (2 vs. 5), but in the noise control. In experiment 3, the Gaussian noise is with mean 10% of the max value of each normalized coordinate (max = 1), and standard deviation of 20% of that max value for the normal distribution, N(0.1, 0.2). Thus, this noise is three times smaller than the highest noise of 30% used in experiment 1, which is much more complex for the human analysis.

The experiment 1 with n = 192 and a high level of noise (30%) points out on the likely *upper bound* of human classification of n-D data using the Radial Coordinates for data modeled as linear hyper-tubes. This *upper bound* is *no greater than 192 dimensions with up to 30% noise*. One of the motivations for the experiment 3 with n = 160 was the failure of Radial Coordinates at n = 192. The decreased dimensions and noise level in experiment 3 was to find out would n = 160 with lower noise be upper bound or not for the Radial Coordinates. The experiment 3 shows that the *upper bound* for human classification on such n-D data is *no less than n = 160 dimensions with up to 20% noise*. Thus the expected *classifiable dimensions are in* [160,192] *dimensions interval* for the Radial Coordinates.

Due to advantages of Star CPC over Radial Coordinates, these limits must be higher for Star CPC and lower for Parallel Coordinates due to higher occlusion in PC. Limits that are more exact are the subject of the future experiments. About 70 respondents participated in the experiment 3, therefore it seems that 160 dimensions can be viewed as a quite firm bound. In contrast, the question that 192 dimensions is the max of the upper limit for Star CPC may need addition studies. Thus, so far the indications are that the upper limit for Star CPC is above n = 192 and it needs to be found in future experiments for linear hyper-tubes. Note that finding bounds for linear-hyper-tubes most likely will be also limits for non-linear hyper-tubes due to their higher complexity.

6.6.2 Application Scope of CPC Stars

As experiment 1 had shown, the application scope of CPC Stars covers tasks with 192 dimensions. While this is a significant progress, the current interests in Big data studies are in larger dimensions and the number of n-D points. However, in many practical diagnostics tasks in medicine and engineering, the dimensions do not exceed 200.

The expansion to dimension *n* up to 1000 can be performed by grouping the coordinates $\{X_i\}$ by 100–150, and representing them by separate or collocated colored stars, and/or mapping some X_i into colors. Lossy reduction of *n* can be applied after visual analysis of these lossless displays, which can reveal the least informative attributes. Another reduction can be based on a priori domain knowledge.

In the mentioned diagnostics tasks, the number of n-D points is often less than 10^3–10^4. In many tasks with millions of records, often a prior knowledge of data and a specific goal allow essential data reduction. The experience shows that visual comparison of thousands of figures to analyze few classes is feasible if the visual representation *effectively applies human shape perception capabilities.* To avoid occlusion each star can be displayed in its own coordinate system in a separate cell.

While this solves the occlusion issue, it requires switching gaze from one star to another one. It takes time, requires memorizing the first star before looking at another one, which complicates the comparison of stars.

One of the solutions for this issue is considering one star as a base and using an animated overlay of other stars with it one after another. The analyst can control a speed of animation. In animation, the color of the overlaid star differs from the color of the base star. The sections of two stars, which are practically identical, can be blinked or shown in a third color. The analyst can use a mouse interaction to indicate that two stars are similar and potentially from the same class. Future experimental studies will be to find most efficient *interactive arrangement.*

6.6.3 Prospects for Higher Data Dimensions

The above advantages of CPC stars versus traditional stars and parallel coordinates are even more essential for data of higher dimensions. We presented these three representations (Figs. 6.19 and 6.20) for musk learning dataset from the UCI machine learning repository. It is an example of very practical design models of drugs and other chemicals without expensive experimental tests, such as clinical trials of the targeted properties. In the data each instance is described by their 170 physical, chemical, structural, etc. properties and its target attribute (musk class or non-musk class).

Although CPC stars show the same information in each cell as the traditional stars, they are better for visual analysis because they have:

- less density of form features,
- bigger sizes, and
- better separability.

In contrast, Parallel Coordinates are unacceptable for such large data dimensions, while the stars above allow comparing data with over a hundred attributes. Open polylines in Parallel Coordinates of the same n-D data points as shown in Fig. 6.20 are practically indistinguishable. These advantages of closed contours are consistent with Gestalt Laws. Therefore, new extensive user studies are not necessary because of extensive previous experiments elsewhere verified Gestalt Laws. These laws are viewed as most universal information about form perception for display choice independently on specific data properties. In addition, we conducted such studies described in this chapter.

6.6.4 Shape Perception Capabilities: Gestalt Law

Gestalt Laws. About century ago, psychologists experimentally revealed fundamental laws [Gestalt Laws (Wertheimer 1944; Elder and Goldberg 2002)] of perception and recognition of figures by a human vision system. According to Gestalt Laws, a figure that *possesses a closure, symmetry, similarity, proximity, and continuity will be detected faster* in the presence of noise.

In the same way its *shape* will be recognized faster and more accurately as well as a *common pattern* of a few figures will be specified better. In concordance with the Gestalt laws, the closed contours such as *stars* in Radial Coordinates show the essential perceptual advantages over lines in the Parallel Coordinates (PC), bar charts, pie charts, etc.

Mapping data vectors into contours allows describing and recognizing very complicated nonlinear structures in a data space. Invariance of shape perception under local affine transformations of image (Wagemans et al. 2000) radically extends these capabilities.

Visualizations with simple connections between data attributes and image features allow effective use of these unique human perceptual capabilities. Polar displays of data vectors (stars), parallel (Cartesian) coordinates, pie- and bar-charts are among visualizations that benefit from these capabilities.

Lack of experimental and theoretical data for display evaluation. Unfortunately, extreme complexity and flexibility of visual shape perception led psychologists to focus on either very common law of vision such as Gestalt law or some basic properties such as perception thresholds.

Overviews (Bertini et al. 2011; Hoffman and Grinstein 2002) and respective publications in last few decades do not expose lossless visualizations for more than 10–20 dimensions intended for effective use of complex shape perception. For lesser dimensions some experimental studies of effectiveness of displaying contours such as stars, pie-charts, bar-graphs, Chernoff faces, etc. have been done e.g., (Elder and Goldberg 2002).

However, usually these visualizations show only specific attributes of given data that are: (1) known to a subject matter expert as important ones, (2) suspected to be important by a researcher, or (3) found by clustering. Therefore, it is difficult to use these results as evaluations of capabilities of the visualization methods or to synthesize better visualizations for other data, especially beyond 20–30 dimensions.

Shape perception features. Humans are able to detect, compare, and describe multiple figures by using hundreds of their local features such as concave, convex, angle, and wave, and combine them into a multilevel hierarchy (Grishin 1982; Grishin et al. 2003).

Each feature itself includes many attributes, e.g., size, orientation, location, and others. A term "holistic picture" denotes an image together with its description that includes image statistics, textures, integral characteristics, shapes, and coloring. Moreover, the holistic concept is appearing at multiple levels of image perception. First, the image is considered as a set of "spot clusters" and relations between them as an overall structure of the image. Then each spot cluster is considered with the same aspects (statistics, textures, integrals, shapes and coloring) where elements are "spots" and the structure represent relations between these "spots". Next, each "spot" is viewed at the holistic level in the same way and at the levels of its elements. At these levels, perceptually important features include symmetry, elongation, orientation, compactness, convexity/concavity, peaks, waves, sharp angle, inside/outside, etc.

6.7 Collaborative Visualization

Visualization of large n-D datasets for pattern discovery can be accomplished collaboratively by splitting a dataset between **collaborating agents**, which can include both humans and software agents. In this case, each agent analyzes and visualizes a subset of data and exchanges findings with other agents.

There are multiple options of for collaborative visualization for knowledge discovery. Below we present 3 major options.

Option 1: Data splitting to support collaboration based on:

- *Location of data* on n-D space (each agent works of the data from a specific location on n-D space produced by data clustering).
- *Class of data* (each agent works only on the data of a specific class or classes),
- *Attributes of data* (each agent works only on the projection of data to the specific subset of attributes.)

Option 2: *Task specialization.*

- Specialization of agents to different visual tasks depending on individual skills and capabilities of agents. In this case, data can be the same for all agents (not split), but *organized and visualized differently*, e.g., different order of the attributes because visualization in parallel coordinates and paired coordinates are sensitive to this change.

Option 3: *Joint work*

- Work without splitting data and tasks between team members. People work as a team on the same task, on the same data and discuss findings. This is a case of the collaborative experiment described in Sect. 6.3.4.

Other options include different combinations of options 1–3. Figures 6.21 and 6.22 illustrate options 1 and 2. Figure 6.23 shows the combination of options 1 and 2, when each agent works on subset of data and own subtask first and then work together exchanging findings using a collaboration platform.

Figures 6.24 and 6.25 illustrate the situation when the base n-D data are the same for all agents, but visualization data (plots) are different and tasks are different too. In Fig. 6.24 each agent needs to evaluate the representation in the respective plot. In Fig. 6.25 each agent needs to evaluate the separation pattern between classes of data.

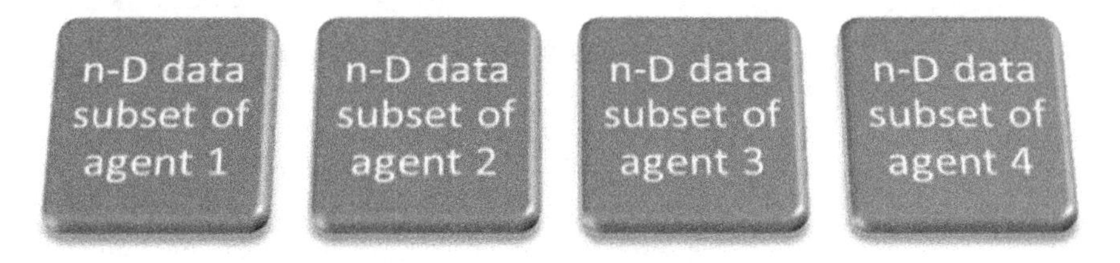

Fig. 6.21 Data splitting for collaborative visualization

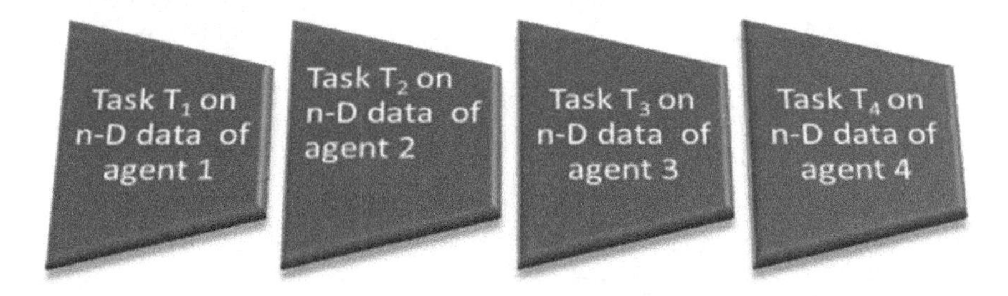

Fig. 6.22 Task splitting for Collaborative Visualization

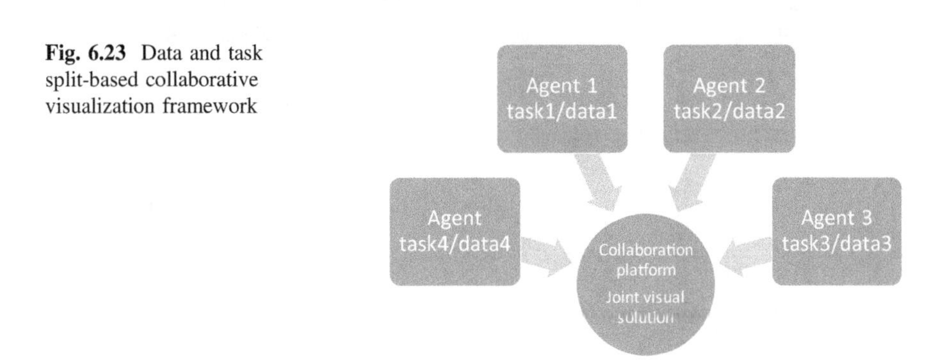

Fig. 6.23 Data and task split-based collaborative visualization framework

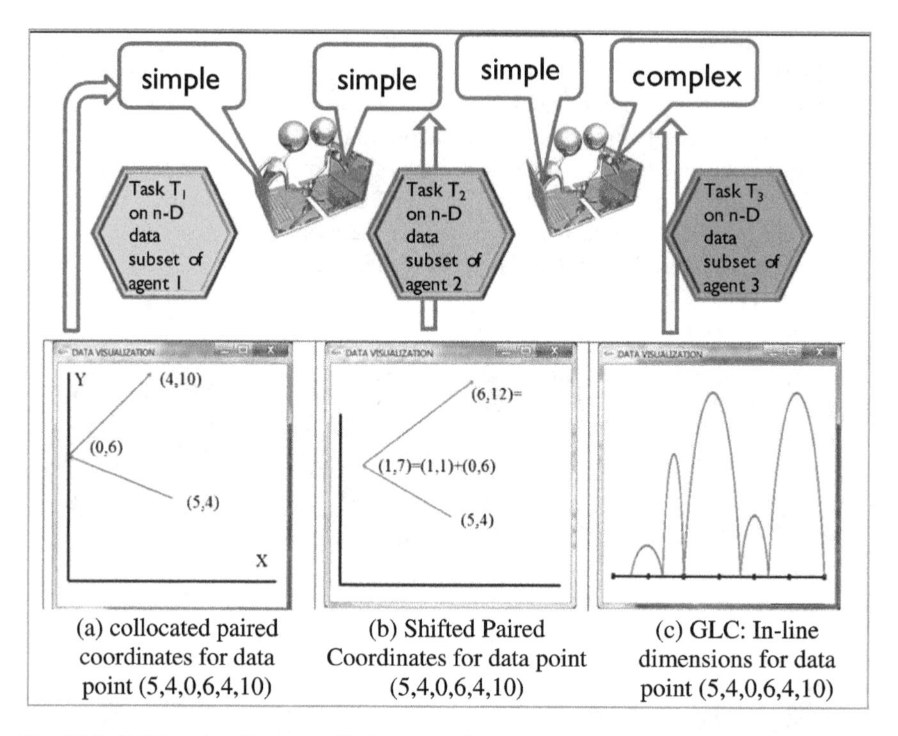

Fig. 6.24 Collaboration diagram with data example 1

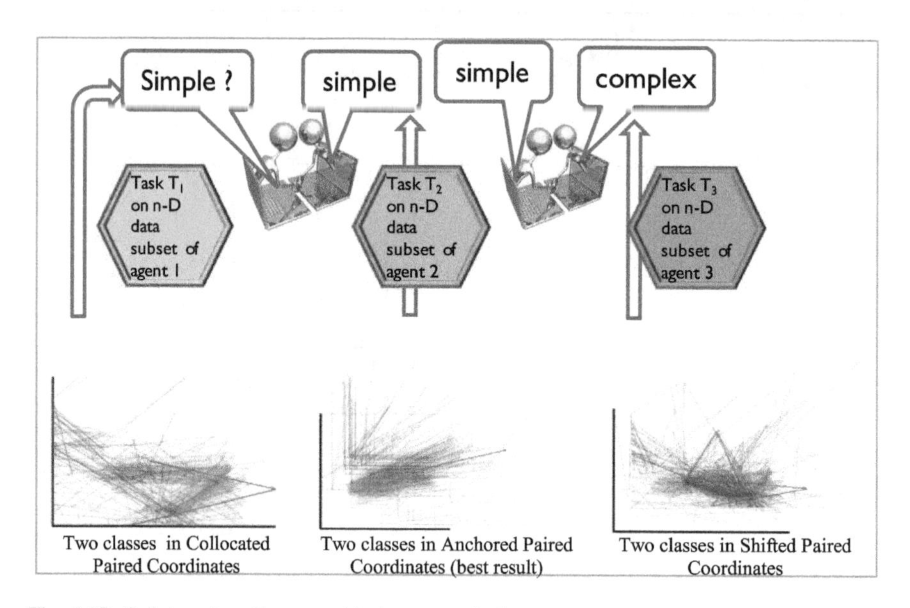

Fig. 6.25 Collaboration diagram with data example 2

6.8 Conclusion

This chapter shows that Paired Closed Contour Coordinates as a part of GLC are capable representing data in 14-D, 48-D, 96-D, 160-D, 170-D, and 192-D where humans are capable discovering features and patterns for classification these high-dimensional data. In particular, the experiment 3 that was involved about 70 participants show the abilities of visual discovery of n-D patterns using GLCs with $n = 160$ dimensions. This shows that this type of GLC is useful at the current stage of its development and is promising for knowledge discovery tasks in visual analytics, visual data mining and machine learning. Advantages of this technique relative to the Parallel Coordinates were shown in experiments described above. This technique can be applied in cooperation with the analytical Data Mining/ Machine Learning methods to decrease the heuristic guesses in selecting a class of Data Mining models.

References

Bertini, E., Tatu, A., Keim, D.: Quality metrics in high-dimensional data visualization: an overview and systematization, IEEE Tr. Vis. Comp. Graph. **17**(12), 2203–2212 (2011)

Elder, J., Goldberg, M.: Ecological statistics of Gestalt laws for the perceptual organization of contours. J. Vis. **2**, 324–353 (2002). http://journalofvision.org/2/4/5/ 324

Grishin: Pictorial Analysis of Experimental Data. Nauka Publishing, Moscow, pp. 1–237 (1982)

Grishin, V., Sula, A., Ulieru, M.: Pictorial analysis: a multi-resolution data visualization approach for monitoring and diagnosis of complex systems. Int. J. Inf. Sci. **152**, 1–24 (2003)

Hoffman, P.E., Grinstein, G.G.: A survey of visualizations for high-dimensional data mining. Inf. Vis. Data Min. Knowl. Discov. 47–82 (2002)

Lichman, M. UCI Machine Learning Repository [http://archive.ics.uci.edu/ml]. Irvine, CA: University of California, School of Information and Computer Science, 2013

Trafimow, D., Marks, M. (eds.): J. Basic Appl. Soc. Psychol. **37**(1), 1–2 (2015)

Trafimow, D., Rice, S.: A test of the null hypothesis significance testing procedure correlation argument. J. Gen. Psychol. **136**(3), 261–270 (2009)

Valentine, J.C., Aloe, A.M., Lau, T.S.: Life after NHST: How to describe your data without "p-ing" everywhere. Basic Appl. Soc. Psychol. **3**;37(5), 260–73 (2015)

Wagemans, J., Van Gool, L., Lamote, C., Foster, D.: Minimal information to determine affine shape equivalence. J. Exp. Psychol. Hum. Percept. Perform. **26**(2), 443–468 (2000)

Wertheimer, M.: Gestalt theory. Social Research **11**, 78–99 (1944)

Chapter 7
Interactive Visual Classification, Clustering and Dimension Reduction with GLC-L

I believe in intuition and inspiration.
It is, strictly speaking, a real factor in scientific research.
Albert Einstein

7.1 Introduction

A representative software system for the interactive visual exploration of multivariate datasets is XmdvTool (2015). It implements well-established algorithms such as parallel coordinates, radial coordinates, and scatter plots with hierarchical organization of attributes (Yang et al. 2003). For a long time, its functionality was concentrated on exploratory manipulation of records in these visualizations. Recently, its focus has been extended to support data mining (version 9.0, 2015), including interactive parameter space exploration for association rules (Lin et al. 2014), interactive pattern exploration in streaming (Yang et al. 2013), and time series (Zhao et al. 2016).

The goal of this chapter is to present a new *interactive visual machine learning system for solving supervised learning classification tasks* based on a GLC-L visualization algorithm and associated interactive and automatic algorithms GLC-IL, GLC-AL and GLC-DRL for discovery of linear and non-linear relations and dimension reduction. Classification and dimension reduction tasks from three domains, image processing, computer-aided medical diagnostics and finance (stock market), are used to illustrate this method.

This chapter is organized as follows. First we presents the approach including the base algorithm GLC-L, its interactive version, the algorithm for automatic discovery of relations combined with interactions, visual structure analysis of classes and generalization of algorithms for non-linear relations. Next, we present the results for five case studies using these algorithms. The discussion and the analysis of the results in comparison with prior results and software implementation follow the results. Then the advantages and benefits of the presented algorithms for multiple domains are summarized.

B. Kovalerchuk, *Visual Knowledge Discovery and Machine Learning*,
Intelligent Systems Reference Library 144,
https://doi.org/10.1007/978-3-319-73040-0_7

7.2 Methods: Linear Dependencies for Classification with Visual Interactive Means

We consider a task of visualizing an n-D linear function $F(\mathbf{x}) = y$, where $\mathbf{x} = (x_1, x_2, \ldots, x_n)$ is an n-D point and y is a scalar,

$$F(\mathbf{x}) = y = c_1x_1 + c_2x_2 + c_3x_3 + \cdots + c_nx_n + c_{n+1}.$$

Such functions play important roles in classification, regression and multi-objective optimization tasks. In regression, $F(\mathbf{x})$ directly serves as a regression function. In classification, $F(\mathbf{x})$ serves as a discriminant function to separate the two classes with a classification rule with a threshold T: if $y < T$ then $\mathbf{x}$ belongs to class 1, else $\mathbf{x}$ belongs to class 2. In multi-objective optimization, $F(\mathbf{x})$ serves as a tradeoff to reconcile n contradictory objective functions with c_i serving as weights for objectives.

7.2.1 *Base GLC-L Algorithm*

This section presents the visualization *algorithm* called *GLC-L* for a linear function. It is used as a base for other algorithms presented in this chapter.

Let $K = (k_1, k_2, \ldots, k_{n+1})$, $k_i = c_i/c_{max}$, where $c_{max} = |\max_{i=1:n+1}(c_i)|$, and

$$G(\mathbf{x}) = k_1x_1 + k_2x_2 + \cdots + k_nx_n + k_{n+1}.$$

Here all k_i are normalized to be in [−1, 1] interval. The following property is true for F and $G: F(\mathbf{x}) < T$ if and only if $G(\mathbf{x}) < T/c_{max}$. Thus, F and G are equivalent linear classification functions. Below we present steps of *GLC-L* algorithm for a given linear function $F(\mathbf{x})$ with coefficients $C = (c_1, c_2, \ldots, c_{n+1})$.

Step 1 *Normalize* $C = (c_1, c_2, \ldots, c_{n+1})$ by creating as set of normalized parameters $K = (k_1, k_2, \ldots, k_{n+1}) : k_i = c_i/c_{max}$. The resulting normalized equation

$$y_n = k_1x_1 + k_2x_2 + \cdots + k_nx_n + k_{n+1}$$

with normalized rule:

if $y_n < T/c_{max}$ then $\mathbf{x}$ belongs to class 1, else $\mathbf{x}$ belongs to class 2,

where y_n is a normalized value, $y_n = F(\mathbf{x})/c_{max}$. Note that for the classification task we can assume $c_{n+1} = 0$ with the same task generality. For regression, we also deal with all data normalized. If actual y_{act} is known, then it is normalized by C_{max} for comparison with y_n, y_{act}/c_{max}.

Step 2 *Compute all angles* $Q_i = arccos(|k_i|)$ of absolute values of k_i and locate coordinates X_1–X_n in accordance with these angles as shown in Fig. 7.1 relative to the horizontal lines. If $k_i < 0$, then coordinate X_i is oriented to the left, otherwise X_i is oriented to the right (see Fig. 7.1). Draw its values of a given n-D point $\mathbf{x} = (x_1, x_2, \ldots, x_n)$, as *vectors* $\mathbf{x}_1, \mathbf{x}_2, \ldots, \mathbf{x}_n$ in respective coordinates X_1–X_n (see Fig. 7.1).

Step 3 *Draw vectors* $\mathbf{x}_1$, $\mathbf{x}_2$,…, $\mathbf{x}_n$ *one after another*, as shown on the left side of Fig. 7.1. Then *project* the last point for $\mathbf{x}_n$ onto the horizontal axis U (see a red dotted line in Fig. 7.1). To simplify, visualization axis U can be collocated with the horizontal lines that define the angles Q_i.

Step 4 Step 4a. For regression and linear optimization tasks, repeat step 3 for all n-D points as shown in the upper part of Figs. 7.2a and 7.3a.
Step 4b. For the two-class classification task, repeat step 3 for all n-D points of classes 1 and 2 drawn in different colors. Move points of class 2 by mirroring them to the bottom with axis U doubled as shown in Fig. 7.2. For more than two classes, Fig. 7.1 is created for each class and m parallel axes U_j are generated next to each other similar to Fig. 7.2. Each axis U_j corresponds to a given class j, where m is the number of classes.
Step 4c. For multi-class classification tasks, conduct step 4b for all n-D points of each pair of classes i and j drawn in different colors, or draw each class against all other classes together.

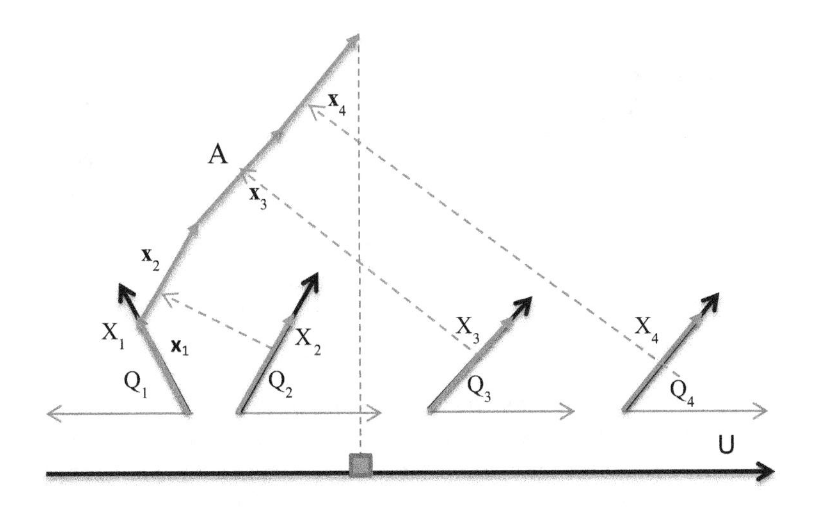

Fig. 7.1 4-D point A – (1, 1, 1, 1) in GLC-L coordinates X_1–X_4 with angles (Q_1, Q_2, Q_3, Q_4) with vectors $\mathbf{x}_i$ shifted to be connected one after another and the end of last vector projected to the black line. X_1 is directed to the left due to negative k_1. Coordinates for negative k_i are always directed to the left

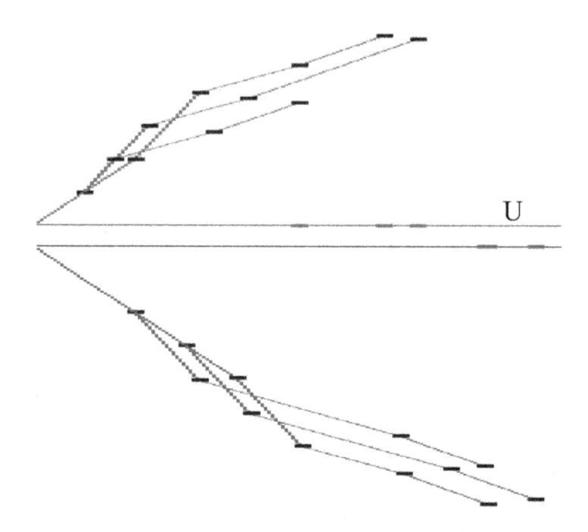

Fig. 7.2 GLC-L algorithm on simulated data. Result with axis X_1 starting at axis U and repeated for the second class below it

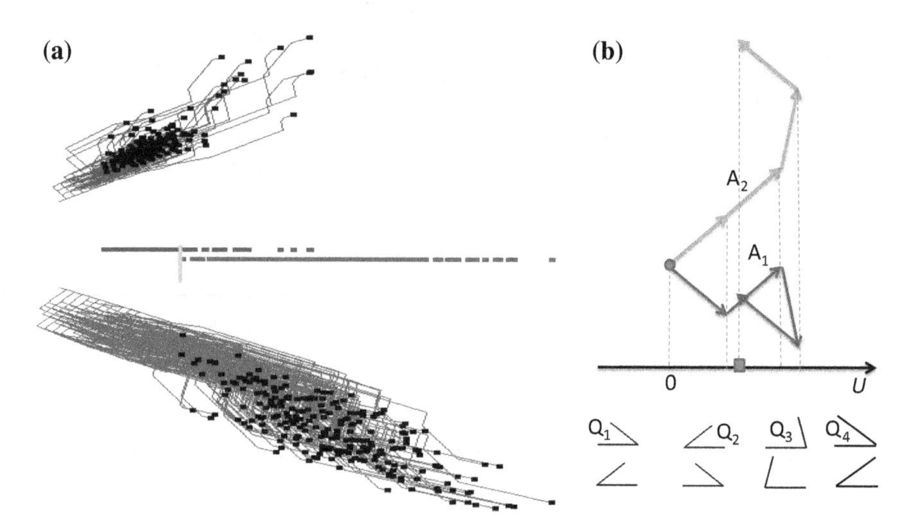

Fig. 7.3 GLC-L algorithm on real and simulated data. **a** Visualized data subset from two classes of Wisconsin breast cancer data from UCI machine learning repository [16]; **b** 4-D point A = (−1, 1, −1, 1) in two representations A_1 and A_2 in GLC-L coordinates X_1–X_4 with angles Q_1–Q_4

This algorithm uses the property that $cos(arccos\ k) = k$ for $k \in [-1, 1]$, i.e., projection of vectors $\mathbf{x}_i$ to axis U will be $k_i x_i$ and with consecutive location of vectors $\mathbf{x}_i$, the projection from the end of the last vector $\mathbf{x}_n$ gives a sum $k_1x_1 + k_2x_2 + \cdots + k_nx_n$ on axis U. It does not include k_{n+1}. To add k_{n+1}, it is sufficient to shift the start point of $\mathbf{x}_1$ on axis U (in Fig. 7.1) by k_{n+1}. Alternatively, for the visual classification task, k_{n+1} can be omitted by subtracting k_{n+1} from the threshold.

Steps 2 and 3 of the algorithm for negative coefficients k_i and negative values x_i can be implemented in two ways. The first way represents a negative value x_i, e.g.,

$x_i = -1$ as a vector $\boldsymbol{x}_i$ that is directed backward relative to the vector that represent $x_i = 1$ on coordinate X_i. As a result, such vectors $\boldsymbol{x}_i$ go down and to the right. See representation A_1 in Fig. 7.3b for point $A = (-1, 1, -1, 1)$ that is also self-crossing. The alternative representation A_2 (also shown in Fig. 7.3b) uses the property that $k_i x_i > 0$ when both k_i and x_i are negative. Such $k_i x_i$ increases the linear function F by the same value as positive k_i and x_i. Therefore, A_2 uses the positive x_i, k_i and the "positive" angle associated with positive k_i. This angle is shown below angle Q_1 in Fig. 7.3b. Thus, for instance, we can use $x_i = 1$, $k_i = 0.5$ instead of $x_i = -1$ and $k_i = -0.5$. An important advantage of A_2 is that it is perceptually simpler than A_1. The visualizations presented in this chapter use A_2 representation.

A linear function of n variables, where all coefficients c_i have similar values, is visualized in GLC-L by a line (graph, path) that is similar to a straight line. In this situation, all attributes bring similar contributions to the discriminant function and all samples of a class form a "strip" that is a simple form GLC-L representation. In general, the term c_{n+1} is included in F due to both mathematical and the application reasons. It allows the coverage of the most general linear relations. If a user has a function with a non-zero c_{n+1}, the algorithm will visualize it. Similarly, if an analytical machine learning method produced such a function, the algorithm will visualize it too. Whether c_{n+1} is a meaningful bias or not in the user's task does not change the classification result. For regression problems, the situation is different; to get the exact meaningful result, c_{n+1} must be added and interpreted by a user. In terms of visualization, it only leads to the scale shift.

7.2.2 *Interactive GLC-L Algorithm*

For the data *classification* task, the interactive algorithm **GLC-IL** is as follows:

- It starts from the results of GLC-L such as shown in Fig. 7.3a.
- Next, a user can interactively slide a yellow bar in Fig. 7.3a to change a classification threshold. The algorithm updates the confusion matrix and the accuracy of classification, and pops it up for the user.
- An appropriate threshold found by a user can be interactively recorded. Then, a user can request an analytical form of the linear discrimination rule be produced and recorded.
- A user sets up two new thresholds if the accuracy is too low with any threshold (see Fig. 7.4a with two green bars). The algorithm retrieves all n-D points with projections that end in the interval between these bars. Next, only these n-D points are visualized (see Fig. 7.4b).
- At this stage of the exploration the user has three options:

  (a) modify interactively the coefficients by rotating the ends of the selected arrows (see Fig. 7.5),

  (b) run an automatic coefficient optimization algorithm GLC-AL described in Sect. 7.2.3,

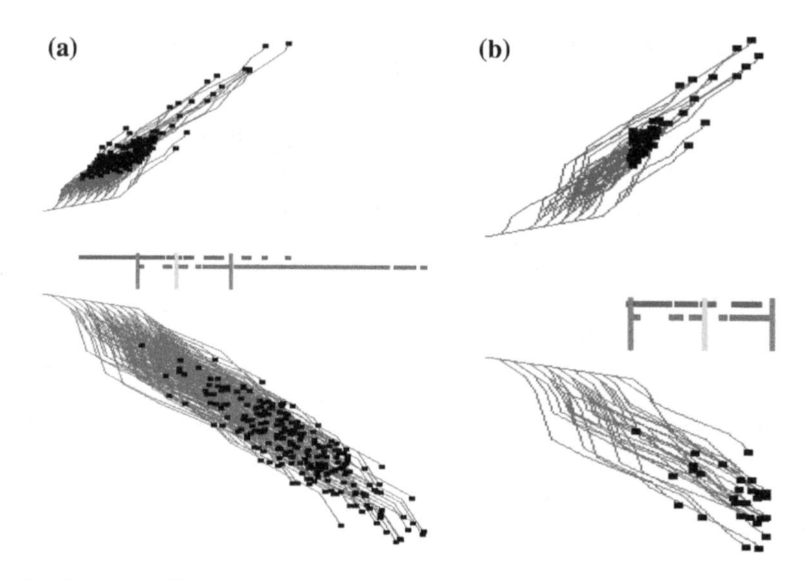

Fig. 7.4 Interactive GLC-L setting with sliding green bars to define the overlap area of two classes for further exploration. **a** Interactive defining of the overlap area of two classes; **b** selected overlapped n-D points

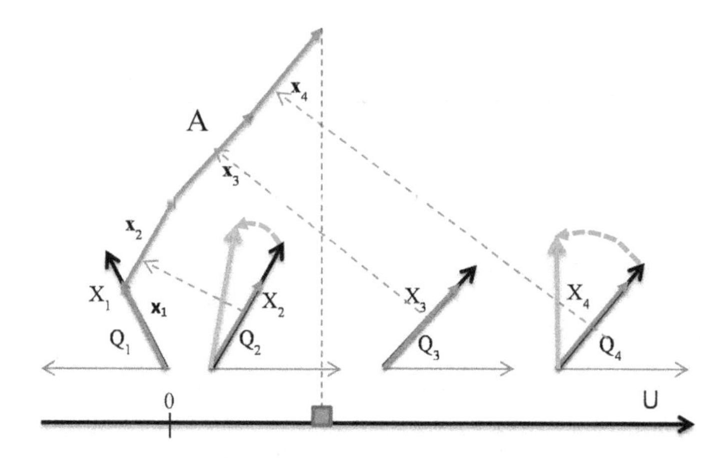

Fig. 7.5 Modifying interactively the coefficients by rotating the ends of selected arrows, X_2 and X_4 are rotated

(c) apply a visual structure analysis of classes.

For *clustering*, the interactive algorithm *GLC-IL* is as follows. A user interactively selects an n-D point of interest *P* by clicking on its 2-D graph (path) *P**. The system will find all graphs *H** that are close to it according to the rule below.

Let $P^* = (p_1, p_2, \ldots, p_n)$ and $H^* = (h_1, h_2, \ldots, h_n)$, where $p_i = (p_{i1}, p_{i2})$ and $h_i = (h_{i1}, h_{i2})$ are 2-D points (nodes of graphs),

T be a threshold that a user can change interactively,

$L(P, T)$ be a set of n-D points that are close to point P with threshold T (i.e., a cluster for P with T),

$L(P,T) = \{H : D(P^*, H^*) \leq T\}$, where $D(P^*, H^*) \leq T \Leftrightarrow \forall i \|p_i - h_i\| < T$, and $\|p_i - h_i\|$ be the Euclidian distance between 2-D points p_i and h_i.

The automatic version of this algorithm searches for the largest T, such that only n-D points of the class, which contains point P, are in $L(P, T)$ assuming that the class labels are known,

$$\max T : \{H \in L(P,T) \Rightarrow H \in \text{Class}(P)\},$$

where (P) is a class that includes n-D point P.

7.2.3 *Algorithm GLC-AL for Automatic Discovery of Relation Combined with Interactions*

The GLC-AL algorithm differs from the Fisher Linear Discrimination Analysis (FDA), Linear SVM, and Logistic Regression algorithms in the criterion used for optimization. The GLC-AL algorithm directly maximizes some value computed from the confusion matrix (typically accuracy), $A = (\text{TP} + \text{TN})/(\text{TP} + \text{TN} + \text{FP} + \text{FN})$, which is equivalent to the optimization criterion used in the linear perceptron (Freund and Schapire 1999) and Neural Networks in general. In contrast, the Logistic Regression minimizes the Log-likelihood (Freedman 2009). The GLC-AL algorithm also allows maximization of the truth positive (TP). Fisher Linear Discrimination Analysis maximizes the ratio of between-class to within-class scatter (Maszczyk and Duch 2008). The Linear SVM algorithm searches for a hyperplane with a large margin of classification, using the regularization and quadratic programming (Cristianini and Shawe-Taylor 2000).

The automatic algorithm GLC-AL is combined with interactive capabilities as described below. The progress in accuracy is shown after every m iterations of optimization, and the user can stop the optimization at any moment to analyze the current result. It also allows interactive change of optimization criterion, say from maximization of accuracy to minimization of False Negatives (FN), which is important in computer-aided cancer diagnostic tasks.

There are several common computation strategies to maximize accuracy A in $[-1, 1]^{n+1}$ space of coefficients k_i. Gradient-based search, random search, genetic and other algorithms are commonly used to make the search feasible.

For the practical implementation, in this study, we used a simple random search algorithm that starts from a randomly generated set of coefficients k_i, computes the accuracy A for this set, then generates another set of coefficients k_i again randomly, computes A for this set, and repeats this process m times. Then the highest value of A is shown to the user to decide if it is satisfactory. This is Step 1 of the algorithm shown below. It is implemented in C++ and linked with OpenGL visualization and interaction program that implements Steps 2–4. A user runs the process m times more if it is not satisfactory. In Sect. 7.4.1, we show that this automatic Step 1 is computationally feasible.

```
    Step 1:
best_coefficients = []
    while n > 0
        coefficients <- random(−1, 1)
        all_lines = 0
        for i data_samples:
                line = 0
                for x data_dimensions:
                        if coefficients[x] < 0:
                            line = line – data_dimensions[x]·cos(acos(coefficients[x]))
                        else:
                            line = line + data_dimensions[x]·cos(acos(coefficients[x]))
                all_lines.append(line)
                //update best_coefficients
                n--
```

Step 2: Projects the end points for the set of coefficients that correspond to the highest A value (in the same way as in Figure 7.5) and prints off the confusion matrix, i.e., for the best separation of the two classes.

Step 3:

Step 3a:

1: User moves around the class separation line.

2: A new confusion matrix is calculated.

Step 3b:

1: User picks the two thresholds to project a subset of the dataset.

2: n-D points of this subset (between the two thresholds) are projected.

3: A new confusion matrix is calculated.

4: User visually discovers patterns from the projection.

Step 4: User can repeat Step 3a or Step 3b to further zoom in on a subset of the projection or go back to Step 1.

Validation Process Typical 10-fold cross validation with 90–10% splits produces *10 different 90–10% splits* of data on the training and validation data. In this study, we used *10 different 70–30% splits* with 70% for the training set and 30% for the validation set in each split. Thus, we have the same 10 tests of accuracy as in the typical cross validation. Note that supervised learning tasks with 70–30% splits are more challenging than the tasks with 90–10% splits.

These 70–30% splits were selected by using permutation of data. The *splitting process* is as follows:

(1) indexing all m given samples from 1 to m, $w = (1,2,\ldots,m)$,
(2) randomly permuting these indexes, and getting a new order of indexes, $\pi(w)$,
(3) picking up first 70% of indexes from $\pi(w)$,

(4) assigning samples with these indexes to be training data,
(5) assigning remaining 30% of samples to be validation data.

This splitting process also can be used for a 90–10% split or other splits.
The total validation process for each set of coefficients K includes:

(i) applying data splitting process,
(ii) computing accuracy A of classification for this K,
(iii) repeating (i) and (ii) t times (each times with different data split),
(iv) computing average of accuracies found in all these runs.

7.2.4 Visual Structure Analysis of Classes

For the visual structure analysis, a user can interactively:

- Select border points of each class, coloring them in different colors.
- Outline classes by constructing an envelope in the form of a convex or a non-convex hull.
- Select most important coordinates by coloring them differently from other coordinates.
- Select misclassified and overlapped cases by coloring them differently from other cases.
- Draw the prevailing direction of the envelope and computing its location and angle.
- Contrast envelopes of difference classes to find the separating features.

7.2.5 Algorithm GLC-DRL for Dimension Reduction

A user can apply the automatic algorithm for dimension reduction anytime a projection is made to remove dimensions that don't contribute much to the overall line in the x direction (angles close to 90°). The contribution of each dimension to the line in the horizontal direction is calculated each time the GLC-AL finds coefficients.

The algorithm for automatic dimension reduction is as follows:

Step 1 Setting up a threshold for the dimensions, which did not contribute to the line significantly in the horizontal projection.
Step 2 Based on the threshold from Step 1, dimensions are removed from the data, and the threshold is incremented by a constant.
Step 3 A new projection is made from the reduced data.
Step 4 A new confusion matrix is calculated

The interactive algorithm for dimension reduction allows a user to pick up any coordinate arrow X_i and remove it by clicking on it, which leads to the zeroing of its projection. See coordinates X_2 and X_7 (in red) in Fig. 7.6b.

The computational algorithm for dimension reduction is as follows.

Step 1 The user visually examines the angles for each dimension, and determines which one is not contributing much to the overall line.

Step 2 The user selects and clicks on the angle from Step 1.

Step 3 The dimension, which has been selected, is removed from the dataset and a new projection is made along with a new confusion matrix. The dimension, which has been removed, is highlighted.

Step 4 Step 4a: The user goes back to Step 1 to reduce further the dimensions
Step 4b: The user selects to find other coefficients with the remaining dimensions for a better projection using the described above automatic algorithm GLC-AL.

7.2.6 Generalization of the Algorithms for Discovering Non-linear Functions and Multiple Classes

Consider a goal of visualizing a function

$$F(\mathbf{x}) = c_{11}x_1 + c_{12}x_1^2 + c_{21}x_2 + c_{22}x_2^2 + c_3x_3 + \ldots + c_nx_n + c_{n+1}$$

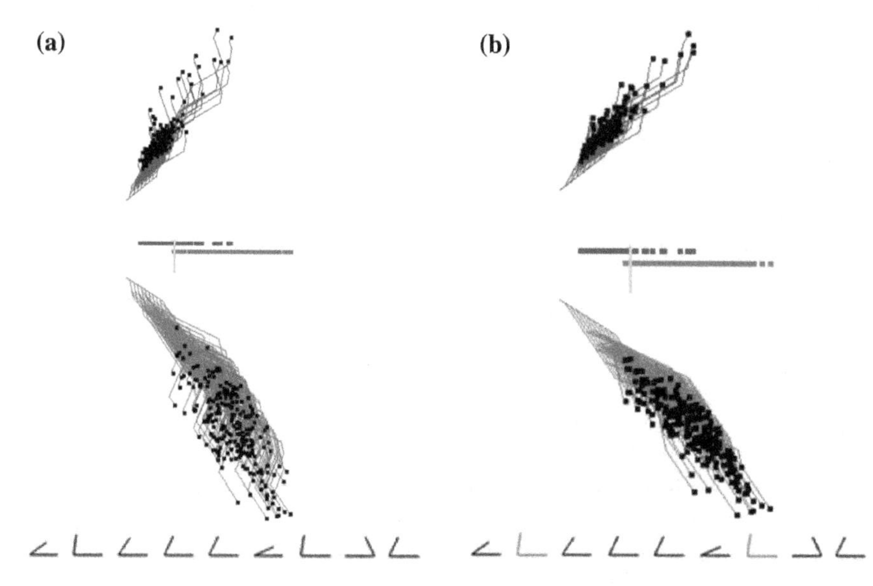

Fig. 7.6 Interactive dimension reduction, angles for each dimension are shown on the bottom. **a** Initial visualization of two classes optimized by GLC-AL algorithm; **b** visualization of two classes after 2nd and 7th dimensions (red) with low contribution (angle about 90°) have been removed

with quadratic components. For this F, the algorithm treats x_i and x_i^2 as two different variables X_{i1} and X_{i2} with the separate coordinate arrows similar to Fig. 7.1. Polynomials of higher order will have more than two such arrows. For a non-polynomial function

$$F(\mathbf{x}) = c_1 f_1(x_1) + c_2 f_2(x_2) + \ldots + c_n f_n(x_n) + c_{n+1},$$

which is a linear combination of non-linear functions f_i, the only modification in GLC-L is the substitution of x_i by $f_i(x_i)$ in the multiplication with angles still defined by the coefficients c_i. The rest of the algorithm is the same.

For the multiple classes the algorithm follows the method used in the multinomial logistic regression by discrimination of one class against all other $k-1$ classes together. Repeating this process k times for each class will give k discrimination functions that allow the discrimination of all classes.

7.3 Case Studies

Below we present the results of five case studies. In the selection of data for these studies, we followed a common practice in the evaluation of new methods—using benchmark data from the repositories with the published accuracy results for alternative methods as a more objective and less biased way than executing alternative methods by ourselves.

We used data from Machine Learning Repository at the University of California Irvine (Lichman 2013), and the Modified National Institute of Standards and Technology (MNIST) set of images of digits (LeCun et al. 2013). In addition, we used S&P 500 data for the period that includes the highly volatile time of Brexit.

7.3.1 Case Study 1

For the first study, *Wisconsin Breast Cancer Diagnostic (WBC) data set* was used (Lichman 2013). It has 11 attributes. The first attribute is the id number which was removed and the last attribute is the class label which was used for classification. These data were donated to the repository in 1992. The samples with missing values were removed, resulting in 444 benign cases and 239 malignant cases.

Figures 7.7 and 7.8 show samples of screenshots where these data are interactively visualized and classified in GLC-L for different linear discrimination functions, providing accuracy over 95%. The malignant cases are drawn in red and benign in blue.

Figures 7.7 and 7.9 show examples of how splitting the data into training and validation affects the accuracy. Figure 7.7 shows results of training on the entire data set, while Fig. 7.9 shows results of training on 70% of the data randomly

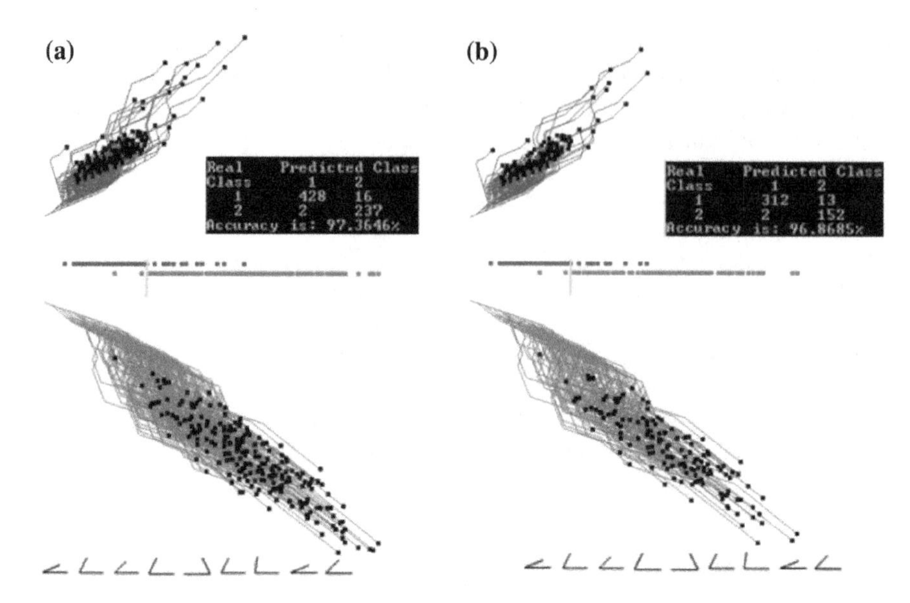

Fig. 7.7 Results for Wisconsin breast cancer data showing the training, validation, and the entire data set when trained on the entire data set. **a** Entire training and validation data set. Best projections of one of the first runs of GLC-AL. Coefficients found on the entire data set; **b** data split into 70/30 (training and validation) showing only 70% of the data, using coefficients and the separation line found on the entire data set in **a**

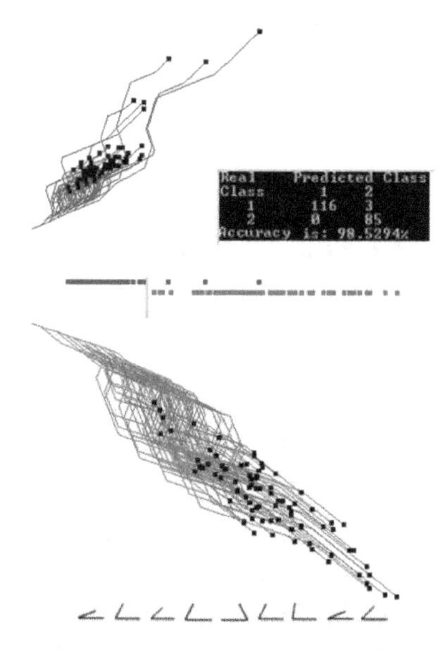

Fig. 7.8 Results for Wisconsin breast cancer data showing the 30% of the entire dataset (validation set). Using the coefficients and the separation line same as in Fig. 7.7a. Accuracy goes up

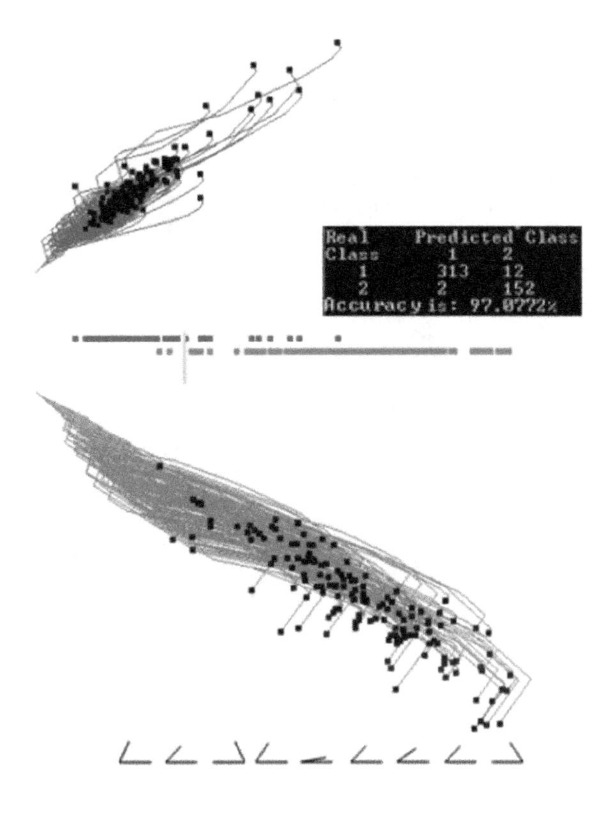

Fig. 7.9 Results for Wisconsin breast cancer data showing the training dataset when trained on the training set. Data are split using 70% (training set) to find coefficients with the projecting training set. Best result from the first runs of GLC-AL

selected. The visual analysis of Fig. 7.9 shows that 70% of data used for training are representative for the validation data too (Figs. 7.10, 7.11).

This is also reflected in similar accuracies of 97.07 and 96.56% on these training and validation data. The next case studies are shown first on the entire data set to understand the whole dataset. Section 7.3.4 presents accuracy on the training and validation data with the same 70/30 split.

Figure 7.12a shows the results for the best linear discrimination function obtained in the first 20 runs of the random search algorithm GLC-AL. The threshold found by this algorithm automatically is shown as a yellow bar. Results for the alternative discriminant functions from multiple runs of the random search by algorithm GLC-AL are shown in Figs. 7.12b, 7.13, 7.14. 7.15 and 7.16.

In these examples, the threshold (yellow bar) is located at the different positions, including the situations, where all malignant cases (red cases) are on the correct side of the threshold, i.e., no misclassification of the malignant cases.

Figures 7.14, 7.15 and 7.16 show the process and results of interactive selecting subsets of cases using two thresholds. This tight threshold interval selects heavily overlapping cases for further detailed analysis and classification. This analysis removed interactively the second dimension with low contribution without decreasing accuracy (see Fig. 7.16).

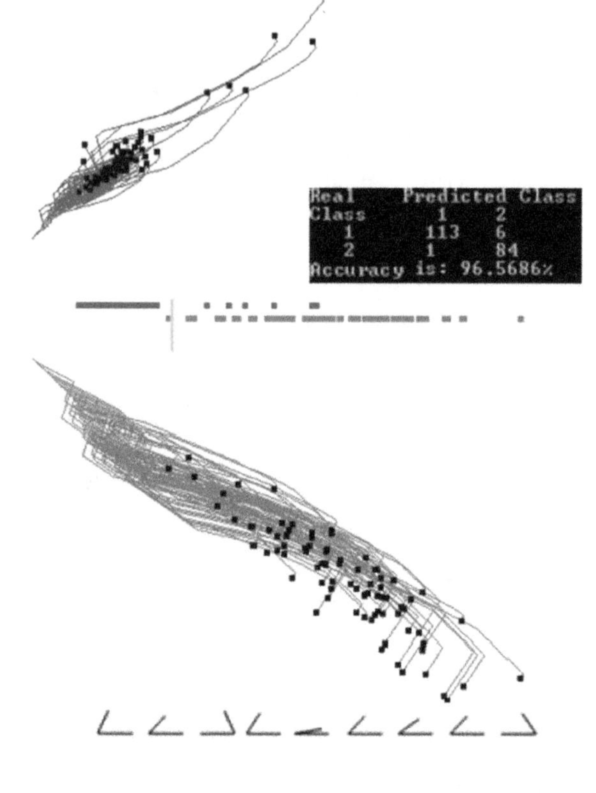

Fig. 7.10 Results for Wisconsin breast cancer data showing the validation dataset (30% of the data) when trained on the training set. Using the coefficients found by the training set in Fig. 7.9 and projecting the validation data set

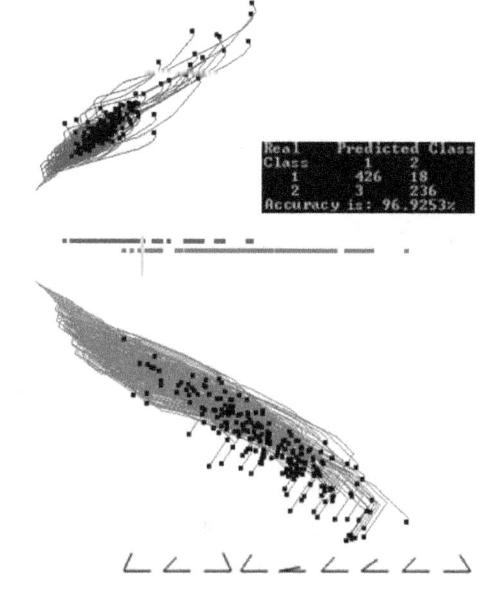

Fig. 7.11 Results for Wisconsin breast cancer data showing the entire data set when trained on the training set. Projecting the entire data set using the coefficients found by the training set in Fig. 7.9

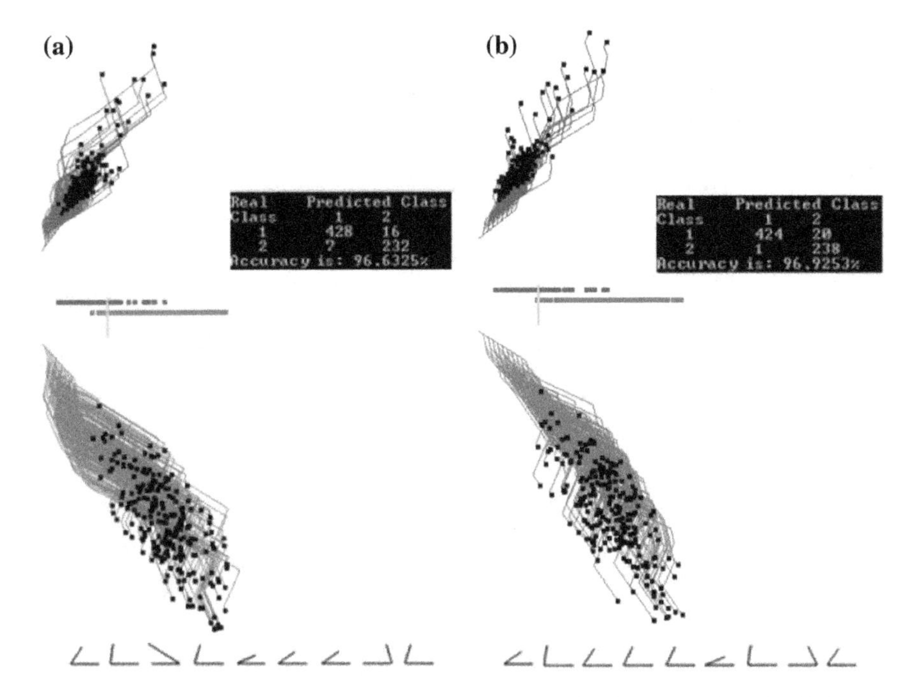

Fig. 7.12 Wisconsin breast cancer data interactively visualized and classified in GLC-L for different linear discrimination functions: **a** using the best function of the first 20 runs of the random search with a threshold found automatically shown as a yellow bar; **b** using an alternative function from the first 20 runs with the threshold (yellow bar) at the positions having only one malignant (red case) on the wrong side and higher overall accuracy than in **a**

7.3.2 Case Study 2

In this study, the *Parkinson's data* set from UCI Machine Learning Repository (Lichmam 2013) were used. This data set, known as Oxford Parkinson's Disease Detection Dataset, was donated to the repository in 2008. The Parkinson's data set has 23 attributes, one of them being status if the person has Parkinson's disease.

The dataset has 195 voice recordings from 31 people of which 23 have Parkinson's disease. There are several recordings from each person. Samples with Parkinson's disease present are colored red in this study. In the data preparation step of this case study, each column was normalized between 0 and 1 separately.

Figures 7.17, 7.18, 7.19, 7.20, 7.21 and 7.22 show examples of how splitting the data into training and validation sets affects the accuracy. Figures 7.17, 7.18 and 7.19 show results of training on the entire dataset, while Fig. 7.20 shows results of training on 70% of the data randomly selected.

The accuracies in Figs. 7.17, 7.18 and 7.19 are 85.19% for whole dataset, 88.32% for the 70% of the dataset and 77.59% for remaining 30% of the dataset. It shows that accuracies depend of subsets used for the splits.

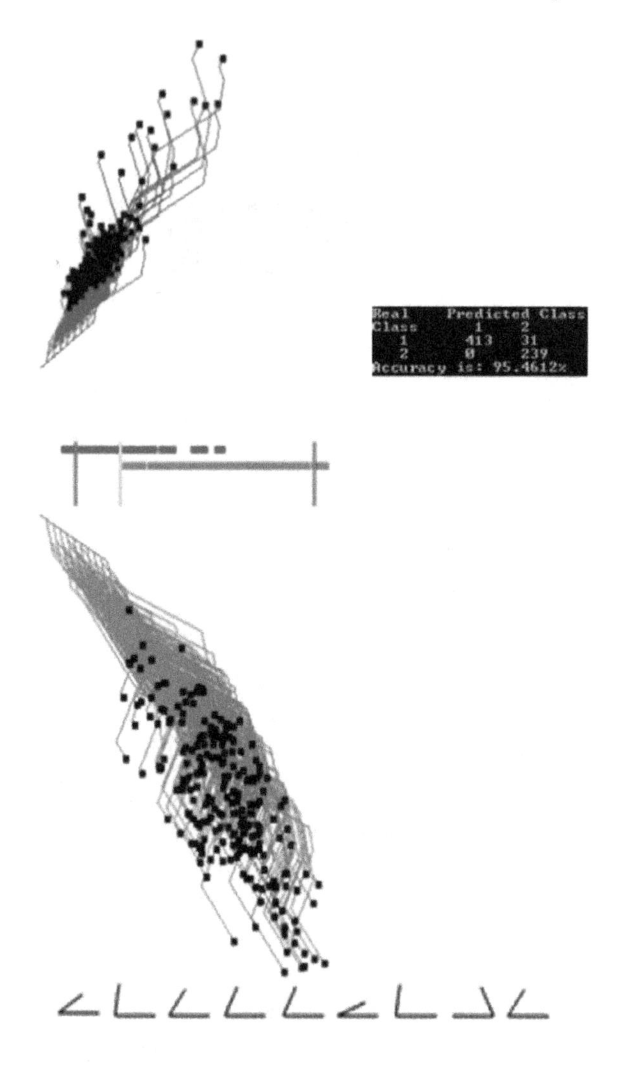

Fig. 7.13 Wisconsin breast cancer data interactively visualized and classified in GLC-L for different linear discrimination functions. Visualization from Fig. 7.12b where the separation threshold is moved to have all malignant (red cases) on the correct side with the tradeoff in the accuracy

The visual analysis of Fig. 7.21 shows that 70% data used for training is also representative for the validation data (30% of data) shown in Fig. 7.21. This is reflected in similar accuracies of 91.24% on training data, 82.75 on validation data and 88.71% overall data shown in Figs. 7.20, 7.21 and 7.22. The rest of illustration for this case study in Fig. 7.22 is for the entire dataset to understand the dataset as a whole. Section 7.4.1 presents accuracy on the training and validation with the same 70/30 split.

The result for the best discrimination function found from the second run of 20 epochs is shown in Fig. 7.23. In Fig. 7.24, five dimensions are removed, some of them are with angles close to 90°. In addition, the separation line threshold is

Fig. 7.14 Wisconsin breast cancer data interactively projecting a selected subset. Two thresholds are set from Fig. 7.13 for selecting overlapping cases

Fig. 7.15 Wisconsin breast cancer data interactively projecting a selected subset. Overlapping cases from the interval between two thresholds from Fig. 7.14

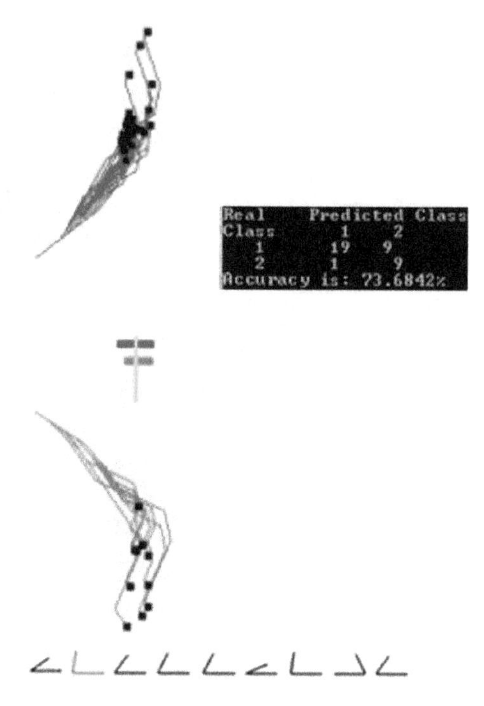

Fig. 7.16 Wisconsin breast cancer data interactively projecting a selected subset. Overlapping cases from the interval between two thresholds, with the second dimension with low contribution removed without decreasing accuracy

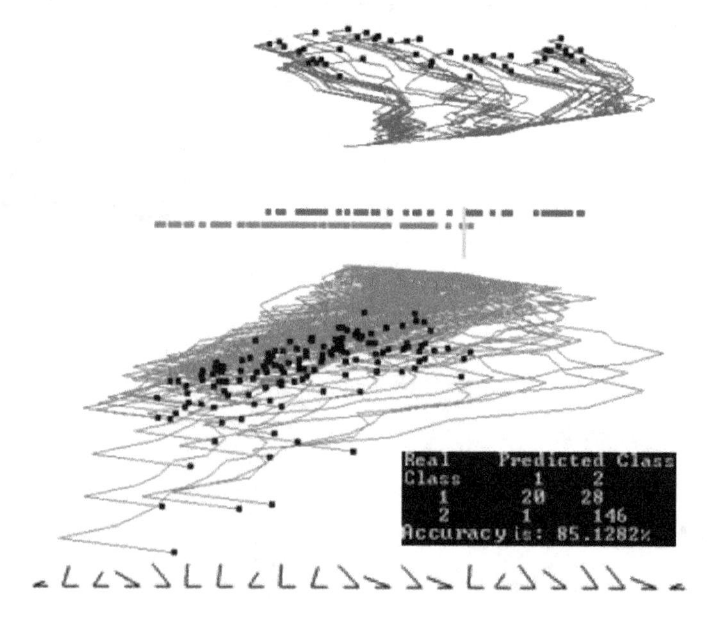

Fig. 7.17 Results with Parkinson's disease data set showing the entire dataset when trained on this dataset. Best projections of one of the first runs of GLC-AL. Coefficients found on the entire dataset

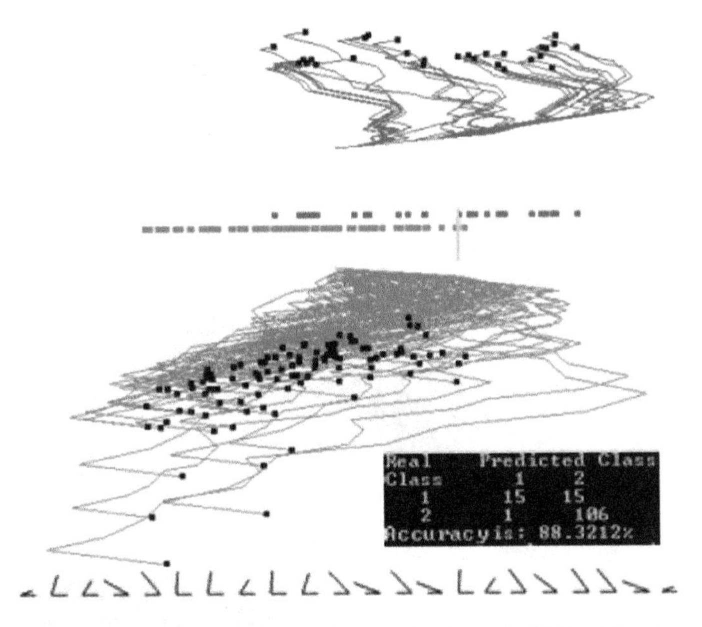

Fig. 7.18 Results with Parkinson's disease data set showing only 70% of the data, using the coefficients and the separation line found on the entire data set in Fig. 7.17

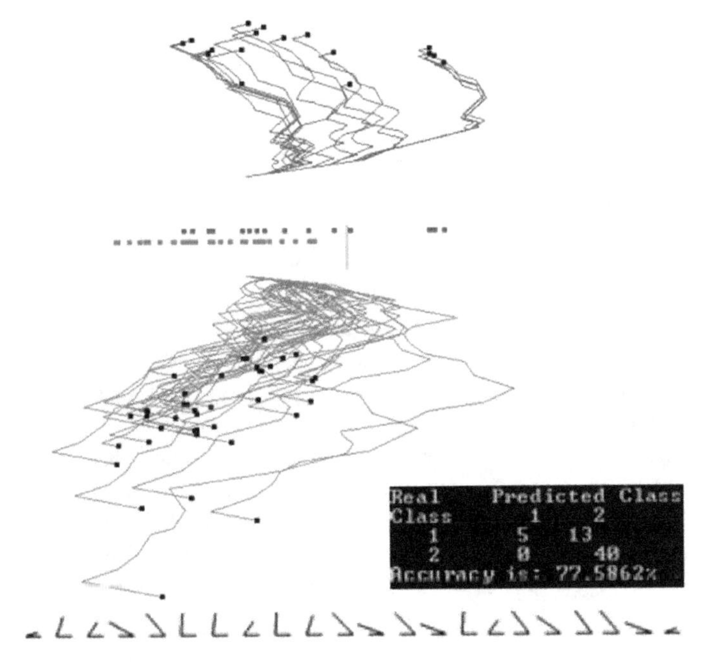

Fig. 7.19 Results with Parkinson's disease data set showing the 30% using coefficients and the separation line the same as in Fig. 7.17. Accuracy goes down

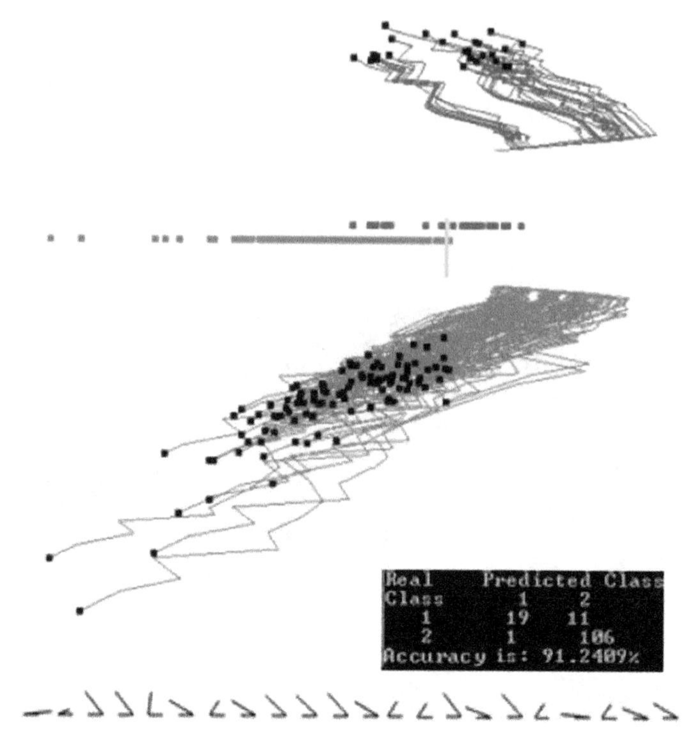

Fig. 7.20 Results with Parkinson's disease data set showing the training dataset (70% of the entire dataset) when trained on the training dataset to find the coefficients. Projecting training set, best from the first runs of GLC-A

moved relative to Fig. 7.23. In Fig. 7.25, the two limits for a subinterval are set to zoom in on the overlapping samples.

In Fig. 7.26, where the subregion is projected and 42 samples are removed, the accuracy only decreases by 4% from 86.15 to 82.35%. Out of those 42 cases, 40 of them are samples of Parkinson's disease (red cases), and only 2 cases are not Parkinson's disease. With such line separation as in Figs. 7.23, 7.24 and 7.25, it is very easy to classify cases with Parkinson's disease from this dataset (high True Positive rate, TP); however, a significant number of cases with no Parkinson's disease are classified incorrectly (high False Positive rate, FP).

This indicates the need for improving FP more exploration, such as preliminary clustering of the data, more iterations to find coefficients, or using non-linear discriminant functions. The first attempt can be a quadratic function that is done by adding a squared coordinate X_i^2 to the list of coordinates without changing the GLC-L algorithm (see Sect. 7.2.6).

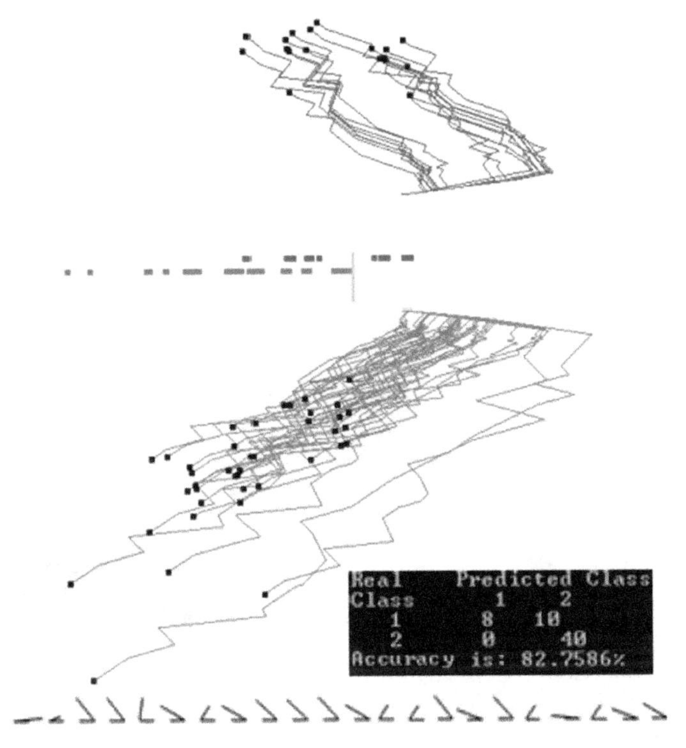

Fig. 7.21 Results with Parkinson's disease data set showing the validation data (30% of the data) using coefficients found by the training set in Fig. 7.20 and projecting the validation dataset

7.3.3 Case Study 3

In this study, a subset of the *Modified National Institute of Standards and Technology (MNIST) database* (LeCun et al. 2013) was used. Images of digit 0 (red) and digit 1 (blue) were used for projection with 900 samples for each digit. In the preprocessing step, each image is cropped to remove the border. The images after cropping were 22 × 22, which is 484 dimensions.

Figures 7.27, 7.28, 7.29, 7.30, 7.31 and 7.32 show examples of how splitting the data into training and validation changes the accuracy. Figures 7.27, 7.28 and 7.29 shows the results of training on the entire data set, while Fig. 7.14 shows the results of training on 70% of the data, which are randomly selected. The visual analysis of Figs. 7.30, 7.31 and 7.32 shows that 70% of the data used for training are also representative for the validation data.

This is also reflected in similar accuracies of 91.58 and 91.44% respectively. The rest of this case study is illustrated on the entire data set to understand the dataset as a whole. Accuracy on the training and validation can be found later in Sect. 7.4.1, where a 70/30 split was also used.

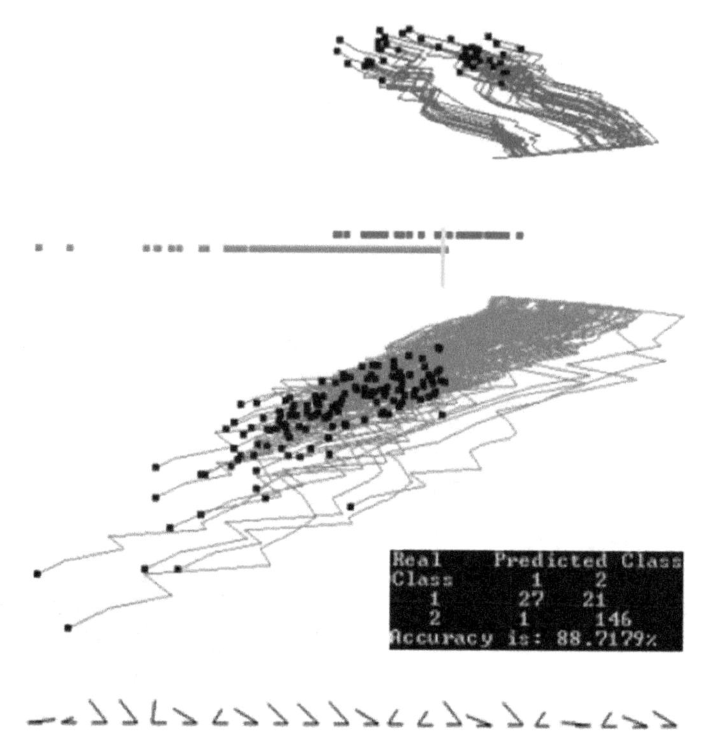

Fig. 7.22 Results with Parkinson's disease data set showing the entire data set when trained on the training set. Projecting the entire data set using coefficients found by the training set in Fig. 7.20

Figure 7.33 shows the results of applying the algorithm GLC-AL to these MNIST images. It is the best discriminant function of the first run of 20 epochs with the accuracy of 95.16%. Figure 7.34 shows the result of applying the automatic algorithm GLC-DRL to these data and the discriminant function. It displays 249 dimensions and removes 235 dimensions, dropping the accuracy only slightly by 0.28%. Figure 7.35 shows the result when a user decided to run the algorithm GLC-DRL a few more times. It removed 393 dimensions, kept and projected the remaining 91 dimensions with the accuracy dropping to 83.77% from 93.84% as shown in Fig. 7.34.

Figure 7.36 shows the result of user interaction with the system by setting up the interval (using two bar thresholds) to select a subset of the data of interest in the overlap of the projections. The selected data are shown in Fig. 7.37. Figure 7.38 shows the results of running GLC-AL algorithm on the subinterval to find a better discriminant function and projection. Accuracy goes up by 5.6% in this subinterval.

Next, the automatic dimension reduction algorithm GLC-DRL is run on these subinterval data, removing the 46 dimensions and keeping and projecting the 45 dimensions with the accuracy going up by 1% (see Fig. 7.39). Figure 7.40 shows

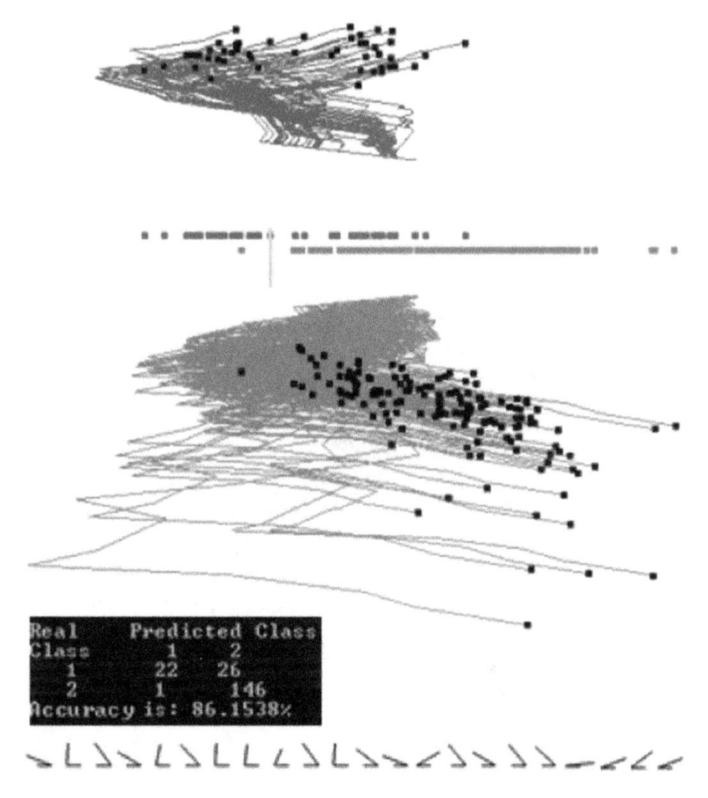

Fig. 7.23 Additional Parkinson's disease experiments. **a** Best projection from the second run of 20 epochs

the result when a user decided to run the algorithm GLC-DRL a few more times on these data, removing 7 more dimensions, and keeping 38 dimensions, with the accuracy gaining 6.8%, and reaching 95.76%.

7.3.4 Case Study 4

Another experiment was done on a *different subset of the MNIST database* to see if any visual information could be extracted on encoded images. For this experiment, the training set consisted of all samples of digit 0 and digit 1 from the training set of MNIST (60,000 images). There was 12,665 samples of digit 0 and digit 1 combined in the training set.

The validation set consisted of all the samples of digit 0 and digit 1 from the validation set of MNIST (10,000 images). There was 2115 samples of digit 0 and digit 1 combined in the validation set. The preprocessing step for the data was the

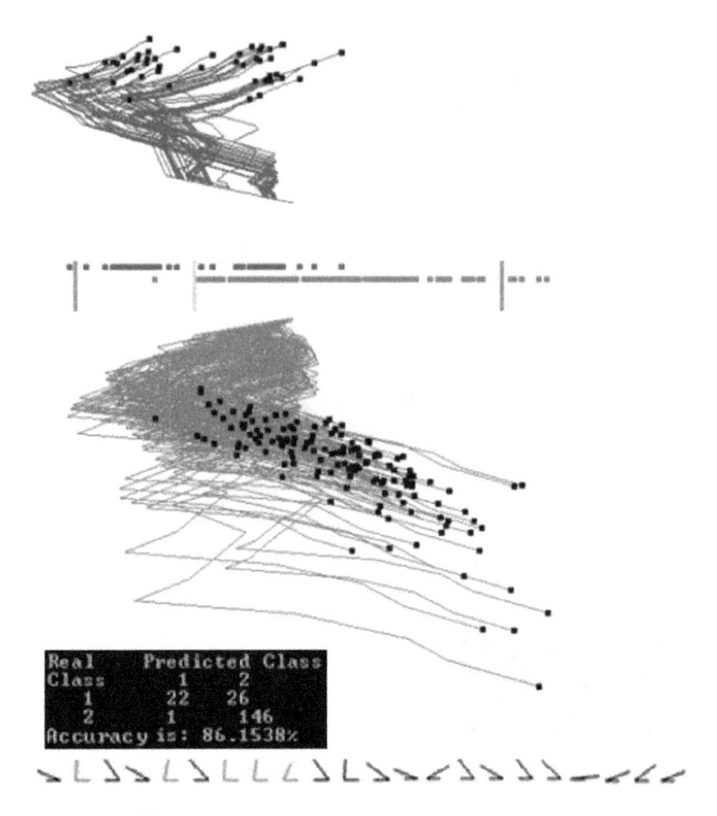

Fig. 7.24 Additional Parkinson's disease experiments. Projection with 5 dimensions removed. Separation line threshold is also moved. Accuracy stays the same

same as for case study 3, where pixel padding was removed resulting in 22 × 22 images.

A Neural Network Autoencoder, which was constructed using Python library Keras (2017), encoded the images from 484 (22 × 22) dimensions to 24 dimensions. François Chollet originally developed the Keras library (Chollet 2015). This study used Keras version 1.0.2 running with python version 2.7.11. The Autoencoder had one hidden layer and was trained on the training set (12,665 samples). Examples of decoded images can be seen in Fig. 7.41. The validation set (2115 images) was passed through the encoder to get its representation in 24 dimensions.

The goal of this case study is to compare side-by-side GLC-L visualization with parallel coordinates. Figures 7.42, 7.43, 7.44 and 7.45 show the comparison of these two visualizations using 24 dimensions found by the Auto encoder among the original 484 dimensions. Figures 7.42 and 7.43 show the difference between the two classes more clearly than Parallel coordinates (PC) in Figs. 7.44 and 7.45. The shapes of the clouds in GLC-L are very different. The red class is elongated and the blue one is more rounded and shifted to the right relative to the red cloud. It also shows the separation threshold between classes and the accuracy of classification.

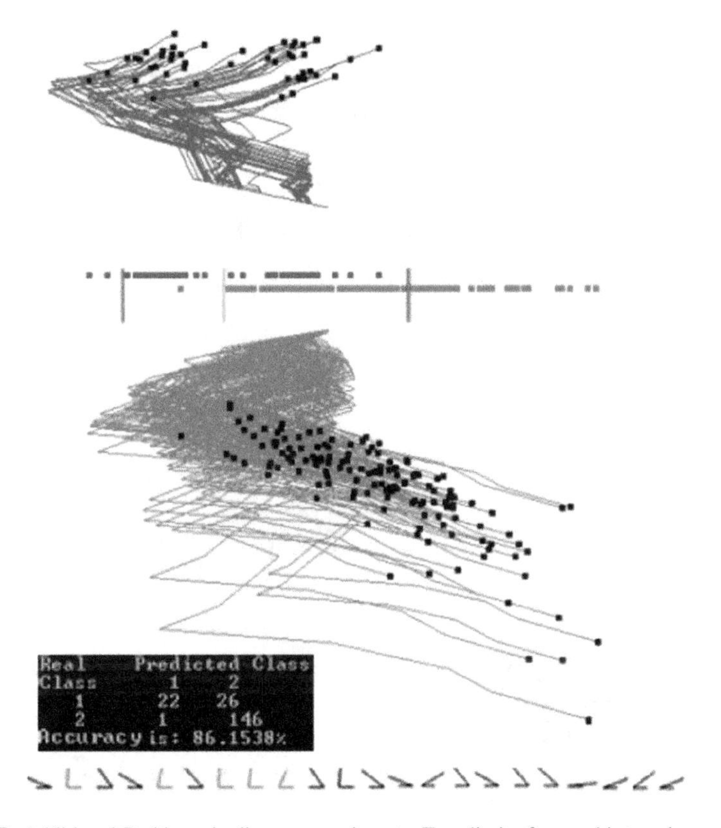

Fig. 7.25 Additional Parkinson's disease experiments. Two limits for a subinterval are set

Parallel coordinates in Figs. 7.44 and 7.45 do not show a separation between classes and accuracy of classification. Only a visual comparison of Figs. 7.44 and 7.45 can be used for finding the features that can help in building a classifier. In particular, the intervals on features 10, 8, 23, 24 are very different, but the overlap is not, allowing the simple visual separation of the classes. Thus, there is no clear way to classify these data using PCs. PCs give only some informal hints for building a classifier using these features.

7.3.5 Case Study 5

This study uses *S&P 500 data* for the first half of 2016 that include highly volatile S&P 500 data at the time of the Brexit vote. S&P 500 lost 75.91 points in one day (from 2113.32 to 2037.41) from 23 June to 24 June 2016 and continued dropping to 27 June to 2000.54 with total loss of 112.78 points since 23 June. The loss of value in these days is 3.59 and 5.34%, respectively.

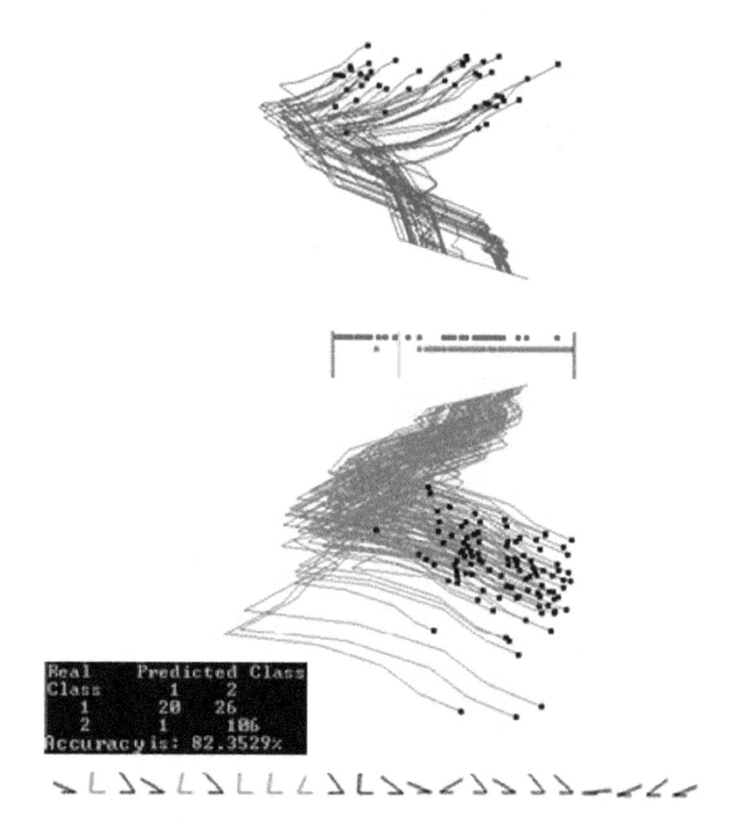

Fig. 7.26 Additional Parkinson's disease experiments. Only cases for the subinterval are projected with the separation line moved. Accuracy drops

The goal of this case study is predicting S&P 500 up/down changes on Fridays knowing S&P 500 values for the previous four days of the week. The day after the Brexit vote is also a Friday. Thus, it is also of interest to see if the method will be able to predict that S&P 500 will go down on that day.

Below we use the following notation to describe the construction of features used for prediction:

$S_1(w)$, $S_2(w)$, $S_3(w)$, $S_4(w)$, and $S_5(w)$ are S&P 500 values for Monday–Friday, respectively, of week w;

$D_i(w) = S_{i+1}(w) - S_i(w)$ are differences in S&P 500 values on adjacent days, $i = 1{:}4$;

Class(w) = 1 (down) if $D_4(w) < 0$,
Class(w) = 2 (up) if $D_4(w) > 0$,
Class(w) = 0 (no change) if $D_4(w) = 0$.

We computed the attributes $D_1(w)$–$D_4(w)$ and Class(w) from the original S&P 500 time series for the trading weeks of the first half of 2016. Attributes D_1–D_3

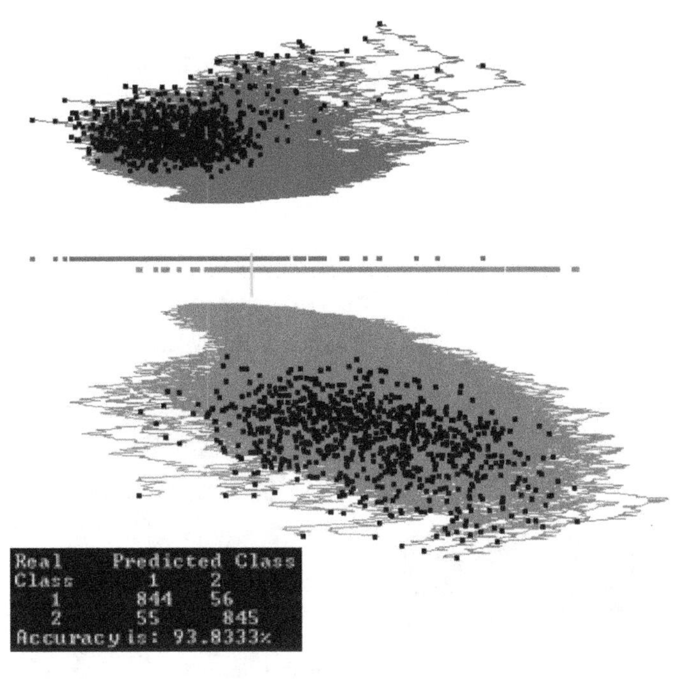

Fig. 7.27 MNIST subset for digits 0 and 1 dataset showing the entire dataset when trained on this dataset. Best projections of one of the first runs of GLC-A. Coefficients found on the entire data set

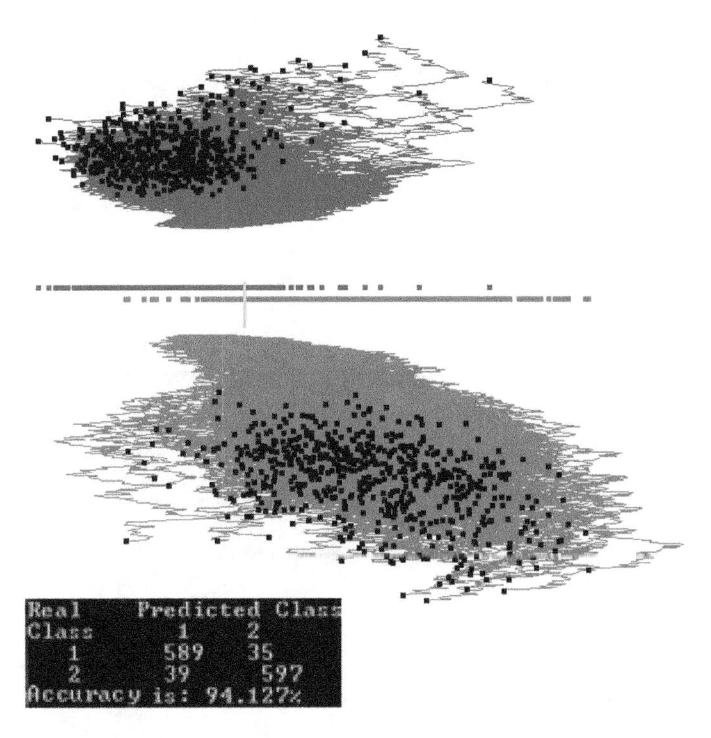

Fig. 7.28 MNIST subset for digits 0 and 1 dataset showing 70% of the data, using coefficients and separation line found on the entire data set in Fig. 7.27

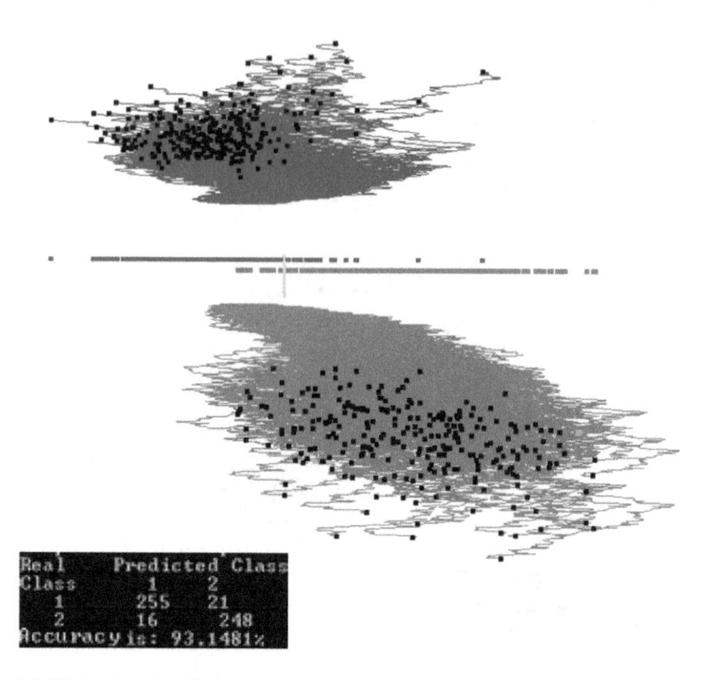

Fig. 7.29 MNIST subset for digits 0 and 1 dataset showing 30% of the data, using coefficients and separation line found on the entire data set in Fig. 7.27

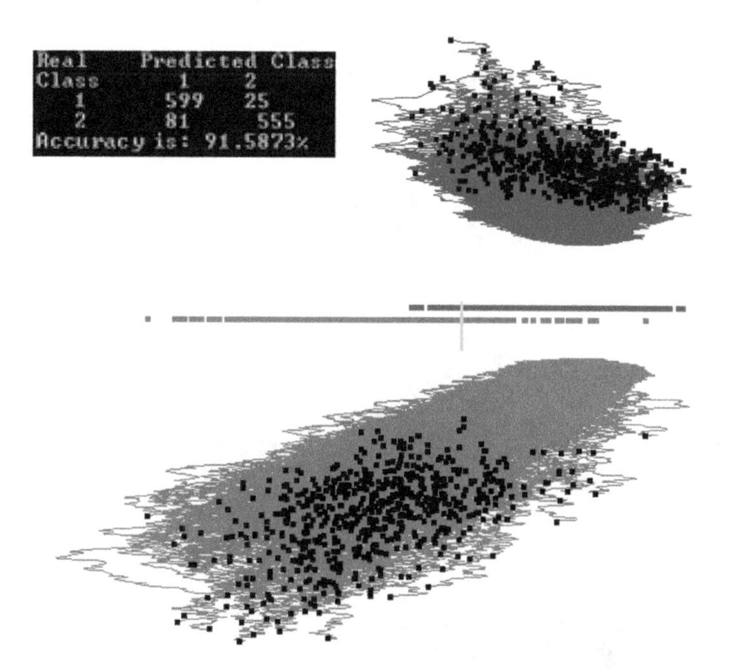

Fig. 7.30 MNIST subset for digits 0 and 1 data set showing training dataset (70% of the whole dataset) when trained on the training dataset to find the coefficients. Projecting training set, best from the first runs of GLC-A

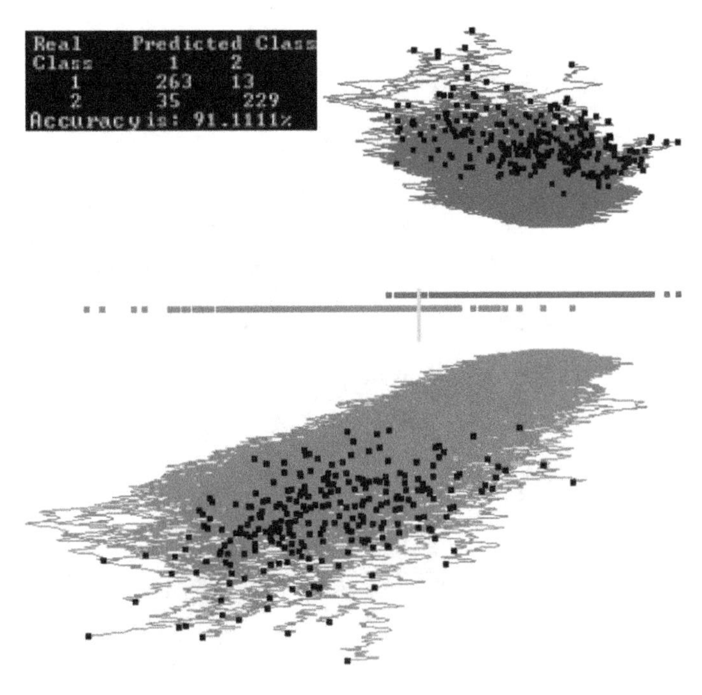

Fig. 7.31 MNIST subset for digits 0 and 1 dataset showing validation dataset (30% of the whole dataset) when trained on the training dataset to find the coefficients. Projecting validation set

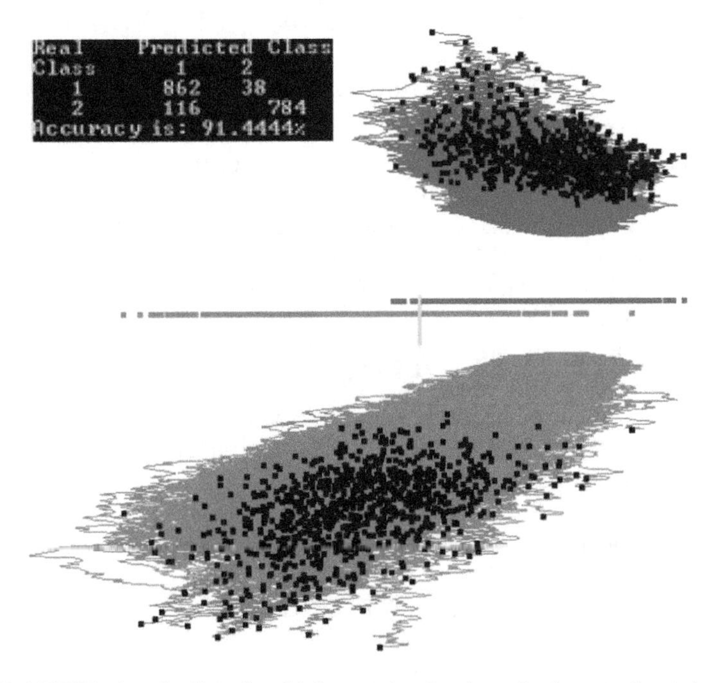

Fig. 7.32 MNIST subset for digits 0 and 1 data set showing the entire data set when trained on the training set. Projecting the entire data set using the coefficients found by the training set

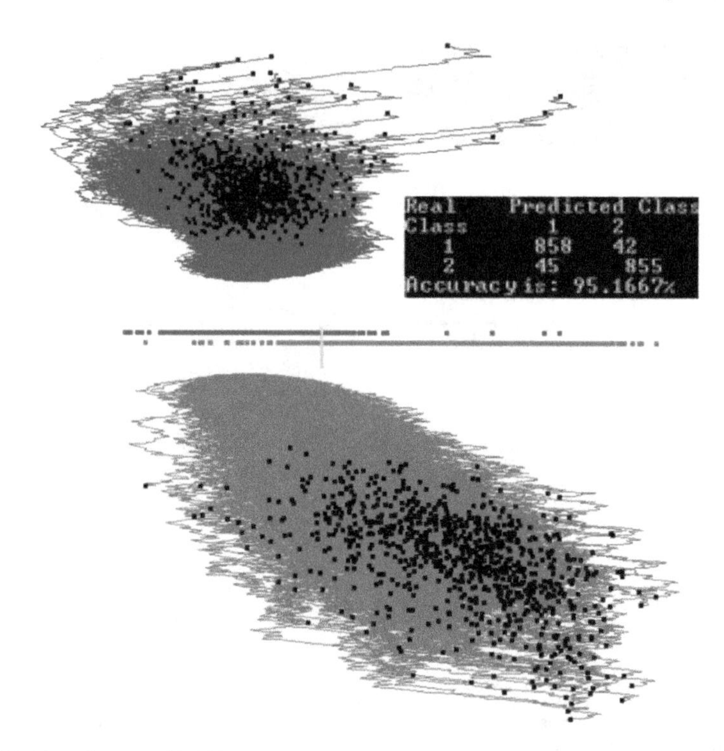

Fig. 7.33 Experiments with 900 samples of MNIST dataset for digits 0 and 1. Results for the best linear discriminant function of the first run of 20 epochs

were used to predict Class (up/down). We excluded incomplete trading weeks, getting 10 "Friday down" weeks, and 12 "Friday up" weeks available for training and validation. We used multiple 60%:40% splits of these data on training and validation due to a small size of these data.

The Brexit week was a part of the validation data set and was never included in the training datasets. Figures 7.46, 7.47, 7.48 and 7.49 show the best results on the training and validation data, which are: 76.92% on the training data, and 77.77% on the validation data.

We attribute the greater accuracy on the validation data to a small dataset. These accuracies are obtained for two different sets of coefficients.

The accuracy of one of the runs was 84.81% on training data, but its accuracy on validation was only 55.3%. The average accuracy on all 10 runs was 77.78% on all training data, and 61.12% on the validation data.

While these accuracies are lower than in the case studies 1–4, they are quite common in such market predictions (see Chap. 8). The accuracy of the down prediction for 24 June (after Brexit) in those 10 runs was correct in 80% of the runs, including the runs shown in Figs. 7.47 and 7.49 as green lines.

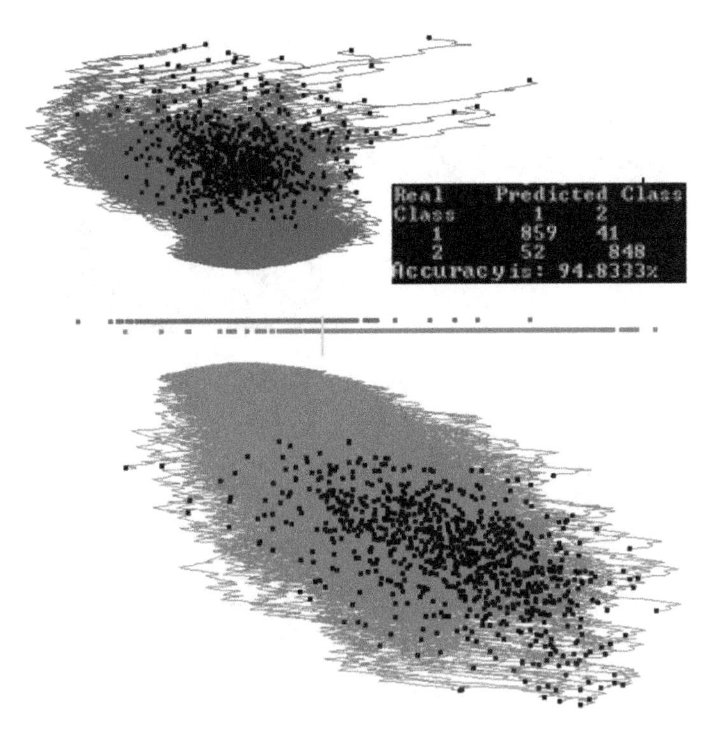

Fig. 7.34 Experiments with 900 samples of MNIST dataset for digits 0 and 1. Results of the automatic dimension reduction displaying 249 dimensions with 235 dimensions removed with the accuracy dropped by 0.28%

7.4 Discussion and Analysis

7.4.1 Software Implementation, Time and Accuracy

All algorithms, including visualization and GUI, were implemented in OpenGL and C++ in the Windows Operating System. A Neural Network Autoencoder was implemented using the Python library Keras (Chollet 2015; Keras 2017). Later, we expect to make the programs publicly available.

Experiments presented in Sect. 7.3 show that 50 iterations produce a nearly optimal solution in terms of accuracy. Increasing the number of iterations to a few hundred did not show a significant improvement in the accuracy after 50 epochs. In the cases where better coefficients were found past 50 epochs, the accuracy increase was less than 1%.

The automatic step 1 of the algorithm GLC-AL in the case studies above had been computationally feasible. For m = 50 (number of iterations) in the case study, it took 3.3 s to get 96.56% accuracy on a PC with 3.2 GHz quad core processor. Running under the same conditions, case study 2 took 14.18 s to get 87.17% accuracy, and for case study 3, it took 234.28 s to get 93.15% accuracy. These

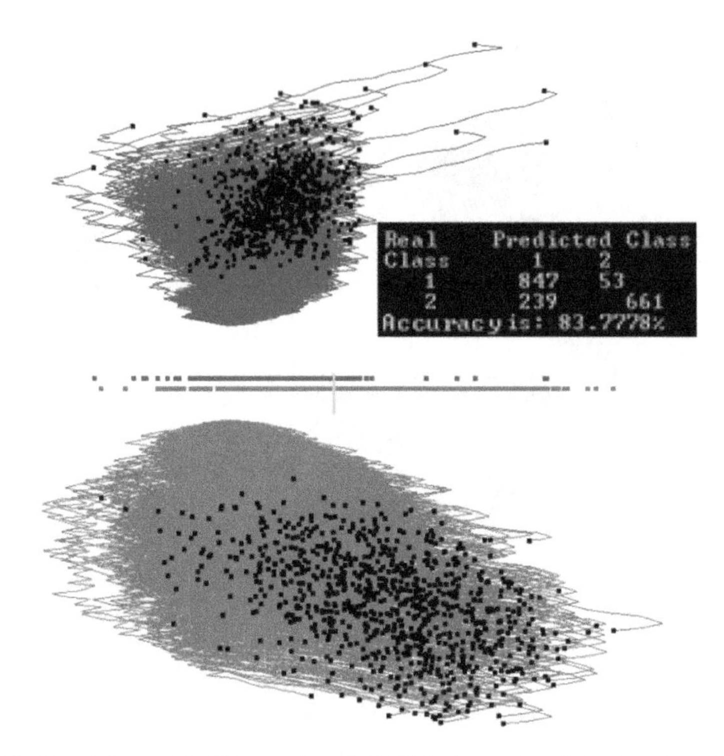

Fig. 7.35 Experiments with 900 samples of MNIST dataset for digits 0 and 1. Automatic dimension reduction, which is run a few more times removing 393 dimensions and keeping 91 dimensions with dropped accuracy

accuracies are on the validation sets for the 70/30 split of data into the training and validation sets. In the future, work on a more structured random search can be implemented to decrease the computation time in high dimensional studies.

Table 7.1 shows the accuracy on training and validation data for the case studies 1–3. We conducted 10 different runs of 50 epochs each with the automatic step 1 of the algorithm GLC-AL. Each run had freshly permutated data, which were split into 70% training and 30% validation as described in Sect. 7.2.3. In some runs for case study 1, the validation accuracy was higher than the training one, which is because the data set is relatively small and the split heavily influences the accuracy.

The training accuracies in Table 7.1 are the highest ones obtained in each run of the 50 epochs. Validation accuracies in this table are computed for the discrimination functions found on respective training data. The variability of accuracy on training data among 50 epochs in each run is quite high. For instance, the accuracy varied from 63.25 to 97.91% with the average of 82.46% in case study 1.

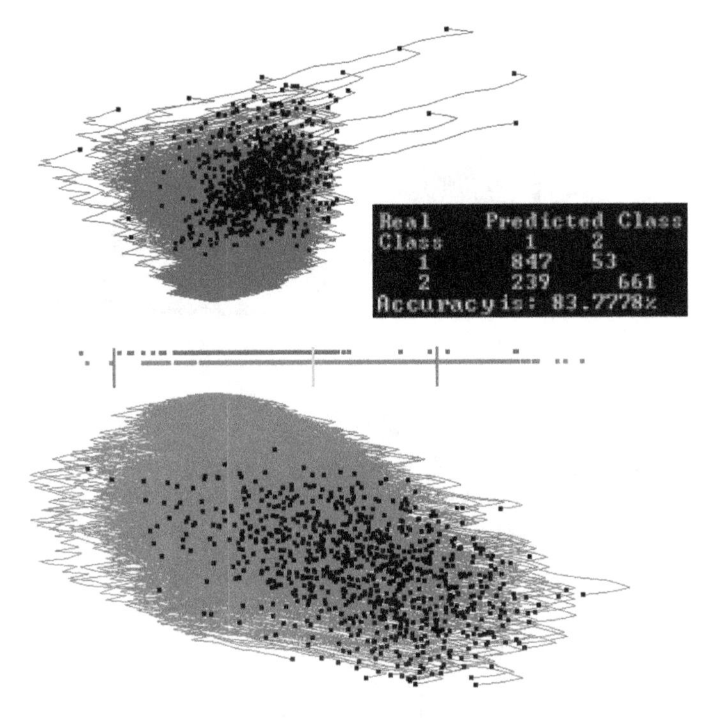

Fig. 7.36 Experiments with 900 samples of MNIST dataset for digits 0 and 1. Thresholds for a subinterval are set (green bars)

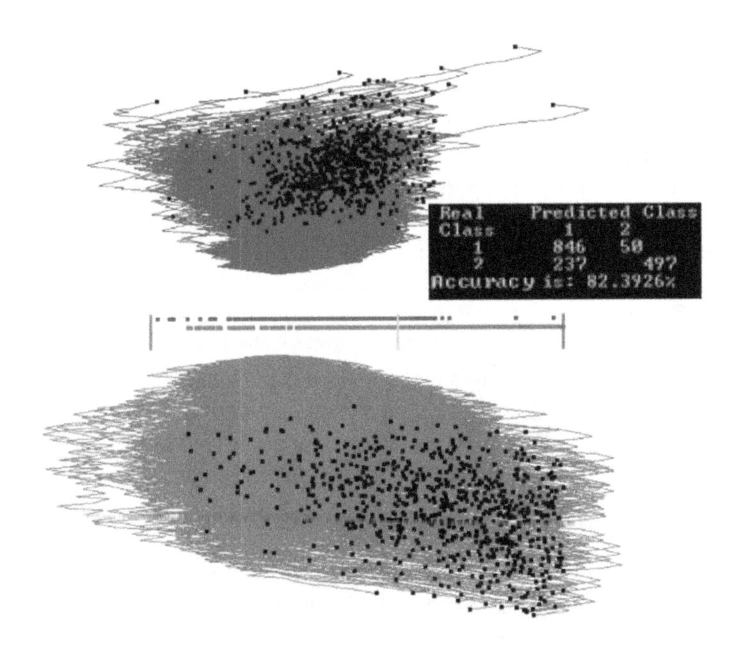

Fig. 7.37 Experiments with a 900 samples of MNIST dataset for digits 0 and 1 with automatic dimension reduction. Data between the two green thresholds are visualized and projected

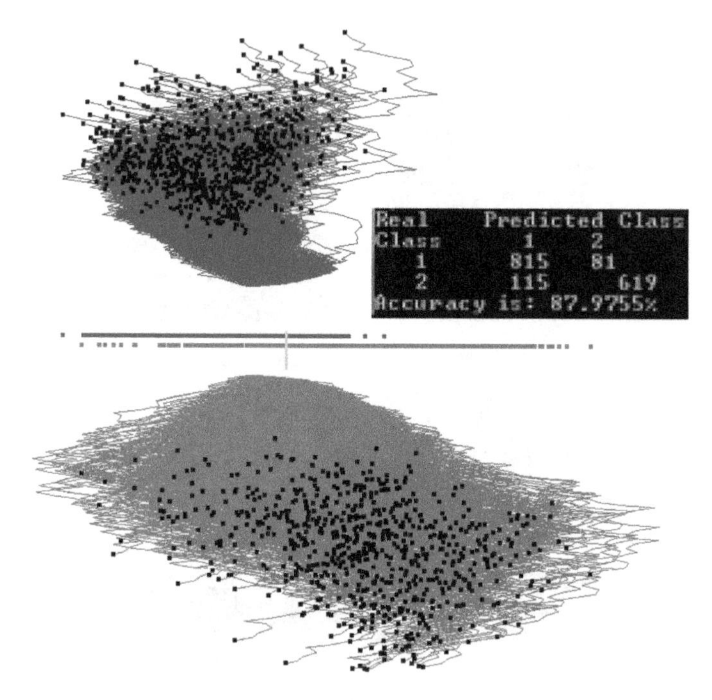

Fig. 7.38 Experiments with a 900 samples of MNIST dataset for digits 0 and 1 with automatic dimension reduction. GLC-AL algorithm on the subinterval to find a better projection. Accuracy goes up by 5.6% in the subregion

7.4.2 Comparison with Other Studies

Case Study 1 For the Wisconsin breast cancer (WBC) data, the best accuracy of *96.995%* is reported in [24] for the 10 fold cross-validation tests using SVM. This number is slightly above our average accuracy of *96.955* on the validation data in Table 1, using the GLC-L algorithm. We use the 70:30 split, which is more challenging, for getting the highly accurate training, than the 10-fold 90:10 split used in that study. The best result on validation is *98.04%,* obtained for the run 8 with 97.49% accuracy on training (see Table 7.1) that is *better than 96.995%* in Salama et al. (2012).

The best results from previous studies collected in Salama et al. (2012) include *96.84%* for SVM-RBF kernel (Aruna et al. 2011) and *96.99%* for SVM in (Christobel and Sivaprakasam 2011). In Salama et al. (2012) the best result by a combination of major machine learning algorithms implemented in Weka (2017) is *97.28%.* The combination includes SVM, C4.5 decision tree, naïve Bayesian classifier and k-Nearest Neighbors algorithms. This result is below the *98.04%* we obtained with GLC-L. It is more important in the cancer studies to control the number of misclassified cancer cases than the number of misclassified benign cases. The total accuracy of classification reported (Salama et al. 2012) does not show

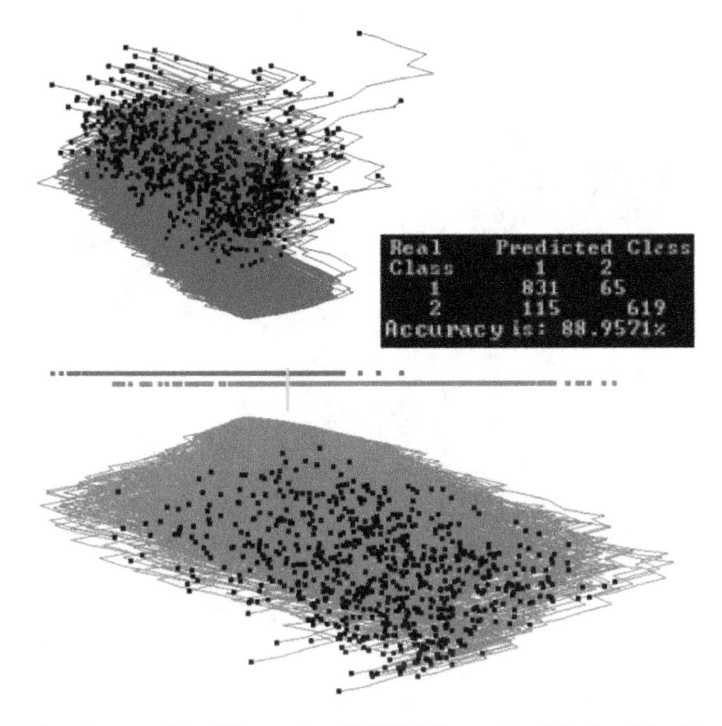

Fig. 7.39 Experiments with a 900 samples of MNIST dataset for digits 0 and 1 with automatic dimension reduction. Result of automatic dimension reduction running a few more times that removes 46 dimensions and keeps 45 dimensions with accuracy going up by 1%

these actual false positive and false negative values, while figures in Sect. 7.3.2 for GLC-L show them along with GLC-L interactive tools to improve the number of misclassified cancer cases.

This comparison shows that the GLC-L algorithm *can compete* with the major machine learning algorithms in accuracy. In addition, the GLC-L algorithm has the following important advantages: it is (i) *simpler* than the major machine learning algorithms, (ii) *visual*, (iii) *interactive*, and (iv) *understandable* by a user without advanced machine learning skills. These features are important for *increasing user confidence* in the learning algorithm outcome.

Case Study 2 In Ramani et al. (2011) 13 Machine Learning algorithms have been applied to these Parkinson's data. The accuracy ranges from 70.77% for Partial Least Square Regression algorithm, 75.38% to ID3 decision tree, to 88.72% for Support Vector Machine, 97.44% for k-Nearest Neighbors and 100% for Random decision tree forest, and average accuracy equal to *89.82%* for all 13 methods.

The best results that we obtained with GLC-L for the same 195 cases are *88.71%* (Fig. 7.22) and *91.24%* (Fig. 7.20). Note that the result in Fig. 7.20 was obtained by using only 70% of 195 cases for training, not all of them. The split for training and validation is not reported in Ramani et al. (2011), making the direct comparison with results from GLC-L difficult. Just by using more training data, the commonly

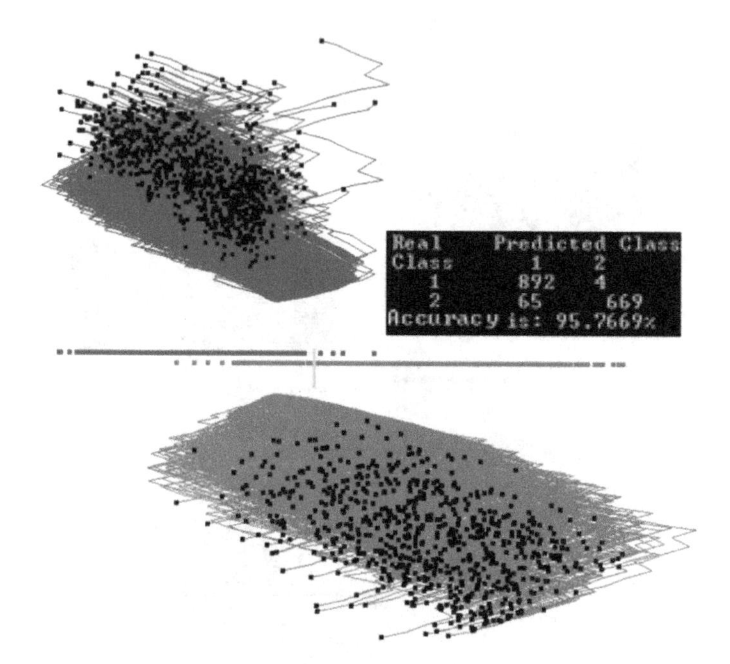

Fig. 7.40 Experiments with a 900 samples of MNIST dataset for digits 0 and 1 with automatic dimension reduction. Result of automatic dimension reduction running a few more times that removes 7 more dimensions and keeps 38 dimensions with accuracy going up 6.8% more

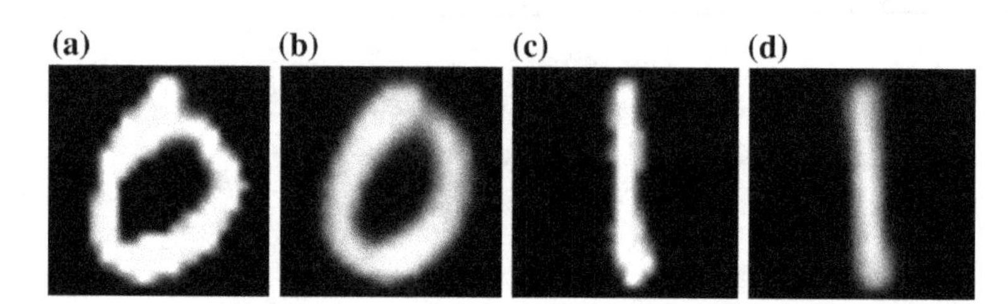

Fig. 7.41 Examples of original and encoded images. **a** Example of a digit after preprocessing; **b** decoded image from 24 values into 484. The same image as in **a**; **c** another example of a digit after preprocessing; **d** decoded image from 24 values into 484. The same image as in **c**

used split 90:10 can produce higher accuracy than a 70:30 split. Therefore, the only conclusion that can be made is that the accuracy results of GLC-L are comparable with average accuracy provided by common analytical machine learning algorithms.

Case Study 3 While we did not conduct the full exploration of MNIST database for all digits, the accuracy with GLC-AL is comparable and higher than the accuracy of other linear classifiers reported in the literature for all digits and whole dataset. Those errors are 7.6% (accuracy 92.4%), for a pairwise linear classifier, and

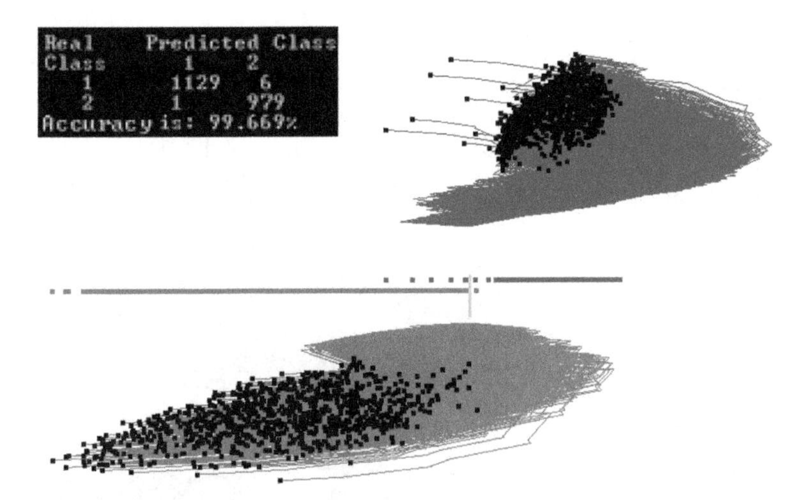

Fig. 7.42 Encoded digit 0 and digit 1 on and GLC-L, using 24 dimensions found by the Autoencoder among 484 dimensions. Results for the best linear discriminant function of the first run of 20 epochs

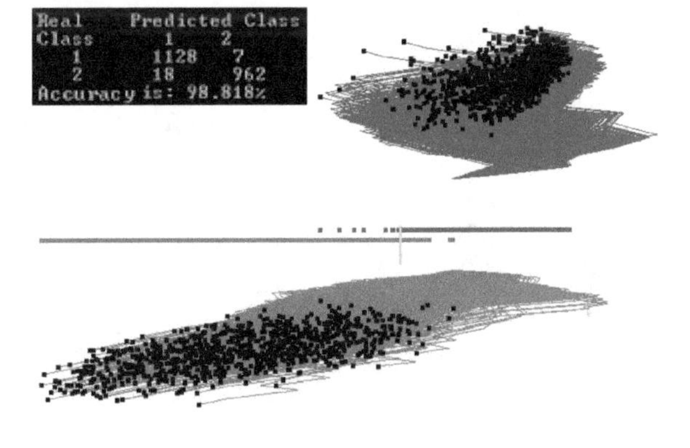

Fig. 7.43 Encoded digit 0 and digit 1 on GLC-L, using 24 dimensions found by the Autoencoder among 484 dimensions. Another run of 20 epochs, best linear discriminant function from this run. Accuracy drops 1%

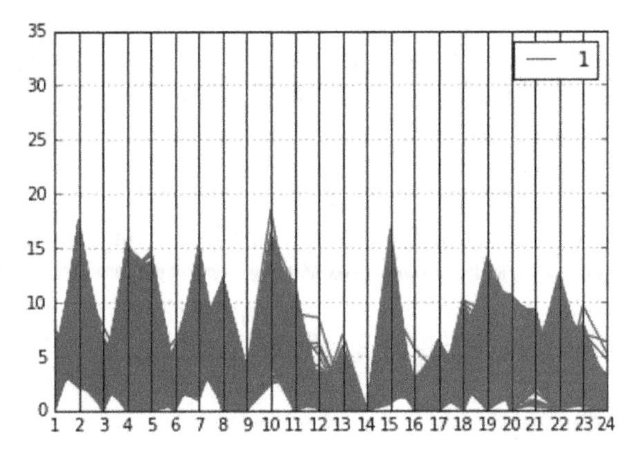

Fig. 7.44 Encoded digit 1 in Parallel Coordinates using 24 dimensions found by the Autoencoder among 484 dimensions. Each vertical line is one of the 24 features scaled in the [0, 35] interval. Digit 1 is visualized on the parallel coordinates

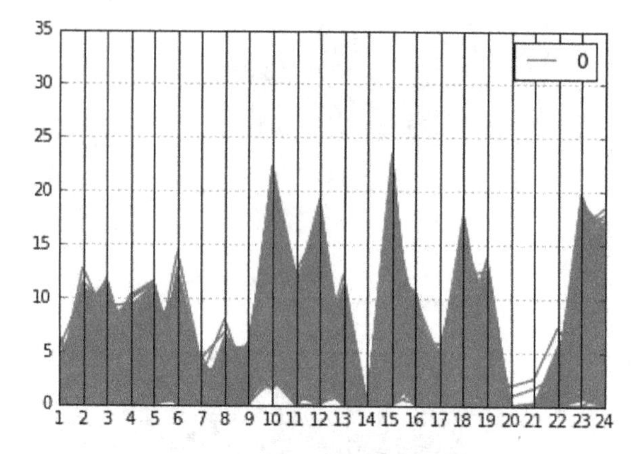

Fig. 7.45 Encoded digit 0 in Parallel Coordinates using 24 dimensions found by the Autoencoder among 484 dimensions. Each vertical line is one of the 24 features scaled in the [0, 35] interval. Digit 0 is visualized in Parallel Coordinates

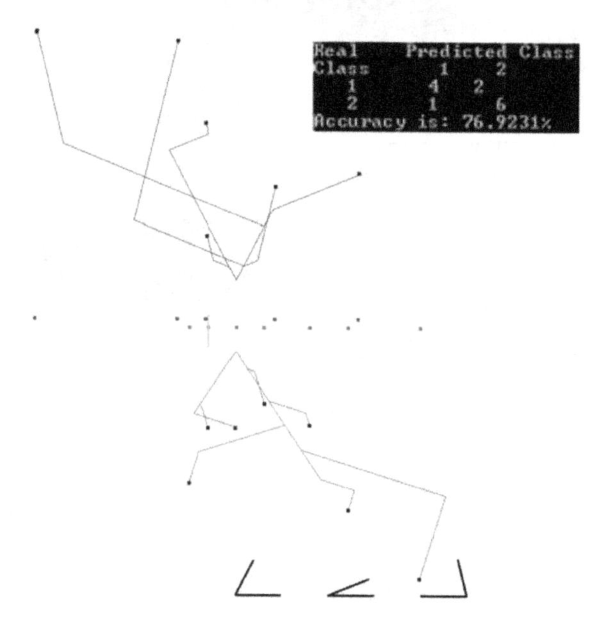

Fig. 7.46 Run 3: data split 60%/40% (training/validation) for the coefficients K = (0.3, 0.6, −0.6). Results on training data (60%) for the coefficients found on these training data

12 and 8.4% (accuracy 88 and 91.6%), for two linear classifiers (1-layer NN) (LeCun et al. 1998). From 1998, dramatic progress was reached in this dataset with non-linear SVM and deep learning algorithms with over 99% accuracy (LeCun et al. 2013). Therefore, future application of the non-linear GLC-L (as outlined in Sect. 7.2.6) also promises higher accuracy.

For **Case study 4**, the comparison with parallel coordinates is presented in Sect. 7.3.4. For **case study 5,** we are not aware of similar experiments, but the accuracy in this case study is comparable with the accuracy of some published stock market predictions in the literature.

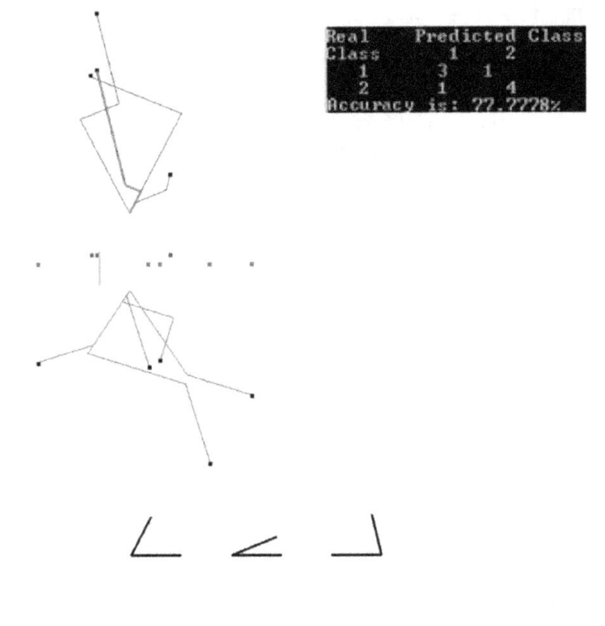

Fig. 7.47 Run 3: data split 60%/40% (training/validation) for the coefficients K = (0.3, 0.6, −0.6). Results on validation data (40%) for coefficients found on the training data. A green line (24 June after Brexit) is correctly classified as S&P 500 down

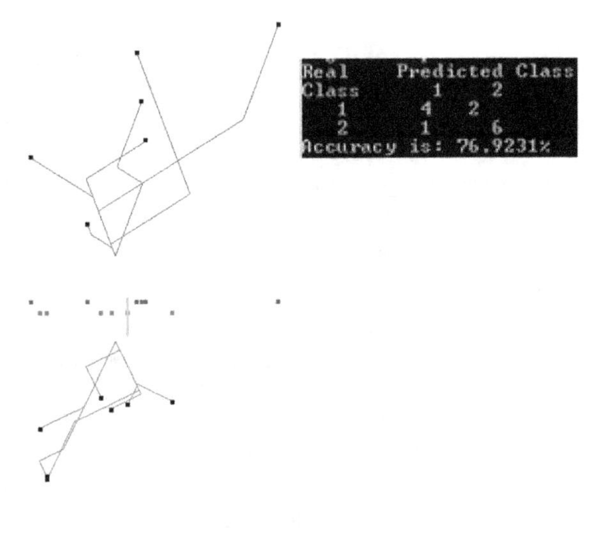

Fig. 7.48 Run 7: data split 60%/40% (training/validation) for coefficients K = (−0.3, −0.8, 0.3). Results on training data (60%) for coefficients found on these training data

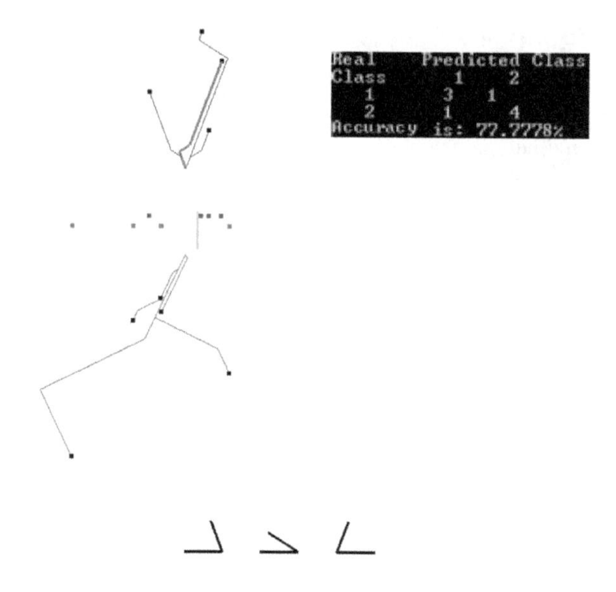

Fig. 7.49 Run 7: data split 60%/40% (training/validation) for coefficients K = (−0.3, −0.8, 0.3). Results on validation data (40%) for a coefficient found on training data. A green line (24 June after Brexit) is correctly classified as S&P 500 down

7.5 Conclusion

This chapter presented GLC-L, GLC-IL, GLC-AL, and GLC-DRL algorithms, and their use for knowledge discovery in solving the machine learning classification tasks in n-D interactively. This chapter presented the five case studies to evaluate these algorithms using the data from three domains: computer-aided medical diagnostics, image processing, and finance (stock market). The utility of our algorithms was illustrated by these empirical studies.

The main advantages and benefits of these algorithms are:

- lossless (reversible) visualization of n-D data in 2-D as graphs,
- absence of the self-crossing of graphs (planar graphs),
- dimension scalability (the case studies had shown the success with 484 dimensions),
- scalability to the number of cases (interactive data clustering increases the number of data cases that can be handled),
- integration of automatic and interactive visual means,
- reached the accuracy on the level of analytical methods,
- opportunities to justify the linear and non-linear models versus guessing a class of predictive models,
- simple to understand for a non-expert in Machine Learning (can be done by a subject matter expert with minimal support from the data scientist),
- supports multi-class classification,
- easy visual metaphor of a linear function, and
- applicable (scalable) for discovering patterns, selecting the data subsets, classifying data, clustering, and dimension reduction.

Table 7.1 Best accuracies of classification on training and validation data with 70–30% data splits

Case study 1 results. Wisconsin Breast Cancer Data			Case study 2 results. Parkinson's Disease			Case study 3 results. MNIST-subset		
Run	Training accuracy (%)	Validation accuracy (%)	Run	Training accuracy (%)	Validation accuracy (%)	Run	Training accuracy (%)	Validation accuracy (%)
1	96.86	96.56	1	89.05	74.13	1	98.17	98.33
2	96.65	97.05	2	85.4	84.48	2	94.28	94.62
3	97.91	96.56	3	84.67	94.83	3	95.07	94.81
4	96.45	96.56	4	84.67	84.48	4	97.22	96.67
5	97.07	96.57	5	85.4	77.58	5	94.52	93.19
6	97.91	96.07	6	84.67	93.1	6	92.85	91.48
7	97.07	96.56	7	84.67	86.2	7	96.03	95.55
8	97.49	98.04	8	87.59	87.93	8	94.76	94.62
9	97.28	98.03	9	86.13	82.76	9	96.11	95.56
10	96.87	97.55	10	83.94	87.93	10	95.43	95.17
Average	97.16	96.95	Average	85.62	85.34	Average	95.44	95.00

In all experiments in this chapter, 50 simulation epochs were sufficient to get acceptable results. These 50 epochs correspond to computing 500 values of the objective function (accuracy), due to ten versions of training and validation data in each epoch. The likely contribution to rapid convergence of data that we used is the possible existence of many "good" discrimination functions that can separate classes in these datasets accurately enough. This includes the situations with a wide margin. In such situations, a "small" number of simulations can find "good" functions. The indirect confirmation of multiplicity of discriminant functions is a variety of machine learning methods that produced quite accurate discrimination on these data that can be found in the literature. These methods include SVM, C4.5 decision tree, naïve Bayesian classifier, k-Nearest Neighbors, and Random forest. The likely contribution of the GLC-AL algorithm is in random generation of vectors of coefficients $\{K\}$. This can quickly cover a *wide range* of K in the hypercube $[-1, 1]^{n+1}$ and capture K that quickly gives high accuracy. Next, a "small" number of simulations is a typical fuzzy set with multiple values. In our experiments, this number was about 50 iterations. Building a full membership function of this fuzzy set is a subject of future studies on multiple different datasets.

Overfitting rejects relevant cases by considering them irrelevant. In contrast, underfitting accepts irrelevant cases, considering them as relevant. The described visual analytics can support improving the *control of underfitting (overgeneralization) and overfitting in learning algorithms* known as *bias-variance dilemma* (Domingos 2000). In this way, it helps selecting a class of machine learning models more rigorously.

A common way to deal with overfitting is adding a *regularization term* to the cost function that penalizes the complexity of the discriminant function, such as requiring its smoothness, certain prior distributions of model parameters, limiting the number of layers, certain group structure, and others (Domingos 2000). Moreover, such cost functions as the least squares can be considered as a form of regularization for the regression task. The regularization approach makes ill-posed problems mathematically *solvable*. However, those extra requirements often are *external* to the user task. For instance, the least-square method may not optimize the error for most important samples because it is not weighted. Thus, it is difficult to justify a regularizer including λ parameter, which controls the importance of the regularization term.

The linear discriminants, considered in this chapter, are among the simplest discriminants with *lower variance* predictions outside training data and do not need to be penalized for complexity. Further simplification of linear functions commonly is done by dimension reduction that is explored in Sect. 7.2.5. Linear discriminants suffer much more from overgeneralization (*higher bias*).

The expected contribution of the visual analytics, discussed in the chapter to deal with the bias-variance dilemma, is not in a direct justification of a regularization term to the cost function. It is in the introduction of another form of a *regularizer outside of the cost function.* In general, it is not required for a regularizer to be a part of the cost function to fulfil its role of *controlling underfitting and overfitting*.

The opportunity for such an outside control can be illustrated by multiple figures in this chapter. For instance, Fig. 7.7a shows that cases of class 1 form an elongated shape, and all cases of class 2 form another elongated shape on the training data. The assumption that the training data are representative for the whole data leads to the expectation that all new data of each class must be *within these elongated shapes of the training data* with some margin. Figure 7.7b shows that this is the case for validation data from Fig. 7.8. The analysis shows that only a few cases of validation data are outside of the convex hulls of the respective elongated shapes of classes 1 and 2 of the training data. This confirms that these training data are representative for the validation data. The linear discriminant in Fig. 7.7a, b, shown as a yellow bar, classifies any case on the left to class 1 and any case on the right to class 2, i.e., significantly underfits (overgeneralizes) elongated shapes of the classes.

Thus, elongated shapes and their convex hull can serve as alternative *regularizers*. We outlined the idea of using convex hulls in Sect. 7.2.4. The actual convex hulls have been constructed in several chapters for several datasets, e.g., Chap. 5 (Sect. 5.6 for Iris data). Additional requirements may include smoothness or simplicity of an envelope at the margin distance μ from the elongated shapes. Here μ serves as a generalization parameter that plays a similar role that λ parameter plays to control the importance of the regularization term within a cost function.

Two arguments for simpler linear classifiers versus non-linear classifiers are presented (Pereira et al. 2009): (1) non-linear classifiers not always provide a significant advantage in performance, and (2) the relationship between features and the prediction can be harder to interpret for non-linear classifiers. Our approach, based on a linear classifier with non-linear constraints in the form of envelopes, takes a middle ground, and thus provides an opportunity to combine advantages from both linear and non-linear classifiers.

While the results are positive, these algorithms can be improved in multiple ways. Future studies are to expand this approach to knowledge discovery in datasets of larger dimensions with the larger number of instances and with heterogeneous attributes. Other opportunities include using GLC-L and related algorithms: (a) as a visual interface to larger repositories: not only data, but models, metadata, images, 3-D scenes and analyses, (b) as a conduit to combine visual and analytical methods to gain more insight.

References

Aruna, S., Rajagopalan, D.S., Nandakishore, L.V.: Knowledge based analysis of various statistical tools in detecting breast cancer. Comput. Sci. Inf. Technol. **2**, 37–45 (2011)

Chollet, F.: Keras. Available online. https://github.com/fchollet/keras (2015). Accessed 14 June 2017

Christobel, A., Sivaprakasam, Y.: An empirical comparison of data mining classification methods. Int. J. Comput. Inf. Syst. **3**, 24–28 (2011)

Cristianini, N., Shawe-Taylor, J.: An Introduction to Support Vector Machines and Other Kernel-Based Learning Methods. Cambridge University Press, Cambridge, UK (2000)

Domingos, P.: A unified bias-variance decomposition. In: Proceedings of the 17th International Conference on Machine Learning, Stanford, CA, USA, 29 June–2 July 2000, pp. 231–238. Morgan Kaufmann, Burlington, MA, USA (2000)

Freedman, D.: Statistical Models: theory and Practice. Cambridge University Press, Cambridge, UK (2009)

Freund, Y., Schapire, R.: Large margin classification using the perceptron algorithm. Mach. Learn. **37**, 277–296 (1999). https://doi.org/10.1023/A:1007662407062

Keras: The Python Deep Learning Library. Available online: http://keras.io. Accessed 14 June 2017

LeCun, Y., Cortes, C., Burges, C.: MNIST Handwritten Digit Database. Available online: http://yann.lecun.com/exdb/mnist/ (2013). Accessed 12 Mar 2017

LeCun, Y., Bottou, L., Bengio, Y., Haffner, P.: Gradient-based learning applied to document recognition. Proc. IEEE **86**, 2278–2324 (1998). https://doi.org/10.1109/5.726791

Lichman, M.: UCI Machine Learning Repository [http://archive.ics.uci.edu/ml]. University of California, School of Information and Computer Science, Irvine, CA (2013)

Lin, X., Mukherji, A., Rundensteiner, E.A., Ward, M.O.: SPIRE: supporting parameter-driven interactive rule mining and exploration. Proc. VLDB Endowment **7**, 1653–1656 (2014)

Maszczyk, T., Duch, W.: Support vector machines for visualization and dimensionality reduction. In: International Conference on Artificial Neural Networks, pp. 346–356. Springer, Berlin/Heidelberg, Germany (2008)

Pereira, F., Mitchell, T., Botvinick, M.: Machine learning classifiers and fMRI: a tutorial overview. NeuroImage **45**(Suppl. 1), S199–S209 (2009)

Ramani, R.G., Sivagami, G.: Parkinson disease classification using data mining algorithms. Int. J. Comput. Appl. **32**, 17–22 (2011)

Salama, G.I., Abdelhalim, M., Zeid, M.A.: Breast cancer diagnosis on three different datasets using multi-classifiers. Breast Cancer (WDBC) **32**, 2 (2012)

Weka 3: Data Mining Software in Java. Available online: http://www.cs.waikato.ac.nz/ml/weka/. Accessed 14 June 2017

XmdvTool Software Package for the Interactive Visual Exploration of Multivariate Data Sets. Version 9.0 Released 31 October 2015. Available online: http://davis.wpi.edu/~xmdv/ (2015). Accessed 24 June 2017

Yang, J., Peng, W., Ward, M.O., Rundensteiner, E.A.: Interactive hierarchical dimension ordering, spacing and filtering for exploration of high dimensional datasets. In: IEEE Symposium on Information Visualization, 2003. INFOVIS 2003. IEEE, pp. 105–112 (2003)

Yang, D., Zhao, K., Hasan, M., Lu, H., Rundensteiner, E., Ward, M.: Mining and linking patterns across live data streams and stream archives. Proc. VLDB Endowment **6**, 1346–1349 (2013)

Zhao, K., Ward, M., Rundensteiner, E., Higgins, H.: MaVis: machine learning aided multi-model framework for time series visual analytics. Electron. Imaging, pp. 1–10. https://doi.org/10.2352/issn.2470-1173.2016.1.vda-493 (2016)

Chapter 8
Knowledge Discovery and Machine Learning for Investment Strategy with CPC

An economist is an expert who will know tomorrow why the things he predicted yesterday didn't happen today.
Laurence J. Peter

8.1 Introduction

Knowledge discovery is an important aspect of human cognition. The advantage of the visual approach is in the opportunity of solving easier perceptual tasks instead of complex cognitive tasks. However for cognitive tasks such as financial investment decision making, this opportunity faces the challenge that financial data are abstract multidimensional and multivariate, i.e., outside of traditional visual perception in 2-D or 3-D world. This chapter presents an approach to find an investment strategy based on pattern discovery in multidimensional space of specifically prepared time series.

Visualization based on the lossless Collocated Paired Coordinates (CPC) defined in Chap. 2 plays an important role in this approach for building the criteria in the multidimensional space for finding an efficient investment strategy. Criteria generated with the CPC approach allow reducing/compressing space using simple directed graphs with the beginnings and the ends located in different time points.

The dedicated subspaces constructed for EUR/USD foreign exchange market time series include characteristics such as moving averages, differences between moving averages, changes in volume, adjusted moving averages known as the Bollinger Band, etc. Extensive simulation studies in learning/testing context are presented below. Effective relations were found for one-hour EURUSD pair for recent and historical data. In this chapter, the method is presented for one-day EURUSD time series in 2-D and 3-D visualization spaces. The main positive result is finding the property in the visualization space that leads to a profitable investment decision (long, short position or nothing). This property is the effective split of a normalized 3-D space on $4 \times 4 \times 4$ cubes. The strategy is ready for implementation in algotrading mode.

B. Kovalerchuk, *Visual Knowledge Discovery and Machine Learning*,
Intelligent Systems Reference Library 144,
https://doi.org/10.1007/978-3-319-73040-0_8

While cognitive algorithms intend to mimic the functioning of the human brain for improving human decision-making, often the scope of mimicking is not obvious. This is evident for the tasks with unclear human decision that must be mimicked. A market investment decision is one example with this difficulty due to complexity and uncertainty of the task and its high dynamics, i.e., a strategy that was correct at time t is not correct at time $t + 1$.

Thus for such tasks, we need two-stage cognitive algorithms:

- mimicking good human decision process at the upper level, and
- mimicking the functioning of the human brain to reproduce that good human decision.

Both stages are active areas of research. This chapter focuses on the first stage for market investment decisions. The concept of dynamic logic for human decisions at the upper level is presented in Kovalerchuk et al. (2012). It includes the sequence of decision spaces and criteria from less specific to more specific.

This concept is applied in this chapter to the development of the investment strategy, where a lossless visual representation of n-D data serves as an initial form of the decision space and more specifics are learned later using a machine learning technique.

Difficulties of defining investment strategy algorithmically are well documented in the literature (Kovalerchuk and Vityaev 2000; Bingham 2014; Li et al. 2009; Martin 2001; Wilinski et al. 2014; Guo et al. 2014; Hoffmann 2014).

Multivariate and multidimensional nature of data complicates both knowledge representation and discovery including:

- identifying a class of predictive models (SVM, regression, ANN, kNN and so on) with associated trading strategies with parameters to be learned, and
- analyzing multidimensional data with a naked eye to stimulate both intuitive discovery of patterns and formal models (Lian et al. 2015; Wichard and Ogorzalek 2004).

The most efficient strategy should take into account the proper balance between both directions of investment (long and short positions) typical for foreign exchange markets (the pair EURUSD belongs to them). The best use of the market potential is reached when numbers of both long and short positions taken are comparable. The methodology presented in this chapter allows the adaptation of the strategy to this symmetry.

The goal is finding visualization-inspired investment strategy using multivariate and multidimensional data. One of the main conclusions is that the new lossless Collocated Paired Coordinates approach is an effective instrument for such inspiration in the synthesis of the investment strategy.

Chapter 2 contains detailed steps of CPC. Below we summarize them:

(1) Representing a normalized to [0, 1] n-D point $\mathbf{x} = (x_1, x_2, \ldots, x_{n-1}, x_n)$, as a set of pairs $(x_1, x_2), \ldots, (x_i, x_{i+1}), \ldots, (x_{n-1}, x_n)$;

(2) Drawing 2-D orthogonal Cartesian coordinates $(\mathrm{X}_1, \mathrm{X}_2), \ldots, (\mathrm{X}_i, \mathrm{X}_{i+1}), \ldots, (\mathrm{X}_{n-1}, \mathrm{X}_n)$

with all odd coordinates collocated on a single horizontal axis X and all even coordinates collocated on a single vertical axis Y;

(3) Drawing each pair (x_i, x_{i+1}) in $(\mathrm{X}_i, \mathrm{X}_{i+1})$;

(4) Connecting pairs by arrows to form a graph $\mathbf{x}^* : (x_1, x_2) \rightarrow (x_3, x_4) \rightarrow \ldots \rightarrow (x_{n-1}, x_n)$.

This graph **x*** represents n-D point **x** in 2-D losslessly, i.e., all values of **x** can be restored. Thus, this visualization is reversible representing all n-D data without loss of them. For the odd n the last pair can be (x_n, x_n) or $(x_n, 0)$.

For 3-D visualization, the pairs are substituted by triples

$$(x_1, x_2, x_3), \ldots, (x_i, x_{i+1}, x_{i+2}), \ldots, (x_{n-2}, x_{n-1}, x_n).$$

If n is not divisible by 3, e.g., $n = 7$ then for $n = 7$ we have triples (x_1, x_2, x_3), (x_4, x_5, x_6), (x_7, x_7, x_7) or triples (x_1, x_2, x_3), (x_4, x_5, x_6), $(x_7, 0, 0)$. If $n = 8$ then we have triples (x_1, x_2, x_3), (x_4, x_5, x_6), (x_7, x_8, x_8) or triples (x_1, x_2, x_3), (x_4, x_5, x_6), $(x_7, x_8, 0)$.

Pairs of variables $(x_i,\ x_{i+1})$ for time series can be sequential pairs of values at time t and t + 1. In this way a 4-D point **x** can be formed as $(v_t,\ y_t, v_{t+1}, y_{t+1})$, where v is volume and y is profit at two consecutive times t and t + 1. Respectively a 4-D point $(v_t,\ y_t, v_{t+1}, y_{t+1})$ will be represented in 2-D as an arrow from 2-D point (v_t, y_t) to another 2-D point (v_{t+1}, y_{t+1}).

This simple graph fully represents 4-D data. It has a clear and simple meaning—the arrow going up and to the right indicates the growth in both profit and volume from time t to t + 1. Similar interpretations have other arrow directions. To make visualization more clear the time pairs starting from odd time t are visualized separately from time pairs starting from even time t. This helps to observe better the beginnings and ends of events.

Similarly, a 6-D point **x** can be formed as $(v_t, d_{MAt}, p_t, d_{MA,t+1}, v_{t+1}, p_{t+1})$, where d_{MA} is the difference between the moving averages for some windows. We represent this 6-D point in 3-D as an arrow from 3-D point (v_t, y_t, d_{MAt}) to 3-D point $(v_{t+1}, d_{MA,t+1}, y_{t+1})$. This simple graph fully represents 6-D data point with a clear meaning—the arrow going up and to the right indicates the growth in all three attributes: profit, d_{MA}, and volume from time t to t + 1.

Several figures such as Figs. 8.5, 8.6, 8.7, 8.8, 8.9, 8.16, 8.17 and 8.18 show and use this visualization. To shorten notation for the spaces in this chapter we will use notation like (Y_r, V_r) instead of $(Y_{rt}, V_{rt}, Y_{r,t+1}, V_{r,t+1})$, and (Y_r, d_{MAr}, V_r), instead of $(Y_{rt}, V_{rt}, d_{MArt}, Y_{r,t+1}, d_{MAr,t+1}, V_{r,t+1})$ where index r stands for normalized profit, volume and d_{MA}. Thus, 2-D and 3-D notations (Y_r, V_r) and (Y_r, d_{MAr}, V_r) will represent 4-D and 6-D spaces, respectively.

The general CPC concept can be applied also more generally for time series of pairs of variables volume v and profit y. Consider consecutive time moments $t, t+1, \ldots, t+k-1$ and a 2 k-D point

$$\mathbf{x} = (v_t, y_t, v_{t+1}, y_{t+1}, v_{t+2}, y_{t+2}, \ldots, v_{t+k-1}, y_{t+k-1}),$$

where v is volume and y is profit as before. This 2k-D point is represented in 2-D as a directed graph $\mathbf{x}^*$ (path that starts in point (v_t, y_t) and ends in (v_{t+k-1}, y_{t+k-1}). This graph fully represents 2k-D data point. In this general case, the path that is going up and to the right indicates the growth in both profit and volume from time t to $t + k$.

Besides CPC the whole class of General Line Coordinates described in previous chapters opens multiple opportunities to represent n-D financial data visually and discovering patterns in these data.

Below we present the stages of the process, a visual method for building an investment strategy using 4-D and 6-D data in 2-D and 3-D spaces, results of 4-D and 6-D pattern discovery in 2-D and 3-D visualization spaces, summarize results and outline the future research.

8.2 Process of Preparing of the Strategy

8.2.1 *Stages of the Process*

The **first stage** of the process is selecting features/indicators from the time series to prepare the variables for creating the multivariate space. The selected features should meet some conditions related to correlation (Hellwig 1969). We selected the features from many well-known indicators such as Bollinger Bands (BB), difference between moving averages (*dMA*), volume (*V*). Some of them were transformed to derivatives of the base indicators such as difference between current and previous Volume (*dV*), current relative value of the observed variable Y_r and others. This catalogue is open for modification and expansion in case of the insufficient accuracy of prediction.

The selected features/indicators are derived from time series with the same frequency of sampling as underlying instrument. In this chapter it is 1 h EURUSD pair. All indicators are normalized to [0, 1], excluding BB due to the nature of this indicator.

The **second stage** is finding good spaces (Y_r, P_r) or (Y_r, P_{r1}, P_{r2}), where Y_r is the main relative normalized outcome in time series, P_{r1}, and P_{r2} are indicators used for space creating, e.g., P_{r1}, P_{r2} belong to $\{B_r, V_r, dV_r, dMA_r, \text{etc}\}$. The relative variables with index r are normalized. Inside the spaces the process is looking for areas or subspaces such as squares, rectangles, cubes or cuboids where the number of profitable events prevails. The event is profitable if it signals to open long or short position in this point of the space and at this time. The areas with big asymmetry between the number of long and short positions are considered as promising from investment strategy point of view. Such asymmetric areas are identified in two steps: (a) by visual and analytical discovery in the CPC space, and (b) by a machine learning process of verification on testing data.

The **third stage** is testing and verifying the strategy by constructing a cumulative profit curve. The curve with a significant profit and a small variance (e.g., with appropriate Calmar ratio (Young 1991) for evaluating risk) is used as a measure of success of the algorithm along with comparison with the result to typical benchmarks.

8.2.2 Variables

Below we list the variables used in this study that include the main outcome and its features/indicators such as based on open, close, high and low values of a variable of trading interest known collectively as a candle or a candlestick.

The *main outcome variable* is $Y_{ri} = (Y_i - Y_{min\,i)}/(Y_{max\,i} - Y_{min\,i})$. This is relative position of Close value $Y(i)$ of current candle normalized to [0, 1] between min and max of last k_b candles, where k_b is a parameter of the strategy (in the first experiment presented below k_b = 120), where i_c is the current value of index i and

$$Y_{max\,i} = max(Y_i);\ \text{for}\ i = i_c - k_b \ldots i_c$$

$$Y_{min\,i} = min(Y_i);\ \text{for}\ i = i_c - k_b \ldots i_c$$

The six indicators are listed below.

(1) *The relative value of position* $Y(i)$ with respect to Bollinger Band is denoted as B_r, $B_r(i) = (Y(i) - B_d(i))/(B_{up}(i) - B_d(i))$. $Y(i)$ can be out of this normalized Bollinger Band range [0, 1]. Here $Y(i)$ is the Close value of i-candle in the time series in i-time, and the Candle is a vector of 4 values known as OHLC,

$$Y_{max}(i) = max(Y(i - k_b : i)),$$

$$Y_{min}(i) = min(Y(i - k_b : i)),$$

k_b is the number of candles back to calculate relative variables;

$$B_{Bd}(i) = M_{Ab}(i) - k_{Bb} \times S_{tb}(i),$$

$M_{Ab}(i) = \text{mean}(Y(i - B_b : i)$ is the average value of last B_b Closes, and $S_{tb}(i) = s_{td}(Y(i - B_b : i)$ is the standard deviation for last B_b-candles.

(2) *Relative Volume*, V_{olr}, is the measure of the position of current Volume between the min and max of Volume within the determined number of last candles (here it is k_b candles). Volume is a measure of how much of a given financial asset has been traded in the current period. In the experiments EURUSD 1 h series was used, thus the period for volume observation is one hour.

(3) *difference between moving averages*, d_{MAr}, where
$d_{MA}(i) = MA_f(i) - MA_s(i)$,
$MA_f(i) = mean(Y(i - f : i - 1))$ is moving average with *f*-parameter

$$MA_s(i) = mean(Y(i - s : i - 1)),$$

f and s are the numbers of candles to calculate moving averages. For the first experiment $f = 5$ ("fast"), and $s = 28$ ("slow"),
$d_{MAr}(i) = (d_{MA}(i) - d_{MAmin}(i))/(d_{MAmax}(i) - d_{MAmin}(i))$ is the relative difference of MA;

$$d_{MAmax}(i) = \max(d_{MA}(i - k_b : i));$$

$$d_{MAmin}(i) = \min(d_{MA}(i - k_b : i)).$$

(4) *Relative return,* r_r,

$$r_r(i) = (r(i) - r_{\min}(i))/(r_{\max}(i) - r_{\min}(i)),$$

where
$r_{max}(i) = max(r(i - k_b : i))$ is return max in last $k_b = 120$ candles,

$$r_{min}(i) = \min(r(i - k_b : i));$$

$$r(i) = Close(i) - Close(i - 1),$$

$Close$ is the last value in i-candle OHLC.

(5) *Relative difference between real value of Y and its prediction,* d_{yr}. Prediction is performed with a linear regression model.

$$d_{yr}(i) = \left(d_y(i) - d_{ymin}(i)\right)/\left(d_{ymax}(i) - d_{ymin}(i)\right),$$

where

$$d_y(i) = Close(i - 1) - y_{next}(i);$$

$y_{next}(i)$ is the prediction in i-Close based on linear regression;

$$d_{ymax}(i) = \max(dy(i - k_b : i));$$

$$d_{ymin}(i) = \min(dy(i - k_b : i)).$$

(6) *First derivative* dV_r *of the Volume, V.* It is also a relative variable:

$$dV_r(i) = (dV(i) - dV_{min}(i))/(dV_{max}(i) - dV_{min}(i)),$$

where
$dV(i) = V_{i-1} - V_{i-2}$, $dV_{max}(i) = max(dV(i - k_b : i)), dV_{min}(i) = \min(dV(i - k_b : i))$.

8.2.3 Analysis

The analysis involves three important parameters: f (numbers of candles to calculate fast moving averages), s (numbers of candles to calculate slow moving averages), and k_b (number of "back" candles). Additional parameters include Stop Loss that is popular in the broker platforms mechanism to avoid big losses. The variable Y_r and every variable from a set $S = \{B_r, V_r, d_{MAr}, r_r, d_{yr}, dV_r\}$ are used to create 2-D spaces such as (B_r, Y_r), (V_r, Y_r) and so on.

Considering the variables as traditional time series allows us to compare each pair (P_r, Y_r) as shown in Fig. 8.1, where P_r is one of the variables from the set S. In Fig. 8.1, it is the first pair.

Figure 8.1 shows a small part of time series (100 one-hour candles of EURUSD). It allows observing a modest correlation with the correlation coefficient equal 0.49 that can be a potential source for a trading strategy rule. Figures 8.2 and 8.3 show far worse situations for finding patterns to construct a trading strategy with correlation coefficients only from 0.055 to 0.15.

Among other indicators from S only dMA_r has a modest correlation coefficient (0.45) with Y_r. See Fig. 8.4. For other pairs the coefficients are much smaller. The visualization of relations of observed variable Y_r with variables V_r or dMA_r in Figs. 8.1 and 8.4 allows getting a preliminary idea for classification of variables on promising and not promising by observing the actual position of correlated variables. However, even for these two selected pairs of variables, it is hard to expect building an efficient trading strategy using these relatively low correlations. Surely, we cannot base high prediction accuracy expectations on these weak correlated variables.

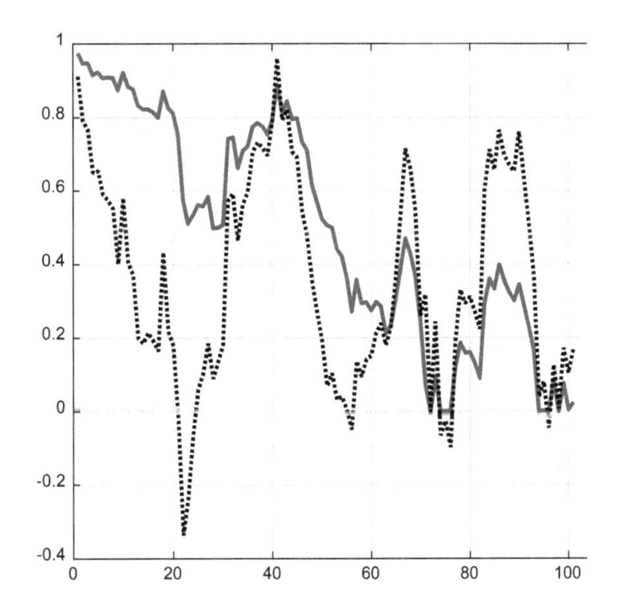

Fig. 8.1 Comparison of two time series: relative outcome *Yr* and relative position of *Y* (*i*) inside Bollinger Band (dotted)

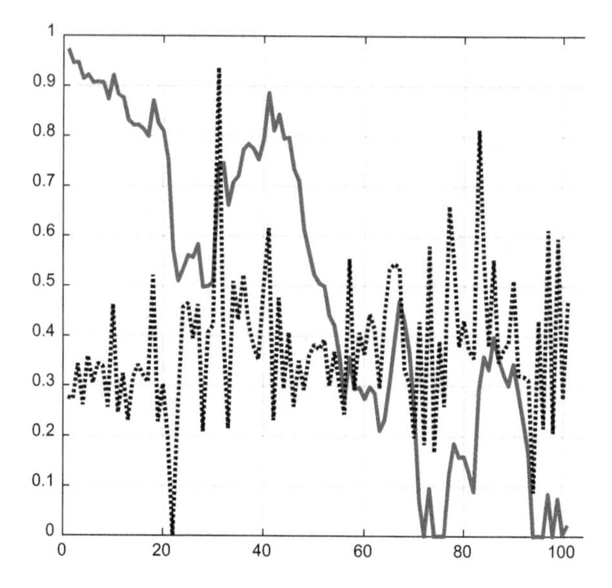

Fig. 8.2 Comparison of two time series: relative outcome *Yr* and relative returns after every *Y*(*i*) (dotted) for one hundred candles

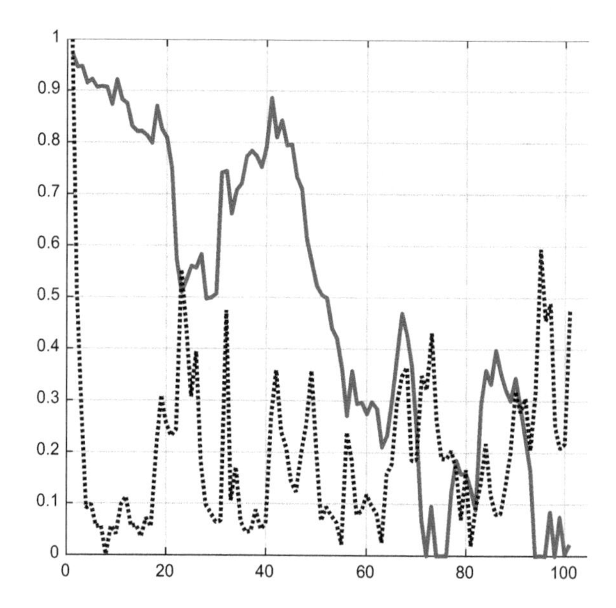

Fig. 8.3 Comparison of two time series: relative outcome *Yr* and relative volume in every one hundred period

In this chapter, we attempt a quite different way using CPC for finding a relation between the selected variables in spite of such relatively low correlation.

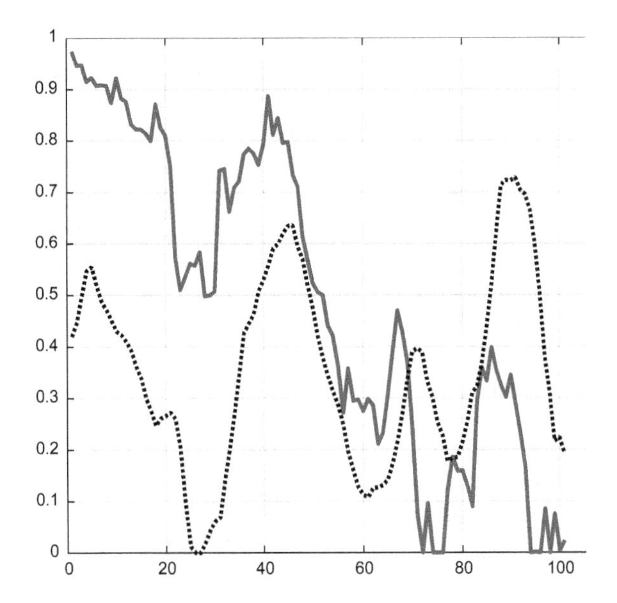

Fig. 8.4 Comparison of two time series: relative outcome *Yr* and relative difference between two moving averages *dMAr* (dotted) for one hundred candles

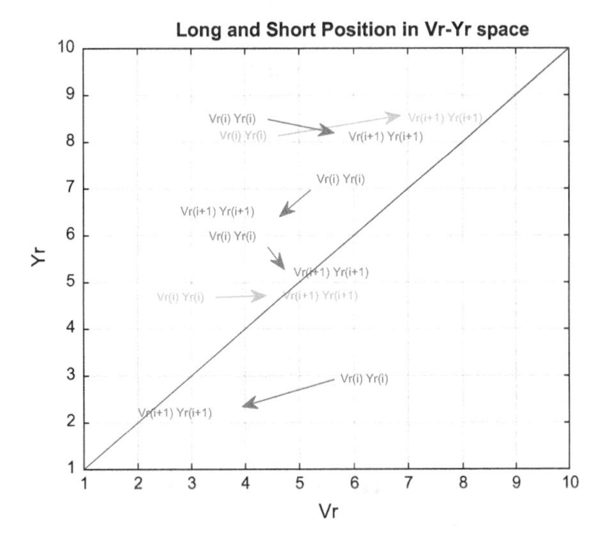

Fig. 8.5 Some examples of arrows which show points of two time series, Y_r and V_r

8.2.4 Collocated Paired Coordinates Approach

Figure 8.5 shows some arrows in (V_r, Y_r) space under the CPC approach. This is also a fragment of two time series such as shown in Fig. 8.3: relative volume V_r and relative main outcome variable Y_r. Figure 8.3 shows time series in a classical form with time as an independent variable. Figure 8.5 has no time axis. The time is represented by arrow direction.

The beginning of the arrow shows the first point in the space (V_{ri}, Y_{ri}), and the head of the arrow shows the next point in the same space (V_{ri+1}, Y_{ri+1}). Figure 8.5 shows the inspiration idea for building a trading strategy in contrast with Fig. 8.3 where we do not see it. Figure 8.5 allows finding the areas with clusters of two kinds of arrows. The arrows for the long positions are solid (green) arrows, for the short positions, are dotted (red). Along the Y_r axis we can observe a type of change in Y in the current candle. Due to the fact that if $Y_{ri+1} > Y_{ri}$ then $Y_{i+1} > Y_i$ the proper decision in i-point would be a long position opening. Otherwise, the best decision in the point would be a short position. Additionally, Fig. 8.5 shows how effective will be a decision in the positions. If the arrows are very horizontal, the profit will be small and more vertical arrows indicate the larger profit.

In comparison with traditional representation in the time series domain, the proposed method brings the additional knowledge about the potential of profit in selected area of parameters (in Fig. 8.5 it is (V_r, Y_r) space).

Figure 8.6 shows multiple arrows for 500 candles period. It shows only candles for even i-candles, denoted as even candles. The arrows for odd candles (odd i) can be visualized in the same way.

This separation simplifies visualization. In contrast, presenting even and odd candles together will show them connected which will mute the visual pattern.

Figure 8.7 shows a part of the set arrows from Fig. 8.6. This set includes only more vertical arrows such that, $|Y_{ri+1} - Y_{ri}| > h = 0.05$. Variable h can be a parameter of future strategy and can have another value.

Figure 8.7 is much more readable and may inspire a trading strategy development more efficiently demonstrating the power of CPC visualization.

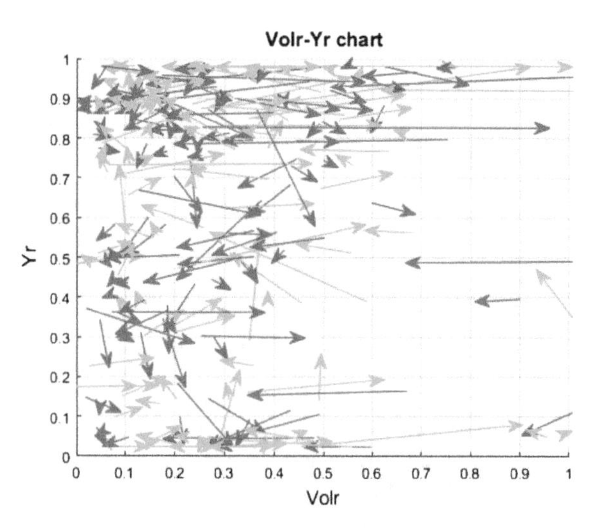

Fig. 8.6 The arrows represent open positions long (green arrow) and short (red arrow) in the space (V_r, Y_r) (relative volume—relative observed variable)

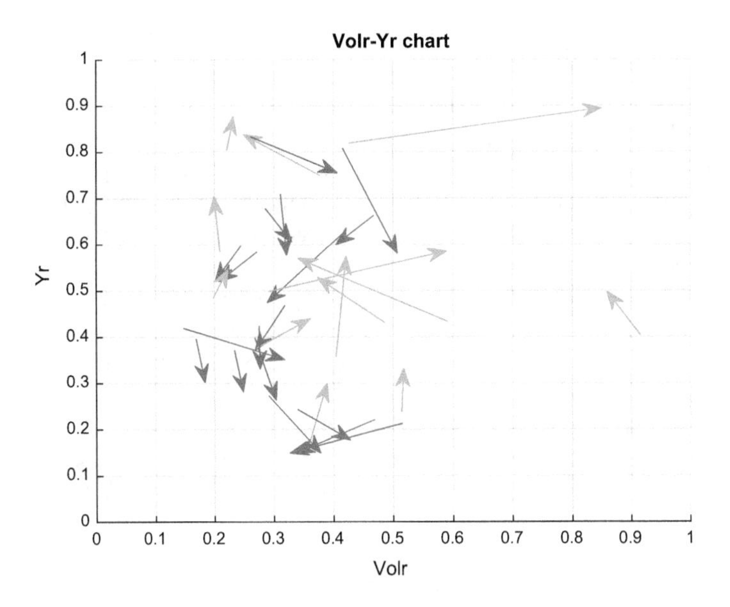

Fig. 8.7 Subset of arrows from Fig. 8.6. Arrows are more vertical

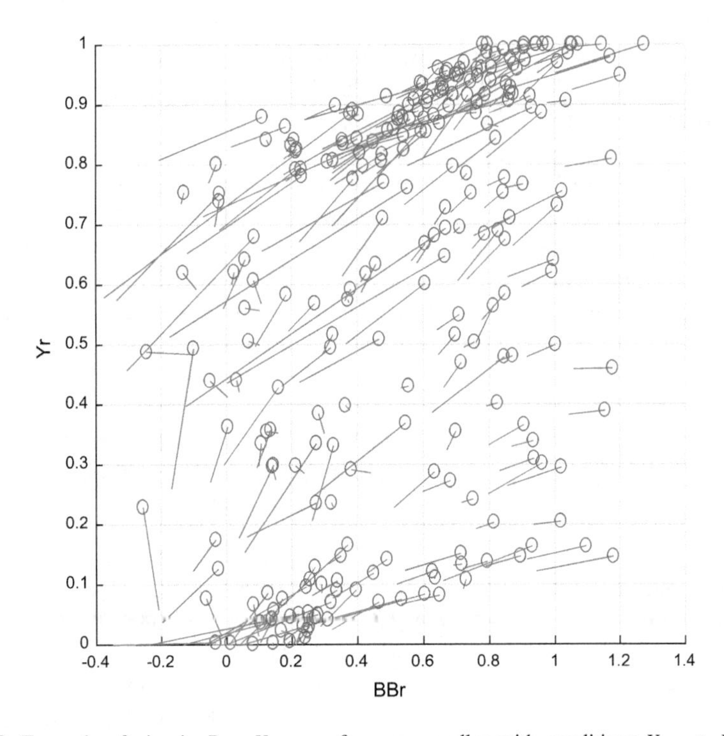

Fig. 8.8 Example of pins in $B_r - Y_r$ space for even candles with conditions $Y_{r\,i-1} > Y_{r\,i}$ with suggestion to open short positions

While arrows indicate well the increase and decrease in values, another notation in CPC space is also beneficial in cases where the beginnings of the CPC sections are more important than ends. In these cases, we use special pins (section with circle at the beginning) instead of arrows.

Figure 8.8 shows the pins in (B_r, Y_r) space with condition $Y_{r\mathrm{i}-1} > Y_{r\mathrm{i}}$ only. It is the condition for opening short positions. It is a result of simulation for 500 candles period for even candles only. Here we see the advantage of the dividing candles to even and odd sets.

Otherwise, we would have the circle in every contact point between two nearest sections and it will be impossible to recognize the end and the beginning of the pins. In this section, the data from the same EURUSD 1 h time series are used for both arrows and pins.

The particular used interval with 500 candles is not specific for the method. The objective is to explain the idea of using CPC on any number of candles and intervals. For another interval with the same or another number of candles, the distribution of pins or arrows will be different but the algorithm will search asymmetric areas in the same way.

The pins with the circles have an important advantage of observing the clusters of the beginnings of the pins (circles) to determine the best places in the space (B_r, Y_r) for opening a long or a short position. For example, Fig. 8.9 shows a rectangle with more empty circles than the filled ones. This leads to the rule: if the next point of time series is exactly in this area, then open a short position.

It gives an idea how to create an investment strategy—looking for areas with big asymmetry between filled and empty circles, i.e., between suggestions of long and short positions. Note that these areas are changing over time and need to be updated regularly. At first glance, arrows in Figs. 8.6 and 8.8 are chaotic without a useful pattern. This situation is quite common in visualization of complex data. It is a strong inspiration for *combining a pure visual technique with analytical computations* for searching the subsets with big asymmetry. In this case, the CPC 2-D or 3-D visual representation opens an opportunity for an *analytical discovery of interesting patterns in this low dimensional CPC visual space*. In general, the combination of analytical and visual means is a core idea of the visual analytics methodology (Keim et al. 2008).

8.3 Visual Method for Building Investment Strategy in 2D Space

The main idea is to find a place in 2-D representation of 4-D (P_r, Y_r) space where asymmetry between number of suggestions to open long position (filled circles) and short positions (empty circles) is high (as in the rectangle in Fig. 8.9). In the experiments, we used a rectangle or a square in 2D space and a cube or a cuboid in 3D space. Consider rectangle $R(i_c, j_c)$, where i_c, j_c are the coordinates of its center with the following properties.

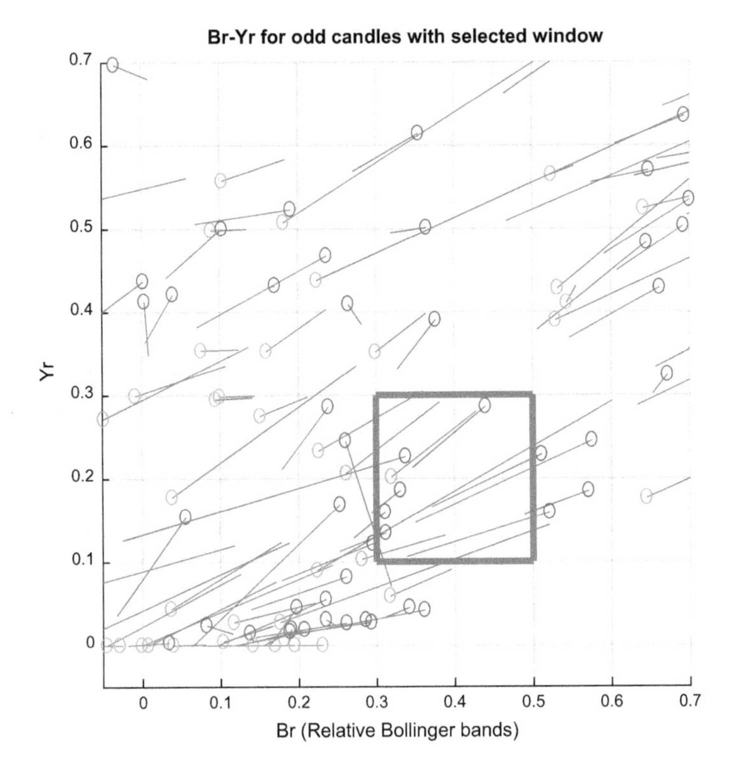

Fig. 8.9 Pins in (*Br*, *Yr*) space for both directions—long (green circles) and short positions (red circles)

(a) Properties for positive *long* positon asymmetry A_1:
Property a1: $(i_c, j_c) = \operatorname{argmax} A_1(i,j)$
where

$$A_l(i,j) = n_l(i,j)/(n_l(i,j) + n_s(i,j)),$$
$$i = i_l,\ i_{l+1}, i_{l+2}, \ldots i_r;\ j = j_b, j_{b+1}, j_{b+2},\ \ldots, j_t$$

$n_1(i,j)$—number of suggestions to open long position in the (i,j) area (number of filled circles);
$n_s(i,j)$—number of suggestions to open short position in the (i,j) area (number of empty circles);
i_1—left barrier for the area; i_r—right barrier for the area; i has increment 0.1;
i_b—bottom barrier for the space; i_t—top barrier for the space; i has increment 0.1.
Property a2: $(n_1 + n_s) > n_{\min}$;
where $n_{\min}$ is the minimal threshold number of the events (circles) in the rectangle. We used $n_{\min} = 5$ in our simulations in the 2D representations to make the conclusions more robust.

(b) Properties for positive *short* positon asymmetry A_s

Property $b1$: $(i_c,j_c) = \text{argmax } A_s(i,j)$

where

$A_s(i,j) = n_s(i,j)/(n_l(i,j) + n_s(i,j)), \quad i = i_l, i_{l+1}, i_{l+2}, \ldots i_r; \, j = j_b, j_{b+1}, j_{b+2}, \ldots, j_t;$
and other notations are the same as in (a).

Property $b2$:

$(n_l + n_s) > n_{\min}$; where $n_{\min}$ is minimal number of the events (circles) inside the rectangle to consider this rectangle; in our simulations in 2D spaces there was $n_{\min} = 5$;

For a rectangle with its center in (i_c,j_c) circle $c(P_{r,}\ Y_{r)}$ belongs to this rectangle when $P_r > i_c$ −w/2 and $P_r < i_c = w/2$, where w = 0.2 is the width of the rectangle; $P_r \in \{B_r, V_r, dV_r, dMA_r, \text{etc}\}$. $Y_r > j_c - h/2$ and $Y_r < j_c = h/2$, where h = 0.2 is the width of the rectangle; $Y_r \cdot [0, 1]$.

While in general the optimal rectangles for long and short positions are different here, we defined a simplified case. The general concept of the first strategy is to find a rectangle with properties 1 and 2 in the learning mode of the algorithm and then test them in the test mode, i.e., test the algorithm when in current candles the parameters of time series c $(P_{r,}Y_r)$ are located in the best areas with coordinates (i_c,j_c).

8.4 Results of Investigation in 2D Space

All the 4-D spaces based on pairs (B_r, Y_r), (dMA_r, Y_r), (V_r, Y_r), (dV_r, Y_r), (r_r, Y_r), (dy_r, Y_r) have been investigated. We considered from 300 to 800 candles for space $(B_{r,}Y_r)$ separately for even and odd candles and found the best area (rectangle or square in the space) relative to the properties 1 and 2 as a pattern for the investment strategy, i.e., with a large asymmetry between suggestions for opening long and short positions.

Figure 8.9 shows the rectangle found for short positions with $A_s = n_s/(n_l + n_s) = 0.83$ and parameters: $(i_c,j_c) = (0.4, 0.2)$, $w = 0.2$; $h = 0.2$; $n_l = 1; n_s = 5$. The found rectangle for long positions has much less $A_l = n_l/(n_l + s_l) = 0.56$ with parameters $(i_c,j_c) = (0.4, 0.2)$, $w = 0.2$, $h = 0.2$, $n_l = 9$; $s_l = 7$.

These rectangles were used 500 times for profit simulation for the even candles in series from 300 to 1300 of candles (learning period) and the same rectangle was used for testing period from 1301 to 2300 candles. For every (B_r, Y_r) that happens in this rectangle a short position was opened. The cumulative profit for short and long sides of the market is shown in Figs. 8.10 and 8.11 for odd candles.

The profit for short positions in Fig. 8.11 is much higher than in Fig. 8.10 for the long positions which is consistent with A_s and A_l values. The cumulative profit is shown in Fig. 8.12 for both sides. The data from 300 to 1300 candles are responsible for learning period, the next candles from 1301 to 2300 for the test period.

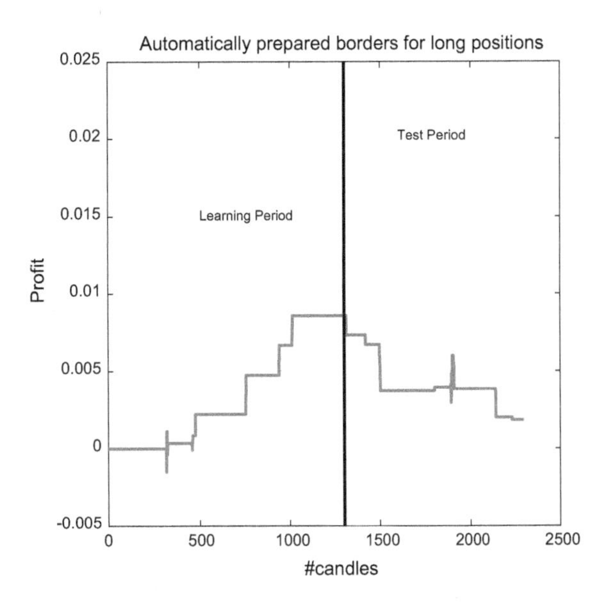

Fig. 8.10 Cumulative profit for long position which were opened based on optimal rectangle in the space (B_r, Y_r)

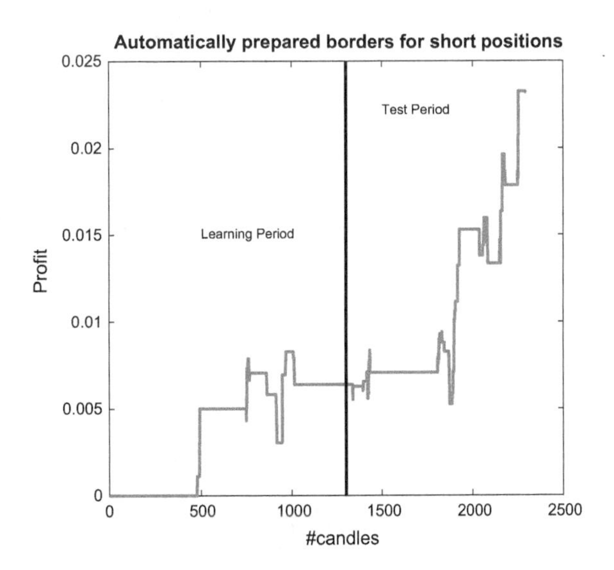

Fig. 8.11 Cumulative profit for short position which were opened based on optimal rectangle in the space (B_r, Y_r)

Usually an optimal relation between the periods is found by the search algorithms (e.g., Wilinski et al. 2014; Wilinski and Zablocki 2015). Typically, it results in the training period that is much longer than the test one.

The first two experiments resulted in a much better split on the training and test periods, which is with the same numbers of candles. The result of initial test guides determining the test period duration. If the result is like in Fig. 8.10 then the reduction of the test period makes sense, otherwise its extension is justified.

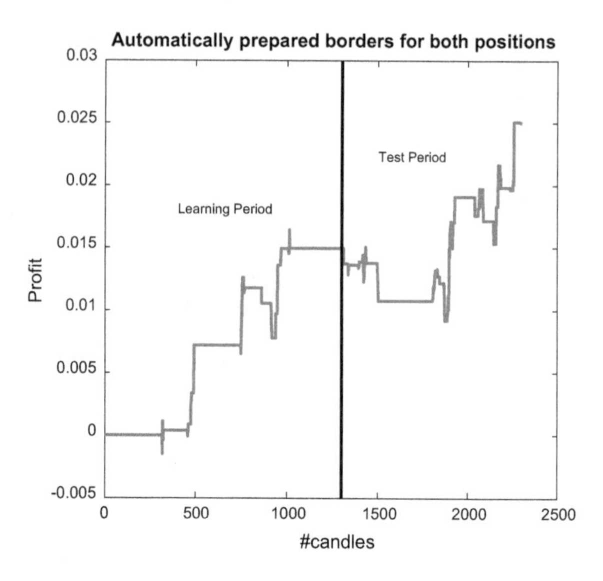

Fig. 8.12 Cumulative profit for strategy in (B_r, Y_r) space for both sides (long and short) of the market

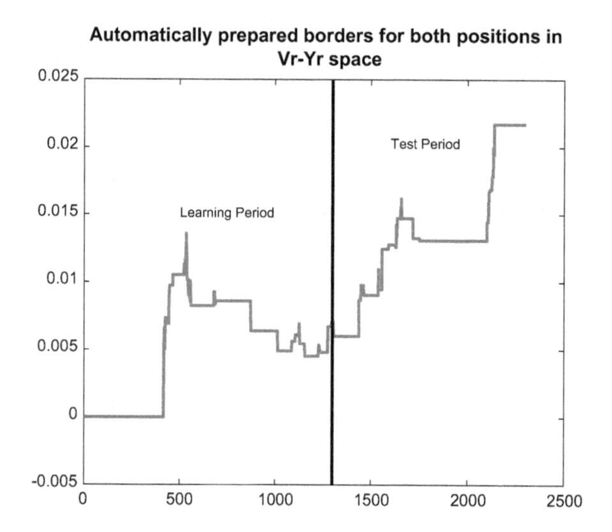

Fig. 8.13 Cumulative profit for strategy in (V_r, Y_r) space for both sides (long and short) of the market

The next simulation experiment was conducted for the space (V_r, Y_r) (relative volume, relative values of observed variable Y). The result with a cumulative curve is shown in Fig. 8.13. We obtained the same positive result in the final point of the time series for the other indicators listed above.

In the next experiment, we joined all rules, for all indicators. The following relative indicators were used: B_r, V_r, dy_r, r_r, dMA_r, dV_r. The result of the simulation is presented in Fig. 8.14. The simulation was performed for arbitrarily determined parameters specified above such as k_b, s, f, SL, $k_w \ldots$ etc.

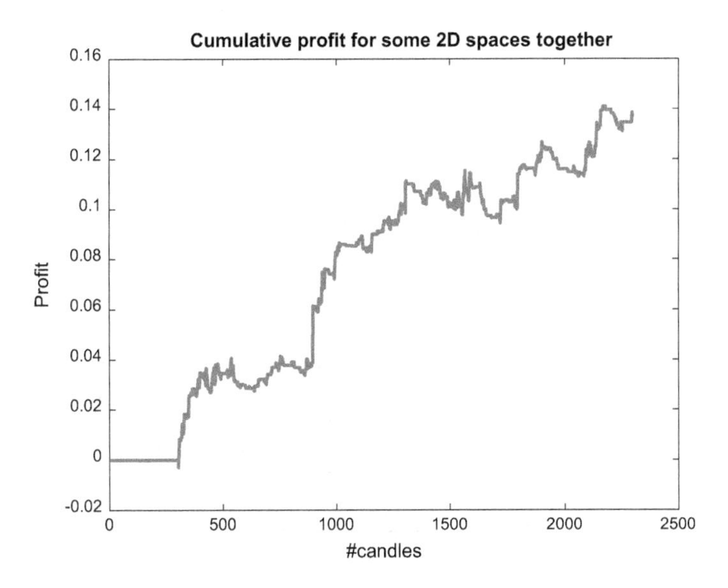

Fig. 8.14 Cumulative profit for some spaces together

The future studies can include looking for optimal values of parameters in a fixed period of time series and for a proper balance between accuracies in learning and testing periods.

Commonly in financial ML, the test period is shorter than the training one (e.g., Wilinski et al. 2014; Wilinski and Zablocki 2015). Figure 8.14 shows significantly increasing profit in the test period with the same length of both training and test periods in this interval of the market data. In another interval of time series, other ratios between length of training and test data need to be learned in the same way

In this study we analyze profit in normalized units: *price interest points* (pip) and PPC (*Profit per candle*). A pip indicates the change in the exchange rate for a currency pair. We assume one pip as 0.0001 USD in the pair EURUSD that is used as a measurement unit of change.

Profit per candle (PPC) represents the difference between values of cumulative profit at the end and the starting point of the period divided by the number of candles in the period. $\text{PPC} = \left(\text{Profit}_{\text{end}} - \text{Profit}_{\text{begin}}\right)/\left(i_{\text{end}} - i_{\text{begin}}\right)$ where $i_{\text{end}}, i_{\text{begin}}$ numbers of considered candles (here—one-hour candles). The result is equal to 1371 pips in 2000 candles with PPC equal to 0.68.

The result in Fig. 8.14 is very positive: the Calmar ratio is equal 6.60 for the learning period and it is 2.94 for the test period from 1300 to 2300 candles. Calmar ratio (Young 1991) is one from many criteria, which evaluate trading quality. It is especially useful for long-term simulations and performances. "Calmar ratio of more than 5 is considered excellent, a ratio of 2–5 is very good, and 1–2 is just good" (Main 2015).

Next in this section, we compare the above result with a common benchmark for the time series of 1 h EURUSD in the same period of 2000 candles from February

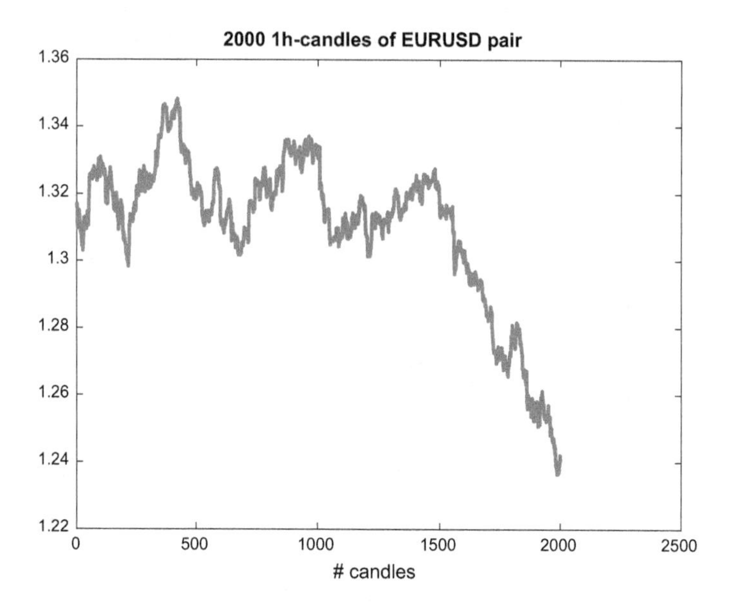

Fig. 8.15 Considered part of 1 h EURUSD time series

2, 2012 to May 31, 2012. The chart of the pair is shown in Fig. 8.15. Another period will be explored in the next section.

The popular benchmark Buy and Hold strategy (only one long position is opened at the beginning) for the considered part of time series provided the negative gain is −0.0771 (loss of 771 pips. The potential maximum possible profit for the 2000 candles is equal to 1.9388 (19388 pips).

It sums up all possible long positions (when $y_{i+1} > y_i$) open for one-hour period and all possible short positions (otherwise). To obtain this absolute result one should have the foreseeing knowledge like God. Note that the result from the first strategy 0.1371 is equal to about 7.1% of this theoretical maximum profit.

The comparison of all considered strategies is presented in Table 8.1. Its columns show results for learning and test periods jointly and the result for test period only. The column with Calmar ratio presents a level of risk in the strategy and

Table 8.1 Comparison of the results of different strategies in 2D spaces

Experiment	*X* axis	*Y* axis	Fin result (pips)	Fin test (pips)	Calmar	PPC
1	B_r	Y_r	250	100	3.43	0.050
2	V_r	Y_r	217	146	2.39	0.073
3	All	Y_r	1371	382	6.60	0.680
4	Pips—b1	Y	−771	−668	–	−0.385
5	Pips—b2	Y	19,388	9674	∞	9.694

column PPC shows Profit per Candle in pips. The column Finresult show the full profit reached in both learning and test periods and in column Fintest shows profit from test period only.

The Table 8.1 allows comparing three considered strategies presented above in figures and two benchmarks. In Table 8.1, "b1" denotes the first benchmark (Buy&Hold); "b2" denotes the second "ideal" benchmark with no losses. For b2, the Calmar ratio is infinity because there is no drawdown in this case. The most interesting result is #3 with all indicators from set S.

The Buy&Hold benchmark (row #4) has no chance to be competitive with our strategies because of decreasing trend in the studied period. Of course one can find another specially prepared period for the Buy&Hold strategy (with increasing trend). In contrast, the strategy proposed in this chapter should work everywhere including periods with both trends.

The second benchmark b2 is the theoretical ideal result, which absolutely no one can reach. The first and the second strategies give us small PPC to be considered as valuable strategies. Later we compare strategy #3 with the results of experiments in 3D spaces presented in the next section.

8.5 Results of Investigation in 3D Space

8.5.1 Strategy Based on Number of Events in Cubes

The next idea is exploring effective areas (cubes and cuboids) in 3D space that will represent 6-D data. Each space includes Y_r (as main axis to observe the outcome of time series) and two indicators from set $S = \{B_r, V_r, dy_r, r_r, dMA_r, dV_r\}$ in consecutive time periods.

The first attempt was made with $(Y_r, V_{r,}dMA_r)$ space. This normalized space was split into $10 \times 10 \times 10 = 1000$ cubes of size $0.1 \times 0.1 \times 0.1$. 500 events have been used to locate "circles" or arrows in these 1000 small cubes for the same period of time series as previously (from 300 to 1300 candles) separately for even and odd candles.

The simulation had shown that many cubes are empty and it makes sense to modify a condition on the minimal number of circles in a cube. In this experiment we started with the limit of 5 circles, $T_{min} = 5$. Figure 8.16 shows two marked cubes (filled circles for suggestions to open long positions and empty circles for short positions) (Fig. 8.17).

The grid resolution (1000 cubes) seems to be too fragmented for a good generalization. The next attempt was made with a new grid $5 \times 5 \times 5 = 125$ cubes of size $0.2 \times 0.2 \times 0.2$. Figure 8.18 shows the two cubes with largest asymmetry factor, which the Matlab program found.

These cubes provide the positive prediction value in terms of the confusion matrix with the accuracy

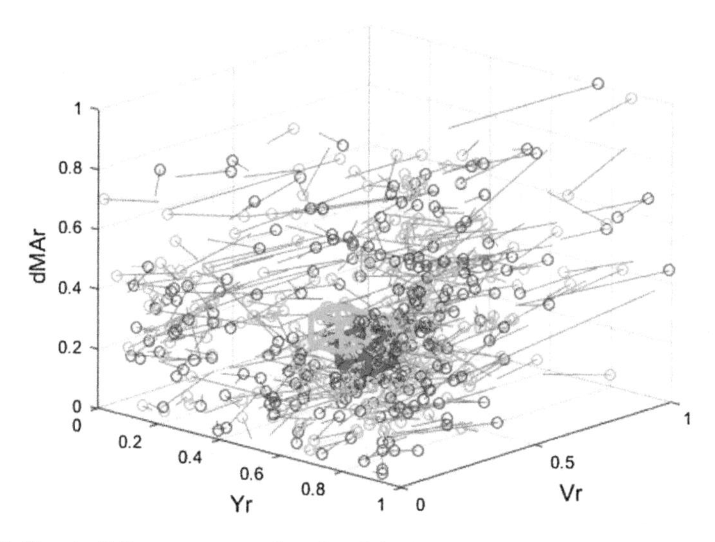

Fig. 8.16 Pins in 3-D space: two cubes found in (Y_r, dMA_r, V_r) space with the maximum asymmetry between long and short positions

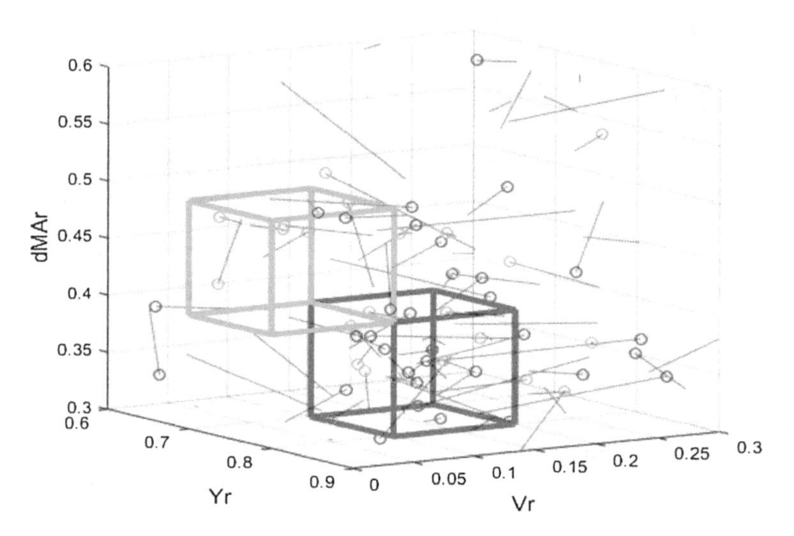

Fig. 8.17 The zoomed cubes with the best asymmetry from Fig. 8.16. The upper cube with green circles is selected for long positions lower cube with red circles is for short positions. For better visibility, the viewpoint is changed from Fig. 8.16

$$\text{TruePositive}/(\text{TruePositive} + \text{FalsePositive})$$

equal to 0.619 (for long positions) and 0.686 (for short positions) with a threshold $T_{min} = 10$ on the number of circles required in the cube .

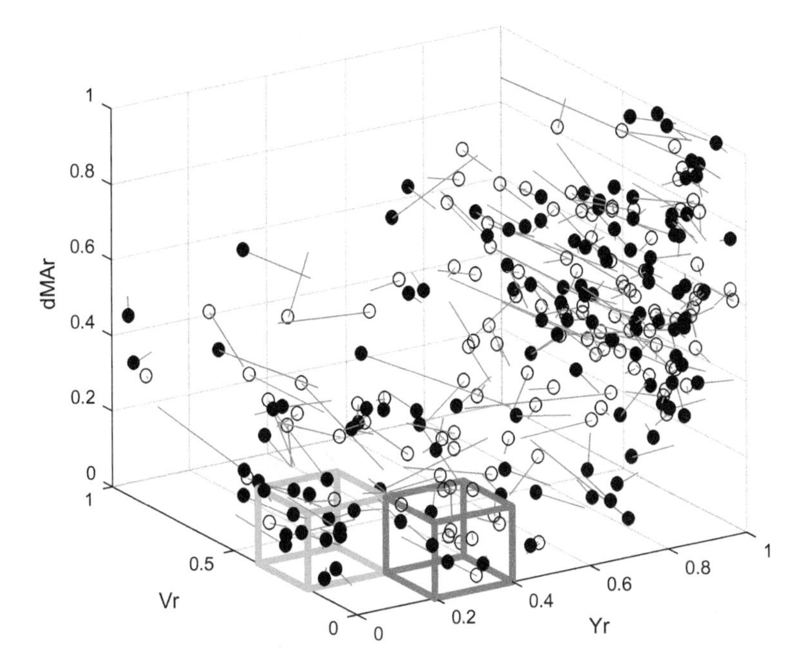

Fig. 8.18 Two determined cubes in Y_r-dMA_r-V_r space with the maximum asymmetry between long and short positions for the new grid resolution

The same algorithm was used in both grid cases ($10 \times 10 \times 10$ and $5 \times 5 \times 5$) to find cubes with maximum of criteria A_l and A_s in cubes with an additional requirement on the minimum number of the circles in the cube to be no less than T_{min}. After these two best cubes have been found in the learning period, the proper positions are opened when subsequent events are located in the cubes. Figure 8.19 presents the cumulative profit for 5000 candles in learning period and 1700 candles during testing. The presented profit is not too rewarding because of small number of positive events.

While formally here Calmar Ratio equals to infinity generally the strategy is extremely careful. For example, we can see here the period of almost 3000 h (between #1200 and #3900) with no trade. The profit of 149 pips in Fig. 8.19 is only 0.022 per candle (compare to Table 8.1) indicates the need for a more dynamic strategy.

8.5.2 Strategy Based on Quality of Events in Cubes

The next approach is an algorithm with different criteria. It uses the sum of returns accumulated in the learning period. Figure 8.6 above shows a lot of arrows with solid stems (long positions) and dashed stems (short positions) which are located

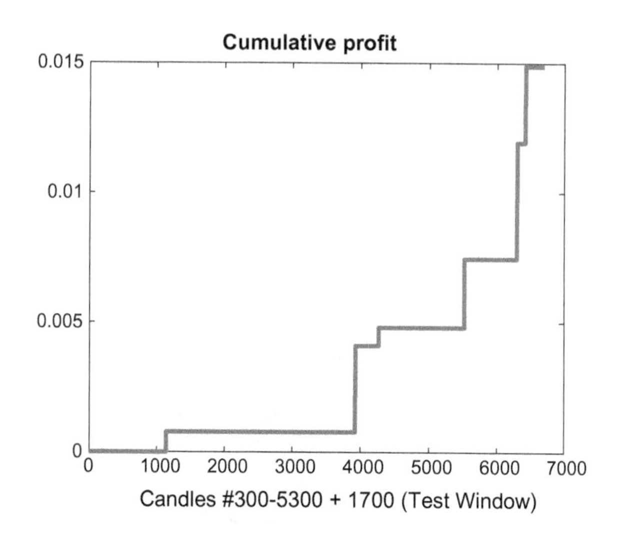

Fig. 8.19 Cumulative profit based on algorithm with maximum level of A_l or A_s

with different angles in the 2D space. The more vertical arrows are also more promising in profit (independently of the direction) due to the design of this space.

Figure 8.7 shows arrows with more vertical stems. Therefore, a new hypothesis for a potential strategy is that the more vertical arrows lead to higher profit. This is reflected in a new criterion for the learning period, which is the value of the projection of the arrows on Y_r axis only.

The projection value is $Y_{ri}-Y_{ri-1}$ that is a normalized value of Y_i-Y_{i-1}. More formally, consider a cube indexed by (k_1, k_2, k_3) in the space (P_1, P_2,Y_r) in a 3D grid, where k_1, k_2 k_3 = 1, 2, …, K. We are interested in maximum of criterion C_l for long positions and C_s for short positions:

$$C_{l\,(k_1,k_2,k_3)} = \sum (Y_{r\,i\,(k_1,k_2,k_3)} - Y_{r\,i-1\,(k_1,k_2,k_3)})$$

when $(Y_{ri}-Y_{ri-1}) > 0$ for all i that belong to a learning period and for all cubes (k_1, k_2, k_3) and

$$C_{s\,(k_1,k_2,k_3)} = \sum (Y_{r\,i\,(k_1,k_2,k_3)} - Y_{r\,i-1\,(k_1,k_2,k_3)})$$

when $(Y_{ri}-Y_{ri-1})<0$ for all i that belong to learning period and for all cubes (k_1, k_2, k_3). Recall that the beginning of the arrow (Y_{ri-1}) belongs to (k_1, k_2, k_3)-cube belongs, not its head. For each learning period, the sums $C_{l\,(k_1,k_2,k_3)}$ and $C_{s\,(k_1,k_2,k_3)}$ are computed in every (k_1, k_2, k_3)-cube.

A new investment strategy is if C_l dominates, then open a long position else open a short position. Figure 8.20 shows the bars that represent the criteria C_l and C_s (by their length) in every cube (in grid 4 × 4 × 4). The bars have different

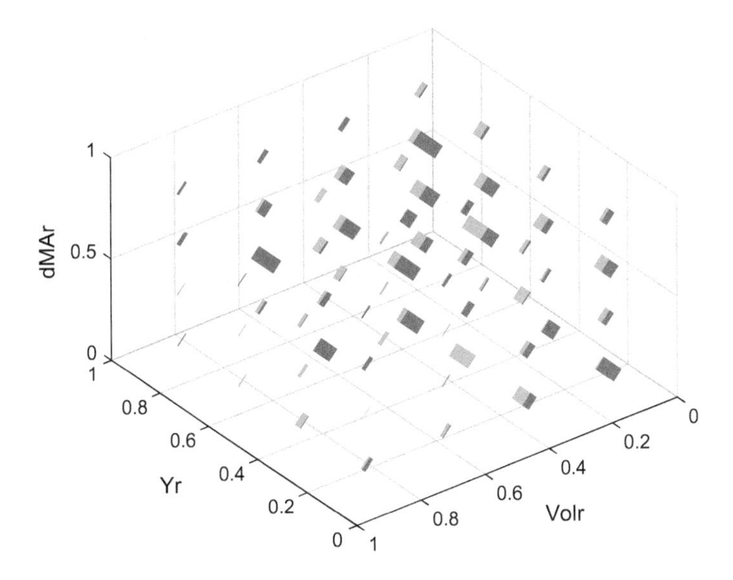

Fig. 8.20 Bars in 3D space after learning period which represent preferences to open long (green) or short (red) position

lengths. A visual strategy can be based on this difference, e.g., if one of the bars is longer than the other one, then open a proper position:

$$C_{l\,(k_1,k_2,k_3)} > C_{s\,(k_1,k_2,k_3)} + d_c \text{ then open long position},$$

$$C_{s\,(k_1,k_2,k_3)} > C_{l\,(k_1,k_2,k_3)} + d_c \text{ then open short position},$$

where d_c is additional difference for C_l and C_s to make the difference more distinct. This strategy has been checked for different values of periods of learning and testing, i.e., bars have been generated for the test data and compared with bars for the training data, because for the test period in every i-candle the (k_1, k_2, k_3) cube can be different.

Parameters (f, s, k_b) have noticeable influence on the result of every simulation. Figure 8.21 shows some curves of cumulative profits as functions of (f, s, k_b). The thicker line is the best one relative to the value of a linear combination of the cumulative profits for learning period and Calmar ratio in the same learning period. For selecting the promising curve, the following criterion is used:

$$C = w_c \cdot Calmar_{learn} + w_p \cdot profit_{learn}$$

where:
w_c is a weight of Calmar component (in these experiments $w_c = 0.3$);
w_p is a weight of profit component (in these experiments $w_p = 100$ for 500 h of learning period);

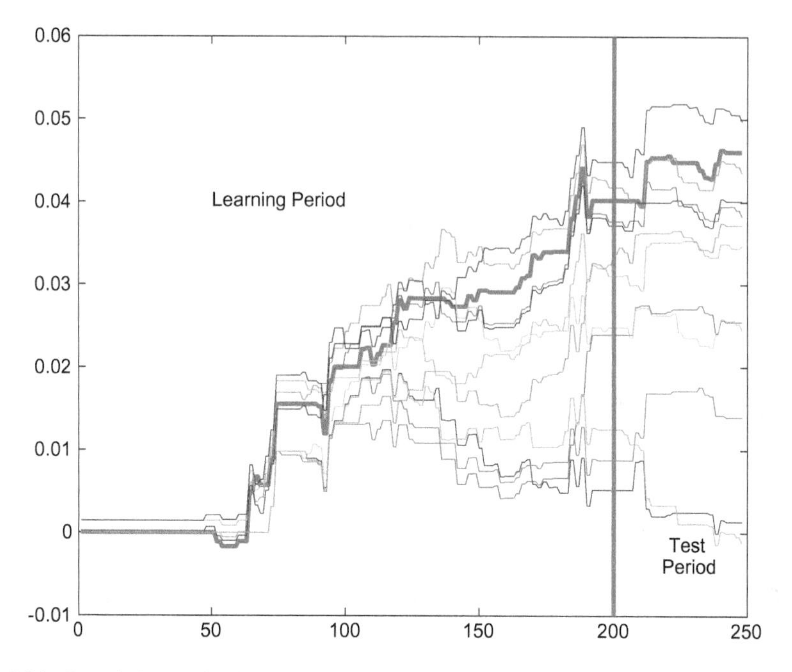

Fig. 8.21 Cumulative profits for different values of parameters03D5 (f, s, k_b)

$Calmar_{learn}$ is the Calmar ratio at the end of learning period;
$profit_{learn}$ is a cumulative profit at the end of the learning period measured as a change in EURUSD rate.

The main idea in constructing the criterion C is to balance Calmar and Profit contributions. We tried to examine some pairs of weights $(w_{c,}\ w_p)$ but this one mentioned above gave us the result we can see below. Here the principles of heuristic were applied—the approach is simple and the results are satisfactory. Figure 8.21 presents the results of simulations for learning periods of 200 candles and test periods of 50 candles.

Figure 8.21 shows that while the thicker line (best in criterion C) provides one of the top results (profits), it does not provide the best cumulative profits at the end of learning period due to the weighting nature of criterion C that also takes into account the Calmar ratio.

Figures 8.22, 8.23, 8.24, 8.25, 8.26 and 8.27 show cumulative profits for different learning and testing periods as a function of 20 shifted datasets (cycles).

The design of these cycles is illustrated in Fig. 8.27. First the best values of parameters (f, s, k_b) were found by optimizing them on training data and then these values were used in the test period which follows the learning period.

The charts in Figs. 8.22, 8.23, 8.24, 8.25 and 8.26, but Fig. 8.25, show very efficient result taking into account the profit-risk relation.

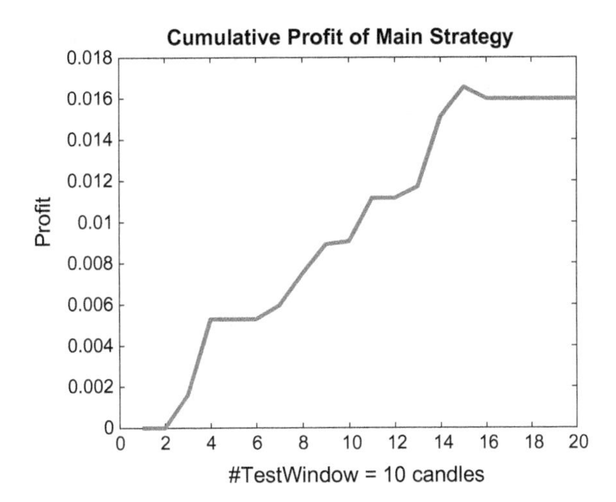

Fig. 8.22 Cumulative profit for main strategy with learning windows of 100 1h-candles and test widows of 10 1h-candles (Calmar ratio ~ 29, PPC = 0.80)

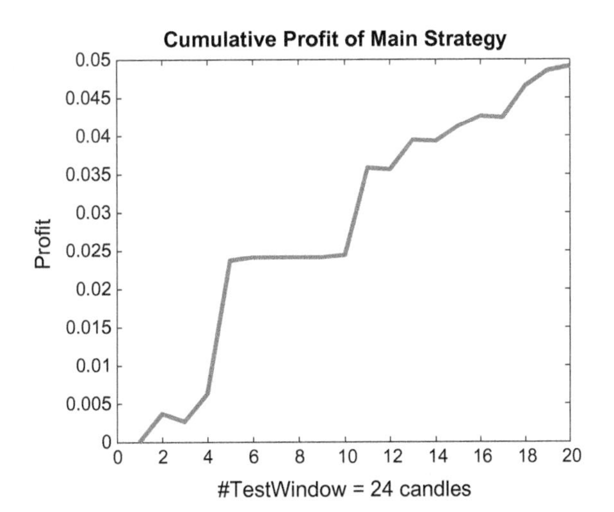

Fig. 8.23 Cumulative profit for main strategy with learning windows of 100 1h-candles and test widows of 24 1h-candles (Calmar ratio ~ 18, PPC = 0.87)

Profit is represented by its PPC value. Practitioners consider the profit at the level of 10–20 pips per day as very high or even unrealistic to be stable (Krutsinger 1997; Pasche 2014). The general conclusion is that the chosen space (Y_r, V_r, dMA_r) is very efficient.

Figures 8.22, 8.23, 8.24, 8.25, 8.26, 8.28 and 8.29 are to show general direction of changes in cumulative profit. It is important that y-axes have different scales and should not be compared directly.

Figure 8.27 shows how a set of learning periods have been generated for selecting a period that provides the best result at its end. At every learning period, multiple simulations were run with the different parameters. The parameters of the

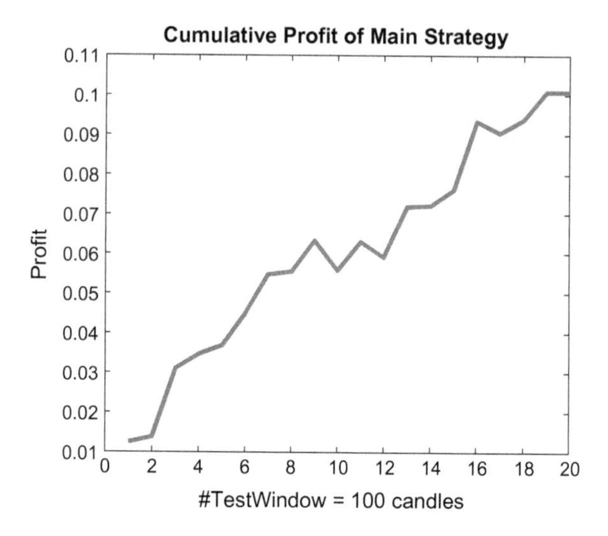

Fig. 8.24 Cumulative profit for main strategy with learning windows of 500 1h-candles and test widows of 100 1h-candles (Calmar ratio ~ 13, PPC = 0.50)

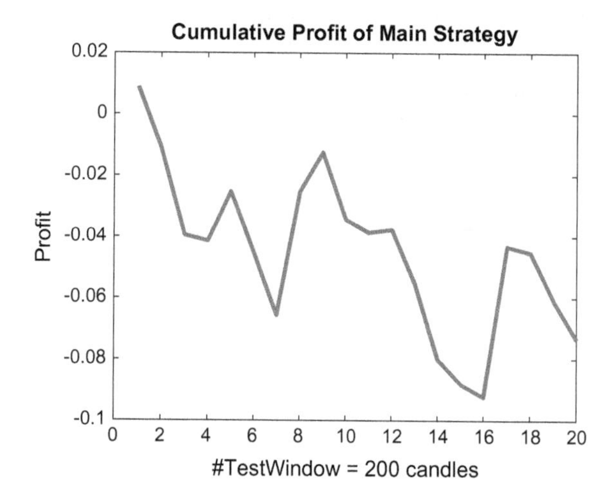

Fig. 8.25 Cumulative profit for main strategy with learning windows of 1000 1 h-candles and test widows of 200 1h-candles (Calmar ratio ~ −0.72. PPC = −0.18)

winning learning period are used in the test period to evaluate the efficiency of the optimized parameters in the investment strategy.

8.5.3 Discussion

The chosen space $(Y_r, V_{r,} dMA_r)$ is the most efficient from explored options from set $S = \{B_r,\ V_r,\ d_{MAr},\ r_r,\ d_{yr}, dV_r\}$. Below we discuss characteristics of this space to clarify the impact of parameters f, s and k_b on profit and risk stabilization. Many

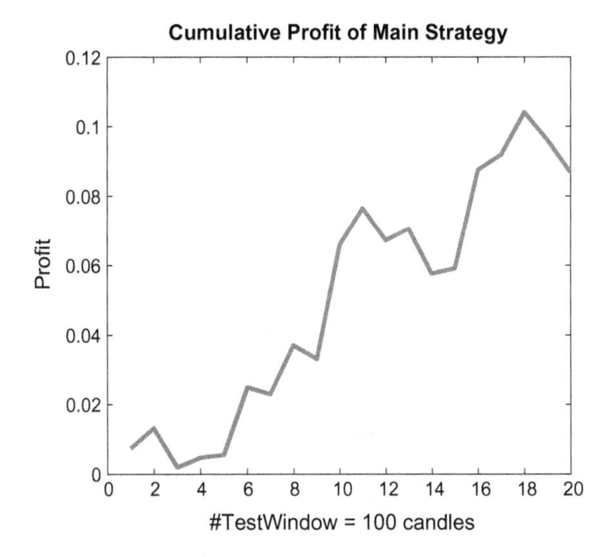

Fig. 8.26 Cumulative profit for main strategy with learning windows of 1000 1h-candles and test widows of 100 1h-candles (Calmar ratio ~ 4.71, PPC = 0.43)

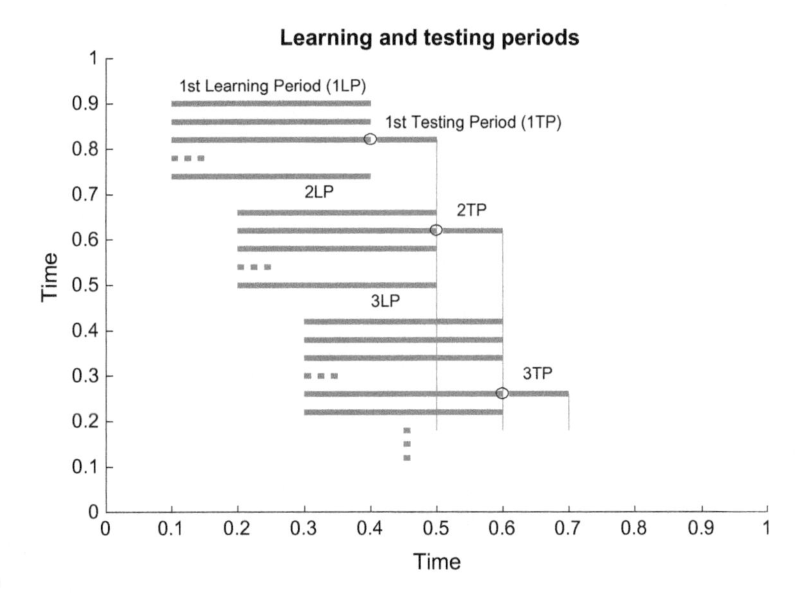

Fig. 8.27 Selection of learning and testing periods

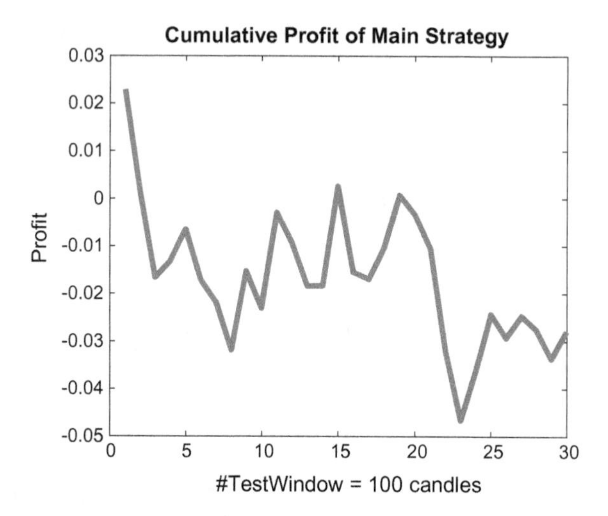

Fig. 8.28 Cumulative profit for main strategy with learning windows of 200 1h-candles and test widows of 100 1h-candles (Calmar ratio = 0.39) for constant parameters, $f = 6$; $s = 20$; $k_b = 150$

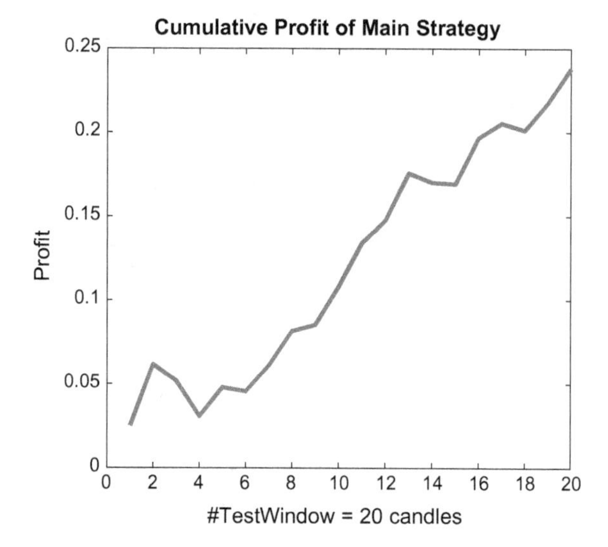

Fig. 8.29 Cumulative profit for the strategy in 1d EURUSD time series with 5-days testing windows and 30-days learning windows in 20 periods, Calmar ratio = 7.69; PPC = 23.83 pips; delta = 0.05

experienced traders try to find the best universal values for their indicators (Krutsinger 1997; Bingham 2014; Fong et al. 2005; Cheng et al. 2012; Mehta and Bhattacharyya 2004; Fong and Yong 2005; Wilinski and Zablocki 2015).

Figure 8.28 shows the cumulative profit for one of our simulations with constant values of these parameters, $f = 6$, $s = 20$, $k_b = 150$ for the pair learning/test periods 200/100 of one-hour candles.

While the chart in Fig. 8.28 starts with values of the parameters that are optimal for the first learning period with high profit value, it shows the significant profit decline for the next periods.

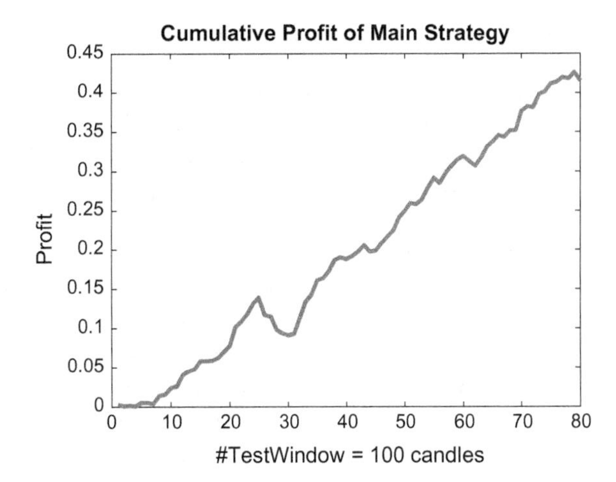

Fig. 8.30 Long-term simulation with new data (up to 2017) with learning period of 500 1h candles and test period of 100 candles. End result (profit): 4151 pips, Calmar ratio = 8.71 and PPC = 0.51)

This leads us to the conclusion that the desire for stable "universal" parameters has no ground. The markets are under permanent volatility.

Multiple figures above present examples of cumulative profit visualizations. Even a novice usually can estimate a quality of different strategies looking at these visualizations, For instance, the advantage of the curve in Fig. 8.24 over the curve in Fig. 8.28 is evident without any formal measurement. Next we explored other aspects of the method by checking the strategy on more recent data with newest candles and different sampling. Figure 8.29 shows simulation with one-day candles.

The result is also very positive. Additionally we can notice that the learning period is equal to about one month and the testing period is a one-week period (five trading days). These are very convenient circumstances for both automatic and manual trading.

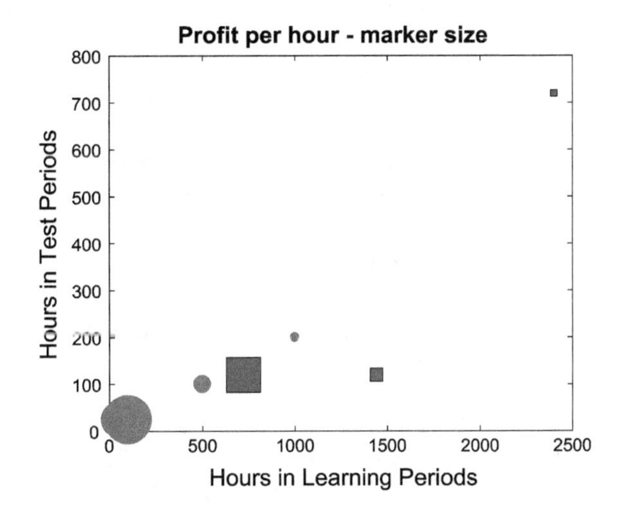

Fig. 8.31 Results of simulations for different ratios of learning/testing periods. The circles present the result for simulations in one-hour time series and the squares present the result for one-day series

The last experiment was carried out on a much larger periods of data of 80 cycles (windows of shifted sets of training and testing data) with each learning period of 500 one-hour candles and each test period of 100 one-hour candles (Fig. 8.30). This Figure shows that the cumulative profit has one period within 25–30 cycles with declining values of profit, but the total results are very positive. While we use measurable objective criteria of success (Calmar ratio and PPC) human perception of success is also important to have confidence in the strategy and the curve in Fig. 8.30 serves this goal.

Figure 8.31 shows some simulation results in a symbolic form of circles and squares of the sizes that depend on PPC.

This figure visualizes the difference in simulation results for one-hour and daily time series, where circles denote one-hour time series and squares denote daily time series. The X coordinates is hours in a learning period and Y coordinate is hours in a test period. The best result for one-hour series (100/24 h ratio between learning and test periods) is a red circle in the left-bottom corner of Fig. 8.31. For the time series of one-day candles (shown as squares) the positive results take place for ratios of days such as 100/24, 50/20, 150/24. In Fig. 8.31 these ratios are shown in hours, e.g., the point 30/5 days is shows as a large blue square for 720/120 h.

8.6 Conclusion

Many parts of the chapter show the inspiring power of CPC visualization in both 2D and 3D spaces of features that represent time series data. This power is coming from the two levels of the approach in searching the best conditions to build the investment strategy. The first level involves examining the best 4D and 6D coordinate systems to build 2D or 3D visualization spaces. The second level involves learning parameters of attributes in each selected space.

A key role of the CPC approach in visualization of 4D and 6D points as arrows in 2D and 3D was in helping to find the best locations (squares in 2D or cubes in 3D) to open long or short positions, respectively. It is shown that the CPC method allows guiding exploration and machine learning for improving the search for the best local combinations of predictive features. For instance, further exploration will benefit from adding Kelly's criterion (Nekrasov 2014) to select the best squares (or cubes) in CPC spaces. This criterion allows taking into account together the quantity (number) of the arrows in the local places and their quality (lengths of projections). This is consistent with our much better results in the 3D distributions for $5 \times 5 \times 5$ or $4 \times 4 \times 4$ grids with a criterion based on length of bars (Fig. 8.20). It allowed us to increase a frequency in simulated trading with a much greater number of open positions in Sect. 8.5, than in the previous method, based on circles in subcubes in Sect. 8.2.

Future challenges include introduction of deeper relations between learning and testing periods beyond the ratio of their lengths, and the implementation of the method in a trading platform and expanding to other investment tasks.

In general, in the investment domain, we want to solve a two-criterion task: max of profit, *p*, and min of risk, *r*. It is obvious that these criteria contradict each other quite often if not all the time. Thus, this is a mathematically ill-posed problem. The common ways in multi-objective optimization to resolve this issue is to find a Pareto frontier or to combine two criteria to one criterion. Calmar ratio implements the second option with an intuitive idea behind it as follows. The component that must be maximized must be a dividend and the component that must be minimized must be a divisor. In this way if we have two pairs (p_1, r_1) and (p_2, r_2) such that $p_1 > p_2$ and $r_1 < r_2$ then Calmar ratio of (p_1, r_1) will be greater than it is for (p_2, r_2). In this case Calmar ratio is consistent with Pareto frontier. However if $p_1 > p_2$ and $r_1 > r_2$ then Calmar ratio of (p_1, r_1) can or cannot be greater than it is for (p_2, r_2). It will depend of actual values of these 4 numbers. In essence, Calmar ratio sets up the order in the Pareto frontier that can be questionable. It can be resolved by analysis of differences of actual pairs (p_i, r_i) from the Pareto frontier. It is commonly done by visualizing the Pareto frontier in this 2-D case of two criteria. In fact, a 2-D case is quite limited and exploration of Pareto frontier in higher dimensions will lead to deeper models. Parallel Coordinated (a special case of GLC) are commonly used to visualize the multidimensional Pareto frontier, but they quickly lead to high occlusion. General Line Coordinates are especially promising way to accomplish such Pareto frontier analysis when the number of criteria is more than two. The issues of Pareto frontier are considered in Chap. 11.

This chapter illustrates the potential of a new emerging joint area of research and application of n-D visual discovery and investment strategies to boost the creativity of both scientists and practitioners.

References

Bingham, N.H.: Modelling and prediction of financial time series. Commun. Stat. Theor. Methods **43**(7), 1351–1361 (2014)

Cheng, C., Xu, W., Wang, J.: A comparison of ensemble methods in financial market prediction. In: Proceedings of the 2012 5th International Joint Conference on Computational Sciences and Optimization, CSO 2012, pp. 755. (2012)

Fong, W.M., Yong, L.H.: Chasing trends: recursive moving average trading rules and internet stocks. J. Empir. Finance **12**(1), 43–76 (2005)

Guo, Z., Wang, H., Liu, Q., Yang, J.: A feature fusion based forecasting model for financial time series. PLoS ONE **9**(6) (2014)

Hellwig, Z.: On the measurement of stochastical dependence. Applicationes Mathematicae **10**(1), 233–247 (1969)

Hoffman, A.J.: Combining different computational techniques in the development of financial prediction models. In: NCTA 2014—Proceedings of the International Conference on Neural Computation Theory and Applications, p. 276

Keim, D., Mansmann, F., Schneidewind, J., Thomas, J., Ziegler, H.: Visual analytics: scope and challenges. In: Visual Data Mining, pp. 76–90. (2008)

Kovalerchuk, B., Vityaev, E.: Data Mining in Finance: Advances in Relational and Hybrid Methods, 325 p. Kluwer/Springer (2000)

Kovalerchuk B., Perlovsky L., Wheeler G.: Modeling of phenomena and dynamic logic of phenomena, J. Appl. Non-classical Logics **22**(1): 51–82 (2012)

Krutsinger, J.: Trading Systems: Secrets of the Masters, p. 242. McGraw-Hill, New York (1997)

Li, X., Deng, Z., Luo, J.: Trading strategy design in financial investment through a turning points prediction scheme. Expert Syst. Appl. **36**(4), 7818–7826 (2009)

Lian, W., Talmon, R., Zaveri, H., Carin, L., Coifman, R.: Multivariate time-series analysis and diffusion maps. Sig. Process. **116**, 13–28 (2015)

Main, R.: Evaluating traders' performers with the Calmar ratio. www.proptradingfutures/the-calamr-ratio/ (2015). Access Jan 2017

Martin, A.D.: Technical trading rules in the spot foreign exchange markets of developing countries. J. Multinational Financ. Manag. **11**(1), 59–68 (2001)

Mehta, K., Bhattacharyya, S.: Adequacy of training data for evolutionary mining of trading rules. Decis. Support Syst. **37**(4), 461–474 (2004)

Nekrasov, V.: Kelly criterion for multivariate portfolios: a model-free approach (30 Sept 2014). Available at SSRN: https://ssrn.com/abstract=2259133. Access Jan 2017

Pasche, R.: How many pips should we target per day? DailyFX, 8 July 2014, www.dailyfx.com (2014). Access Jan 2017

Wichard, J. D., & Ogorzalek, M. (2004, July). Time series prediction with ensemble models. In Neural Networks, 2004. Proceedings. 2004 IEEE International Joint Conference on (Vol. 2, pp. 1625-1630). IEEE

Wilinski, A., Zablocki, M.: The investment strategy based on the difference of moving averages with parameters adapted by machine learning. In: Advanced in Intelligent Systems and Computing, vol. 342, pp. 207–227. Springer, Cham, Heidelberg, New York (2015)

Wilinski, A., Bera, A., Nowicki, W., Błaszynski, P.: Study on the effectiveness of the investment strategy based on a classifier with rules adapted by machine learning. ISRN Artif. Intell. **2014**:10 pages (Article ID 451849) (2014). https://doi.org/10.1155/2014/451849

Young, W.T.: Calmar ratio: a smoother tool. Futures (magazine), October 1991. (1991)

Chapter 9
Visual Text Mining: Discovery of Incongruity in Humor Modeling

All intellectual labor is inherently humorous.
George Bernard Shaw

9.1 Introduction

Garden path jokes. This chapter presents a visual text mining approach to modeling humor within text. It includes algorithms for visualizing and discovering shifts in text interpretation as intelligent agents parse meaning from garden path jokes. Garden path jokes (Dynel 2012) can occur when a reader's initial interpretation of an ambiguous text turns out to be incorrect often triggering a humorous response.

We describe three successful approaches to text visualization conducive to identifying distinguishing features given humorous and non-humorous texts. These visualization methods include Collocated Paired Coordinates defined in Chap. 2, Heat maps, and two-dimensional Boolean plots (Kovalerchuk and Delizy 2005; Kovalerchuk et al. 2012).

This methodology and tools offer a new approach to testing and generating hypotheses related to theories of humor as well as other phenomena involving incongruity-resolution and shifts in interpretation including non-verbal humor.

While many theories of humor agree that humor often involved the detection of incongruities and their resolution, the details remain vague and there is no agreed upon theoretical framework, which describes how these incongruities are formed and detected by intelligent agents (Ritchie 1999).

Humor modeling and natural language understanding. While visual text/data mining and machine learning have been extensively used in many domains (Kovalerchuk and Schwing 2005; Simov et al. 2008), the modeling humor is a new area for these methods. The presented approach visualizes shifts in meaning assignment over time as jokes are processed to deeper understand the specific mechanisms underlying humor.

Furthermore, these approaches can be used to model and detect other forms of humor, in particular sequential physical humor in nonverbal settings and other phenomena involving shifts of interpretation. In general, visualization and visual

B. Kovalerchuk, *Visual Knowledge Discovery and Machine Learning*,
Intelligent Systems Reference Library 144,
https://doi.org/10.1007/978-3-319-73040-0_9

text mining give us more tools for detecting features associated with various natural language phenomena. A major challenge in natural language understanding and humor modeling is automated recognition of *shifts in the* meaning for an ambiguous word within the text.

Correlation-based measure. The visualization approaches presented in this chapter use *correlation-based measures* to assign the meaning of ambiguous words. These measures incorporate:

- the *context* of the ambiguous word in different parts of a surface level text and
- *relations* associated with different meanings of that word as defined in an *ontology* at a deeper level.

These measures represented as *meaning correlation scores* capture opposing meanings of the part of the joke. They form a 4-D space of measures with each joke represented as a 4-D point in that space. Then these 4-D points are visualized in CPCs, heat maps and Boolean plots.

Visualization. The CPC lets us visually see shifts of meaning in jokes when compared with non-jokes. The heat maps color code the differences of meaning given different time steps. The resulting heat maps of jokes distinguishable from that of non-jokes with respect to these meaning correlation differences. Finally, the third Boolean visualization displays in two dimensions an entire model 4-D space. It consists of Boolean vectors that describe meaning correlation over time. The set of jokes and non-jokes, plotted in this space, allows an analyst to see the *boundary* between what is a joke and a non-joke.

To show the power of this visual approach we compare the results with traditional analytical data mining/machine learning approaches on the same data with the same features.

This chapter includes the construction of an informal *ontology* using *web mining* to identify semantic relations to visualize jokes and non-jokes. While improvements can be made, the results are encouraging.

Related Work. Both Computational Humor and Text Visualization as fields have seen extensive activity lately, but tend to work on separate topics. Computational Humor deals a lot with the modeling and detection of incongruities within text and many attempts have recently been made attempting to detect or generate jokes using computers (Labutov and Lipson 2012; Mihalcea et al. 2010; Petrovic and Matthews 2013; Ritchie 2003; Taylor and Mazlack 2007; Valitutti et al. 2013; Taylor and Raskin 2012), but no attempt focused on visualization has been made. On the other hand, studies on Text Visualization tend to focus on other topics such as identifying the central topic within a text.

9.2 Incongruity Resolution Theory of Humor and Garden Path Jokes

In this chapter, we use the Fishtank joke '*Two fish are in a tank. One looks to the other and asks: How do you drive this thing*?' to illustrate the approach. Many predominant theories of verbal humor state that humor is triggered by the detection and resolution of incongruities (Ritchie 1999; Schultz and Horibe 1974). The dictionary defines 'incongruous' as lacking harmony of parts, discordant, or inconsistent in nature. During the parsing of a text incongruities form when a reader's interpretation of some concept conflicts with the other possible interpretations, as the text is read.

Below we focus on a particular humor subtype where there is a shift from some interpretation to an opposing one. Dynel calls these jokes "garden path" jokes using the garden path metaphor of being misled (Dynel 2012), while other theorists use the terminology of 'forced reinterpretation' and 'frame shifting' (Ritchie 1999). These jokes are *sequential* in nature and describe a certain pattern of incongruity and resolution. With a garden path joke, readers first establish some interpretation A as they read the first part of a joke, called the *setup*. Then, given new evidence included in the second part, called the *punchline*, readers must discard this interpretation and establish a new interpretation B. The fishtank joke above displays such joke, and an incongruity. The reader initially interprets the *tank* to be an *aquarium*, but given additional information, the alternative meaning of a *vehicle* becomes possible and probable.

Incongruities often arise when 'opposing' or 'mutually exclusive' elements simultaneously occur. Different word meanings oppose when a tank is a vehicle, it is not an aquarium. Opposition occurs in many other areas, e.g., when something is hot it is not cold. To model this we visualize changes of correlation in the context established for mutually exclusive meanings. Similar meanings have similar contexts according to the distributional hypothesis (McDonald and Ramscar 2001; Sahlgren 2008). Next we discuss the approach to establish meaning of words and how we are identifying correlation between the meaning representations.

Dynamic model Ritchie (2014) emphasized the importance of adding time to models of jokes making them dynamic. A deeper semantic analysis is required to be able to model jokes adequately (Taylor and Raskin 2013). Garden path jokes consist of two parts that we denote P_1 and P_2. Respectively we denote the text with two parts as $T = (P_1, P_2)$. Model M described in Table 9.1 involves six consecutive moments t_1–t_6 that involve parts consecutive parts P_1 and P_2.

Table 9.1 Incongruity process for model M

Time	Description
t_1	Agent G (human or a software agent) reads the first part of the text P_1 and concludes (at a *surface* level) that P_1 is a usual text
t_2	Agent G reads the second part of the text P_2 and concludes that (at a *surface* level) that P_2 is a usual text
t_3	Agent G starts to analyze a *relation* between P_1 and P_2 (at a *deeper* semantic level). Agent G retrieves semantic features (words, phrases) $F(P_1)$ of P_1
t_4	Agent G retrieves semantic features (words, phrases) $F(P_2)$ of P_2
t_5	Agent G compares (correlates) features $F(P_1)$ with P_2 and features $F(P_2)$ with P_1 finding significant differences in meaning (*incongruity*)
t_6	Agent G reevaluate usuality of P_2 taking into account these correlations and concludes that P_2 is unusual

9.3 Establishing Meanings and Meaning Correlations

We chose a vector representation of meaning, based on the *frequency* at which words occur in the context of some target word. This is a common approach taken by a number of researchers in the past for dealing with meaning (Mikolov et al. 2013). These vectors of word associations form an informal ontology describing entities and their relations. The material used to build these vectors was retrieved via a *web search* in line with our previous work (Galitsky and Kovalerchuk 2014).

9.3.1 Vectors of Word Association Frequencies Using Web Mining

Below we consider some ambiguous word *A* with a number of possible meanings $AM_1 \ldots AM_n$ and different parts $P_1, \ldots, P_m$ of some text containing the ambiguous word *A*. For each *meaning AMx* we establish a *set of disambiguating keywords K* (*AMx*), which uniquely identify that meaning. While we hand-chose our keyword sets these can be automated using a variety of resources such as Wordnet.

The keywords *K*(*AMx*) are used as a *query* for a search engine (e.g., Google) to retrieve the top *n* documents. Let *D*(*q*, *n*) be a *search function* which retrieves *n* documents relevant to some query *q*. The resulting document set for some meaning *AMx* is thus designated *D*(*K*(*AMx*), *n*). For short we will also write *D*(*x*,*n*) for this set of documents when the word *A* is clear from the context.

Next, we compute frequencies of all words occurring within distance *j* of *A* given the document set *D*(*x*,*n*). Let *F* (*A*, *j*, *x*, *n*) be a *vector of word frequencies*, where *F* is a function that returns a vector of word frequencies. We use the approach known as *term-frequency times inverse document frequency* approach (*TF·IDF*) (Rajaraman and Ullman 2011) for identifying word frequencies *F*. The value of *F* for a given term (word) *w* in the document set *D*(*x*,*n*) is as follows:

$$F_w = TF_{wD} \times IDF_w$$

where

$TF_{wD} = f_{wD} / max_k f_{kD}$,

f_{wD} is the number of occurrences of term (word) *w* in document set *D(x,n)*,

$max_k f_{kD}$ is the max of the number of occurrences of any word in the document set *D(x,n)*,

$IDF_w = log_2(n/n_w)$,

n_w is the number of times the term *w* appears in *D(x,n)*.

The terms with the highest *TF·IDF* score are often the terms that best characterize the topic of the document (Rajaraman and Ullman 2011).

F (*A*, *j*,*x*, *n*) represents the meaning for *AMx* as a set of word association frequencies, or in other words its *contexts*. These frequencies are ordered by the *lexicographic order* of the words. Note that we include the frequency of the given word A itself though the variants without it being explored too.

In a similar fashion we establish *semantics* for the ambiguous word A given the different parts $P_1, \ldots, P_m$ of some text containing A. We denote them as $F(A, j, P_1, n), \ldots, F(A, j, Pm, n)$ where $F(A, j, P_i, n)$ is a vector of frequencies of P_i within distance j from A in the top n documents, which contain a selected part of P_i or a whole P_i.

9.3.2 Correlation Coefficients and Differences

Correlation coefficients. Denote the meaning of A given a search result for a phrase P_i that contains A as AP_i. We are interested in correlation of AP_i with each of its meanings $AM_1, \ldots, AM_n$.

The first meaning AP_i is formalized by a vector of frequencies $F(A,j,P_i,n)$. Respectively, the meanings AM_x are formalized as vectors of frequencies $F(A,j,x, n)$. Having these vectors, we can compute the correlation coefficients between them

$$C(F(A, j, P_i, n), F(A, j, x, n))$$

Each joke in our dataset is a two-part joke with parts P_1 and P_2 in which two meanings are invoked as vectors $F(A, j, P_1, n)$ and $F(A, j, P_2, n)$. Given two meanings AMx and AMy of the ambiguous word A and some statement with parts P_1 and P_2 that refer to A, we calculate the following correlation scores.

Given P_1 (part one of the given text):

$C_{1x} = C(F(A, j, P_1, n), F(A, j,x, n))$ is a correlation of meaning AP_1 with meaning AMx,

$C_{1y} = C(F(A, j, P_1, n), F(A, j, y, n))$ is a correlation of meaning AP_1 with meaning AMy.

Given P_2 (part two of the given text):

$C_{2x} = C(F(A, j, P_2, n), F(A, j, x\ n))$ is a correlation of meaning AP_2 with meaning AMx,

$C_{2y} = C(F(A, j, P_2, n), F(A, j, y, n))$ is a correlation of meaning AP_2 with meaning AMy.

Differences of correlation coefficients. Finally, we calculate differences between the correlation coefficients, which are useful for joke classification as they describe correlation movement patterns. For example, the difference between C_{1x} and C_{1y} tells us which meaning x or y has greater correlation with part P_1 of the joke. Similarly, the difference between C_{2x} and C_{2y} tells us which meaning x or y has greater correlation with part 2. Thus, if $C_{1x} - C_{1y} > 0$ then the meaning x is more relevant to P_1.

On the other hand, the difference between C_{1x} and C_{2x} tells us if a correlation of meaning x has increased or decreased when moving from part 1 to part 2. of some text. If $C_{1x} - C_{2x} > 0$ then the correlation of meaning x has decreased as the text is read in, while if $C_{1x} - C_{2x} < 0$ then it has increased. Such changes indicate the shift in the meaning and incongruity.

Features from correlation coefficients. We define four Boolean variables $u_1 - u_4$ using these differences:

$$
\begin{aligned}
u_1 &= 1, \mathit{if}\ C_{1x} > C_{1v}, \text{else}\ u_1 = 0 \\
u_2 &= 1, \mathit{if}\ C_{1x} > C_{2y}, \text{else}\ u_1 = 0 \\
u_3 &= 1, \mathit{if}\ C_{1v} < C_{2v}, \text{else}\ u_3 = 0 \\
u_4 &= 1, \mathit{if}\ C_{2x} < C_{2v}, \text{else}\ u_4 = 0
\end{aligned}
$$

Example In order to concentrate on the issue at hand, i.e. modeling incongruity, many jokes are simplified. Respectively, we consider a distilled version of the two-part garden path joke J with parts

$$P_1 = \text{‘fish in tank'}, P_2 = \text{‘they drive the tank'}$$

that include the ambiguous word A = ‘*tank*’.

Let $tankM_1$ and $tankM_2$ be the two meanings invoked at different points while reading J,

$$tankM_x = \text{‘aquarium'}, tankM_y = \text{‘vehicle'}$$

Let $K(tankM_x)$ = [“aquarium”, “tank”] and

$$K(tankM_y) = [\text{“vehicle"}, \text{“panzer"}, \text{“tank"}].$$

. Then we compute meaning vectors F for different meanings of ‘tank’ using data from four web searches for P_1, P_2, $K(tankMx)$ and $K(tankMy)$ and the correlation coefficients between these meaning vectors.

Meaning correlation coefficients given P_2:

$$
\begin{aligned}
C_{1x} = C(&F(\text{tank}, 5, \text{fish in a tank}, 10), \\
&F(\text{tank}, 5, \{\text{aquarium}, \text{tank}\}, 10)) = 0.824
\end{aligned}
$$

$$
\begin{aligned}
C_{1y} = C(&F(\text{tank}, 5, \text{fish in a tank}, 10), \\
&F(\text{tank}, 5, \{\text{vehicle}, \text{panzer}, \text{tank}\}, 10)) = 0.333
\end{aligned}
$$

Meaning correlation coefficients given P_2:

$$
\begin{aligned}
C_{2x} = C(&F(\text{tank}, 5, \text{drive the tank}, 10), \\
&F(\text{tank}, 5, \{\text{aquarium}, \text{tank}\}, 10)) = 0.389
\end{aligned}
$$

$$
\begin{aligned}
C_{2x} = C(&F(\text{tank}, 5, \text{drive the tank}, 10), \\
&F(\text{tank}, 5, \{\text{aquarium}, \text{tank}\}, 10)) = 0.389
\end{aligned}
$$

Over the course of a garden path joke there should be a switch in dominant meaning. Given part 1, correlation with meaning x, C_{1x} should be greater than C_{1y} and given part 2 C_{2y} correlation with meaning y should be greater than C_{2x}. The correlation coefficients above capture these shifts, 0.824 > 0.333 and 0.573 > 0.389. The visual analysis of this example is presented later in next sections.

9.4 Dataset Used in Visualizations

Two-part jokes of garden path form that contain lexical ambiguities have been collected and converted into a simple form by hand, as we want to model incongruity rather than focusing on other issues related to parsing text.

Algorithmically selecting relevant parts of text P_1 and P_2 from longer texts that contain a lot of additional material is a valid approach, but outside the scope of this study. Therefore, "Two fish are in tank" becomes "a fish in a tank." as the number of fish has little to do with the lexical ambiguity involved in the incongruity we model.

For each joke, a non-joke of similar form was created. It contains the same first part, but a different non-humorous second part with minimal change, usually only a noun or verb, to preserve the structure of the statement. The following are some examples of jokes and non-jokes contained in the dataset.P_1: Two fish are in a tank. P_2: They drive the tank.P_1: Two fish are in a tank.P_2: They swim in the tank.

Meaning x search query: 'Aquarium tank'

Meaning y search query: 'Panzer tank'P_1: No charge said the bartender.P_2: To the neutron.P_1: No charge said the bartender.P_2: To the customer.

Meaning x search query: 'Cost charge'

Meaning y search query: 'Electron charge.'

9.5 Visualization 1: Collocated Paired Coordinates

CPC setup. The first visualization method used for humor data is Collocated Paired Coordinates described in Chap. 2, and illustrated in several consecutive chapters. Each text from the dataset with two parts P_1 and P_2 is represented as a 4-D point $\mathbf{z} = (z_1, z_2, z_3, z_4)$ in the same way as in the fishtank example above. Then odd coordinates X_1 and X_3 are mapped to the Cartesian coordinate X and even coordinates X_2 and X_4 are mapped to the Cartesian coordinate Y for visualization in 2-D.

In CPC each 4-D point is a directed graph (arrow) drawn in (X, Y) coordinates with start node (z_1, z_2) and end node (z_3, z_4). Respectively, the X-axis shows the correlations with the first meaning and the Y-axis shows the correlation with the second meaning. This allows us to visualize correlation patterns.

The arrow from point (z_1, z_2) to point (z_3, z_4) represents the time of parsing and understanding the text. At the beginning, a reader:

(1) reads part P_1,
(2) correlates P_1 with two opposing meanings x and y of word A from P_1, and
(3) selects the most correlated meaning of A from x and y with larger C value. Next, the reader does the same steps (1)–(3) for part P_2.

A garden path jokes involves a shift from one meaning to another one, e.g., from x in P_1 to y in P_2 or vice versa. This means that these jokes should form a line (arrow) moving away from one axis and towards another as the meaning correlation score C for one meaning lessens and other meaning increases.

Results. In visualization in Fig. 9.1, the X-axis represents coordinate X_1 that corresponds to the highest correlations computed for P_1. The Y-axis represents coordinate X_2 that shows the lowest correlations computed for P_1. Formally, this means for each joke $\mathbf{z} = (z_1, z_2 . z_3, z_4)$ that

$$z_1 = \max(C_{x1}, C_{y1}), \ z_2 = \min(C_{x1}, C_{y1}) \tag{9.1}$$

$$\text{If } C_{x1} = \max(C_{x1}, C_{y1}) \text{ then } (z_3 = C_{x2}, \ z_4 = Cy_2), \text{ else } (z_3 = Cy_2, \ Z_4 = C_{x2}) \tag{9.2}$$

Thus, the arrows should all move in the same direction as a meaning shift occurs.

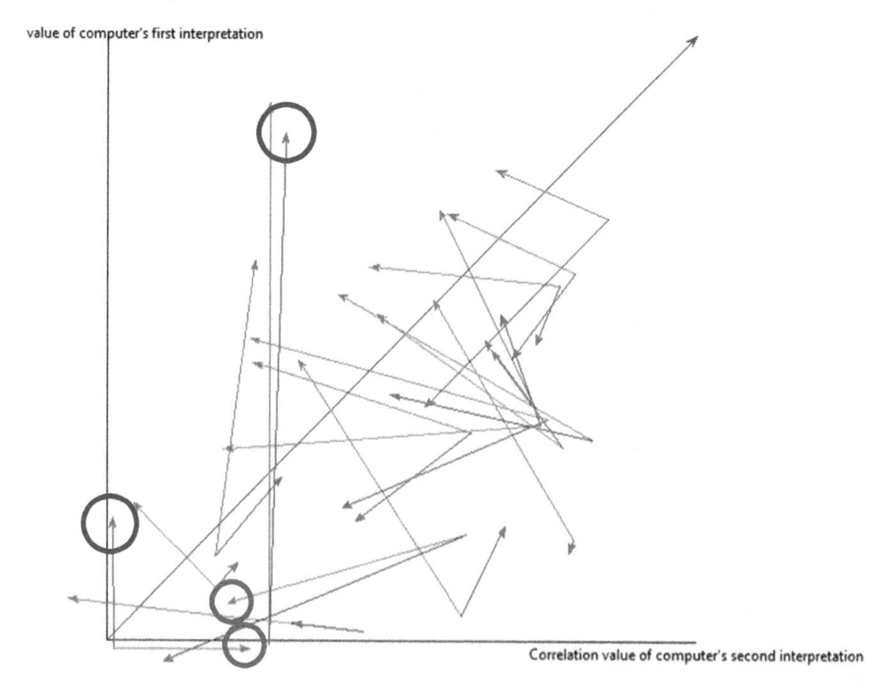

Fig. 9.1 Collocated Paired Coordinate plot of meaning context correlation over time. The set of jokes and non-jokes plotted as meaning correlation over time using collocated paired coordinates

In Fig. 9.1, the arrows are green, if the text is humorous, and red if not. Figure 9.1 shows that most jokes involve a shift away from correlation with the highest meaning in P_1 towards the lowest meaning in P_2 of the joke, while two jokes do not match this pattern. In Fig. 9.1, the *dominant pattern* for green arrows, which represent jokes is starting below the diagonal line, and ending above it. We record this pattern formally as a **classification rule R1**:

$$\text{R1}: \textit{If } z_1 > z_2 \ \& \ z_3 < z_4, \text{then } \mathbf{z} \text{ is a joke, else } \mathbf{z} \text{ is a non-joke.}$$

Visually this rule means that the first point (z_1, z_2) must be below the blue diagonal line shown in Fig. 9.1 and the second point (z_3, z_4) must be above this diagonal line. The two green arrows in Fig. 9.1 that do not follow this pattern are marked by blue circles around their end points. Both of them start and end below the diagonal line. Two red arrows also do not follow rule R1. They are marked by circles too.

All four deviations are likely a result of getting the irrelevant documents from the web search due poor choice in keywords or semantic noise, which do not match the human preferred meaning. Methods such as latent semantic analysis (Dumais 2004) may help with this.

As we see in Fig. 9.1, the pattern of the red arrows is different. They tend to stay on the same lower side of the diagonal line, as there is no meaning change. The confusion matrix for rule R1 is in Table 9.2 The total accuracy of Rule R1 is equal to 88.23%, based on the confusion matrix in Table 9.2.

We can also only look at the meaning correlation coefficients (z_3, z_4) given P_2, which clearly shows for jokes that $z_3 < z_4$, i.e., there is higher correlation with the meaning in z_4, which opposes some meaning that was initially established in z_3. In other words, most of the jokes end above the diagonal line; while only two non-jokes end above the diagonal. We can record this visual discovery as a rule R2:

$$\text{R2}: \text{If } z_3 < z_4 \text{ then } \mathbf{z} \text{ is a joke, else } \mathbf{z} \text{ is a non-joke.}$$

The confusion matrix and accuracy for rule R2 is the same as for rule R1.

Comparison with decision tree. To compare the visual results with traditional analytical data mining/machine learning approaches the C4.5 decision tree algorithm (Quinlan 1993) was used. The C4.5 decision tree algorithm produces a model indicating the same key features that involve changes in meaning correlation as the visual approach with CPC shows.

Table 9.2 Confusion matrix for visual classification rule R1

	Predicted joke	Predicted non-joke
Actual jokes 17	15	2
Actual non-jokes 17	2	15

The resulting C4.5 rule R3 is

$$\text{If } z_3 < z_4 < 0.0075 \text{ then } \mathbf{z} \text{ is a joke, else } \mathbf{z} \text{ is a non-joke.}$$

The confusion matrix and accuracy for rule R3 is the same as for rules R1 and R2 presented above, because 0.0075 is very close to 0 that is the case on rules R1 and R2. Rules R2 and R3 are very similar and result in one key splitting feature, which is the same that we found through the visual data mining process.

Rules R3 and R2 are as accurate as rule R1 but, at first glance, are simpler than R1, because they only use $z_3 < z_4$. However, they use z_1 and z_2 implicitly due to dependencies presented in formulas (9.1) and (9.2) above.

9.6 Visualization 2: Heat Maps

In the CPC visualization above, we saw a shift from one meaning correlation being higher to the opposite. To test this intuition we use heat maps based on differences in correlation coefficient values given the different meanings and different parts of text. This allows identifying potential features that distinguish jokes from non-jokes, assisting in model discovery. The heat map visualization process includes:

- Organizing the dataset from correlation coefficient differences $\Delta_1 = C_{1y} - C_{2x}$, and $\Delta_2 = C_{2x} - C_{2y}$ along with classification label of being a joke or not,
- Color coding the correlation score differences based on values,
- Sorting the rows into groups by classification that is into two groups of joke and non-joke,
- Identifying regions of the heat map where there is a distinguishable difference between the joke and non-joke sections in terms of color.

Figure 9.2 shows the resulting heat map for correlation differences. While this heat map only uses three colors for color coding by value, clearly we can identify areas where the joke dataset differs from the non-joke dataset. In Fig. 9.2 the second column shows $C_{2x} - C_{2y}$, i.e., the difference between correlation with meaning x and meaning y given the second part of the joke. If this value is negative then meaning y is greater given P_2, if it is positive then meaning x remains dominant. While we already expected this to happen, the heat map allows us to identify this value automatically as being a distinguishing feature between classes.

Fig. 9.2 Heat map for correlation differences

	A	B	C	D
1	name	C1y-C2x	C2x-C2y	class
2	webJoke1	-0.384	-0.012	joke
3	soapJoke1	-0.135	-0.084	joke
4	mouseJoke1	0.039	-0.119	joke
5	terminalJoke1	0.03	-0.129	joke
6	framedJoke1	0.084	-0.138	joke
7	balanceJoke1	-0.2824	-0.1406	joke
8	freeJoke1	0.548	-0.142	joke
9	dogJoke1	0.144	-0.174	joke
10	fishJoke1	-0.056	-0.184	joke
11	chargeJoke1	0.0455	-0.1855	joke
12	chopJoke1	0.096	-0.211	joke
13	potatoJoke1	0.124	-0.256	joke
14	houseJoke1	0.0256	-0.2606	joke
15	bankJoke1	-0.114	-0.375	joke
16	virusJoke1	-0.176	-0.153	joke
17	wavesJoke1	0.158	-0.118	joke
18	catJoke1	-0.283	-0.61	joke
19	webNonjoke1	-0.615	0.639	nonjoke
20	balanceNonJol	-0.635	0.485	nonjoke
21	framedNonjoke	-0.299	0.285	nonjoke
22	dogNonjoke1	-0.139	0.24	nonjoke
23	chopNonJoke1	-0.076	0.224	nonjoke
24	terminalNonjok	-0.08	0.222	nonjoke
25	houseNonjoke	-0.192	0.195	nonjoke
26	potatoNonjoke	-0.03	0.179	nonjoke
27	mouseNonjoke	0.159	0.154	nonjoke
28	soapNonJoke1	-0.318	0.141	nonjoke
29	chargeNonJok	-0.137	0.095	nonjoke
30	fishNonJoke1	-0.146	0.073	nonjoke
31	bankNonjoke1	-0.16	0.027	nonjoke
32	freeNonjoke1	0.645	-0.13	nonjoke
33	catNonjoke1	-0.309	-0.534	nonjoke
34	wavesNonjoke	-0.311	0.128	nonjoke
35	virusNonjoke1	-0.266	0.17	nonJoke

9.7 Visualization 3: Model Space Using Monotone Boolean Chains

In this section, we represent the difference between garden path jokes and non-jokes in 4-D Boolean space. This 4-D Boolean space is visualized in 2-D using the method from (Kovalerchuk and Delizy 2005). In Fig. 9.3 each 4-D Boolean vector is shown as a binary sequence without comma separation. These Boolean vectors

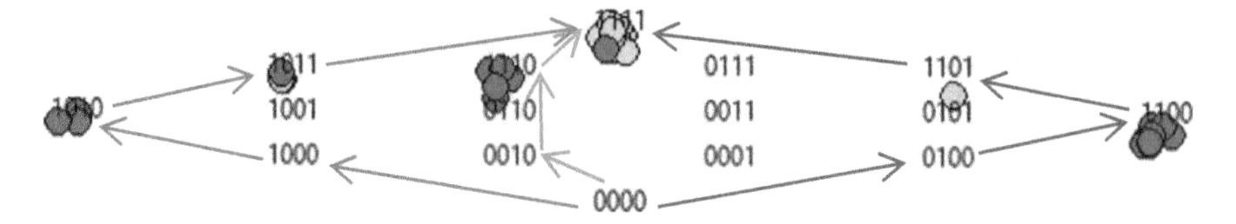

Fig. 9.3 Plot of Monotone Boolean space of jokes (green dots) and non-jokes (red dots)

form monotonically increasing chains, i.e., each succeeding vector in the chain is the same as the last, except it has an additional bit set to one.

Each chain describes the change in features. The chains altogether represent the entire model space of 2^4 Boolean vectors based on the Boolean four features u_1, u_2, u_3, u_4 derived from the meaning correlation coefficient differences described in Sect. 9.3.2.

The steps of this visualization process are:

- Establishing a 4-D Boolean vector (u_1, u_2, u_3, u_4) for each joke/non-joke as described in Sect. 9.3.2;
- Establishing and visualizing 2-D Boolean space of chains in accordance with (Kovalerchuk and Delizy 2005) as explained below;
- Plotting vectors established for jokes as green dots and as red dots for non-jokes on the Boolean plot;

Figure 9.3 shows the resulting visualization using the same dataset, which was used in the two other visualizations. Figure 9.3 shows the full 4-D space, which consists of 2^4 = 16 Boolean vectors with the smallest vector (0,0,0,0) at the bottom and the largest vector (1,1,1,1) at the top. The vectors with single "1" form the second layer, vectors with two "1" the third (middle) layer, and the vectors with three "1" form the forth layer. Together these five layers are called the *4-D Multiple Disk Form* (MDF) (Kovalerchuk, Delizy, 2005). This is a special 4-D case of the general MDF, which is constructed in (Kovalerchuk and Delizy 2005).

Each joke and non-joke is encoded as a Boolean vector (u_1, u_2, u_3, u_4) and is placed as a dot in the respective Boolean vector in Fig. 9.3. Here jokes are green dots and non-jokes are red dots. Figure 9.3 shows that only two vectors (1011) and (1111) that have has mixed content with one joke in each of them. All other boxes represent a single class.

Just visual observation of Fig. 9.3 allows formulating a simple rule, which we denote as rule R4:

$$\text{R4} : \text{If}(\mathbf{w} \geq (1,0,1,1) \vee \mathbf{w} \geq (0,1,0,1) \text{ then } \mathbf{w} \text{ is joke else } \mathbf{w} \text{ is a non-joke.}$$

It is based on noticing that green dots are located only in three vectors. This rule expresses this property—if **w** is in one of them, then it is a joke. The accuracy of this rule is 94.12% (32 out of 34 cases are correctly classified) in contrast with 88.23% for previous rules above. This rule also covers vector (1101) by monotone generalization while this vector is not present in the given dataset. Why is accuracy of R4 higher than for Rules R1–R3? In fact those rules use only one or two out of four values u_1, u_2, u_3 and u_4, but rule R4 uses all 4 values, i.e., captures more complex relations as patterns.

The vectors in Fig. 9.3 have a hierarchical structure that allows outlining the border between jokes and non-jokes. Consider the increasing chain of Boolean vectors outlined by the blue arrows in Fig. 9.3:

$$(0000),\ (1000),\ (1010),\ (1011),\ (1111)$$

Rule R4 established the border between jokes and non-jokes between vectors (1010) and (1011) in this chain, i.e., in R4 vector (1010) is a non-joke, and (1011) is a joke. Both errors of rule R4 are in this chain.

Now consider another increasing chain of Boolean vectors outlined by the violet arrows in Fig. 9.3:

$$(0000),\ (0100),\ (1100),\ (1101),\ (1111)$$

Rule R4 established the border between jokes and non-jokes between vectors (1100) and (1101), i.e., in R4 vector (1100) is a non-joke and (1101) is a joke. Only one error of rule R4 is in this chain.

In both chains $u_4 = 1$ plays a critical role for the text to become a joke in transition from the third vector with $u_4 = 0$ to fourth vector with $u_4 = 1$.

The analysis of the third chain (0000), (0010), (0110), (1110), (1111) (shown by orange arrow) again shows the critical role of $u_4 = 1$ in transition from the forth non-joke vector (1110) to the fifth vector (1101) to become a joke.

After discovering visually such an important role of $u_4 = 1$ in Fig. 9.3 we can take a close look at the meaning of u_4. It is defined in Sect. 9.3.2 as follows: $u_4 = 1$ if $C_{2x} < C_{2y}$, i.e., correlation with meaning y is greater than meaning x.

This is rule R2 discovered with CPCs in Sect. 9.5. However, the Boolean chains shows that we cannot use $u_4 = 1$ alone. The transition to jokes depends on values of other u_i in different combinations of u_i in the Boolean vector depending on the chain. This confirms the insufficiency of rules R3 and R4 for modeling of jokes that we pointed out above.

Figure 9.3 allows generalizing rule R4 to construct a more complete border between jokes and non-jokes as set of pairs based on the discovered role of $u_4 = 1$:

$$(1010) - (1011);\ (1000) - (1011);\ (1110) - (1111);$$
$$(0000) - (0001),\ (0100) - (1001);\ (1100) - (1101)$$

where the first vector in the pair is a non-joke and the second is joke. However, we can make some judgements about them. This generalized rule can be tested on new data by getting more garden path joke and non-jokes.

The visually discovered fact that most of the jokes and non-jokes concentrated in few Boolean vectors on Fig. 9.3 is the insight for further exploration. The existence of the vectors with the mixture of cases is another visual insight likely pointing to the need for more features beyond u_1-u_4, to separate such cases.

9.8 Conclusion

Incongruities and their resolution appear in many other places where classification occurs though. The incongruity visual analysis approaches can be used to detect and classify other forms of humor including physical humor (Nijholt 2014).

In general, incongruities can arise and are resolved where classification occurs based on multiple sources of evidence, for example, where multiple sensors are used, or where a sensor takes readings at multiple steps in time. These incongruity visualization should be able to identity some of these nonverbal and non-humorous 'mistakes' and their resolution which can be useful for a number of tasks from process control to sensor management.

Overall, the results from this study show that visualization can be used as a valid strategy for approaching the modeling and detection of humor within text. This chapter presented three approaches, which were all successful in enabling a person to identify key features that distinguish humorous and non-humorous garden path jokes. One future direction is to use these visualization techniques on other joke types to see what they would look like in terms of patterns of meaning correlation over time.

These techniques can potentially be used to visualize many other forms of incongruity within texts such as:

- shifting within product review sets,
- paradigm level formation of incongruity and resolution within academic document sets over time,
- the writings of a bipolar patient who might shift from one opposing emotion to the other in a cyclic fashion,
- identify non-verbal incongruities, and their resolution such as non-verbal humor, and
- other phenomena involving opposing states and patterns of shifting.

In addition, these visualization techniques allow plotting many examples at once, which is beneficial for analysis of natural language texts.

References

Dumais, S.T.: Latent semantic analysis. Ann. Rev. Inf. Sci. Technol. **38**(1), 188–230 (2004)

Dynel, M.: Garden paths, red lights and crossroads. Israeli J. Hum. Res. **1**(1), 289–320 (2012)

Galitsky, B., Kovalerchuk, B.: Improving web search relevance with learning structure of domain concepts. In: Aleskerov, F. Goldengorin, B. Pardalos, P.M. (eds.) Clusters, Orders, and Trees: Methods and Applications, pp. 311–363. Springer, Berlin (2014)

Kovalerchuk, B., Delizy, F.: Visual data mining using monotone Boolean functions. In: Kovalerchuk, B., Schwing, J. (eds.) Visual and Spatial Analysis, pp. 387–406. Springer, Dordrecht (2005)

Kovalerchuk, B., Schwing, J. (eds.): Visual and spatial analysis: advances in data mining, reasoning, and problem solving. Springer Science & Business Media (2005)

Kovalerchuk, B., Delizy, F., Riggs, L., Vityaev E.: Visual data mining and discovery with binarized vectors. In: Data Mining: Foundations and Intelligent Paradigms 2012 (pp. 135–156). Springer Berlin Heidelberg.

Labutov, I., Lipson, H.: Humor as circuits in semantic networks. In: Proceedings of the 50th Annual Meeting of the Association for Computational Linguistics: Short Chapters-Volume 2, pp. 150–155. Association for Computational Linguistics (2012)

McDonald, S., Ramscar, M.: Testing the distributional hypothesis: the influence of context on judgements of semantic similarity. In: Proceedings of the 23rd Annual Conference of the Cognitive Science Society, pp 611–616 (2001)

Mihalcea, R., Strapparava, C., Pulman, S.: Computational models for incongruity detection in humour. In: Gelbukh, A. (ed.) CICLing 2010. LNCS, vol. 6008, pp. 364–374. Springer, Heidelberg (2010). https://doi.org/10.1007/978-3-642-12116-6 30

Mikolov, T., Sutskever, I., Chen, K., Corrado, G.S., Dean, J.: Distributed representations of words and phrases and their compositionality. In: Burges, C., Bottou, L., Welling, M., Ghahramani, Z., Weinberger, K. (eds.) Advances in Neural Information Processing Systems 26, pp. 3111–3119. Curran Associates, Inc. (2013)

Nijholt, A.: Towards humor modelling and facilitation in smart environments. In: Proceedings of the 5th International Conference on Applied Human Factors and Ergonomics AHFE 2014 (2014)

Petrovic, S., Matthews, D.: Unsupervised joke generation from big data. In: ACL (2), pp. 228–232. Citeseer (2013)

Quinlan, J.R.: C4.5: Programs for Machine Learning. Morgan Kaufmann Publishers (1993)

Rajaraman, A., Ullman, J.D.: Mining of Massive Datasets. Cambridge University Press (2011). http://i.stanford.edu/~ullman/mmds/ch1.pdf

Ritchie, G.: Developing the incongruity-resolution theory. University of Edinburgh, UK (1999)

Ritchie, G.: Logic and reasoning in jokes. Europ. J. Hum. Res. **2**(1), 50–60 (2014)

Ritchie, G.: The jape riddle generator: technical specification. Institute for Communicating and Collaborative Systems (2003)

Sahlgren, M.: The distributional hypothesis. Rivista di Linguistica (Italian Journal of Linguistics) **20**(1), 33–53 (2008)

Schultz, T.R., Horibe, F.: Development of the appreciation of verbal jokes. Dev. Psychol. **10**(1), 13 (1974)

Simov S., Bohlen M., Mazeika A. (eds.): Visual Data Mining. Springer, (2008). doi:10.1007/978-3-540-71080-61

Taylor, J.M., Mazlack, L.J.: An investigation into computational recognition of children's jokes. In: Proceedings of the National Conference on Artificial Intelligence, vol. 22, no. 2, p. 1904. Menlo Park, CA; Cambridge, MA; London; AAAI Press; MIT Press (2007)

Taylor, J.M., Raskin, V.: On the transdisciplinary field of humor research. J. Integr. Des. Process Sci. **16**(3), 133–48 (2012)

Taylor, J.M., Raskin, V.: Towards the cognitive informatics of natural language: the case of computational humor. Int. J. Cogn. Inform. Nat. Intell. (IJCINI) **7**(3), 25–45 (2013)

Valitutti, A., Toivonen, H., Doucet, A., Toivanen, J.M.: Let everything turn well in your wife: generation of adult humor using lexical constraints. ACL **2**, 243–248 (2013)

Chapter 10
Enhancing Evaluation of Machine Learning Algorithms with Visual Means

Science can progress on the basis of error as long as it is not trivial.

Albert Einstein

10.1 Introduction

10.1.1 Preliminaries

Previous chapters demonstrated the ways of visual discovery of patterns using different General Line Coordinates. This chapter demonstrates the hybrid visual and analytical way to enhance the estimation of *accuracy* and *errors* of machine leaning discovery. It focuses on improvement of k-fold cross validation. It provides: (1) a justification for the worst case estimates using the Shannon Function, (2) hybrid visual and analytical ways to get these estimates, and (3) illustrative case studies. The visual means include the point-to-point and GLC point-to-graph mappings of the n-D data to 2-D.

The algorithm of k-fold Cross Validation (CV) is a common tool actively used to evaluate and compare machine learning algorithms. However, it has several important deficiencies documented in the literature along with its advantages. The advantages of quick computations are also a source of its major deficiency. It tests only *a small fraction of all the possible splits of data* on training and testing data leaving untested many difficult for prediction splits. The associated difficulties include bias in estimated average error rate and its variance, the large variance of the estimated average error, and possible irrelevance of the estimated average error to the problem of the user.

The improvement of the cross validation described below combines visual and analytical means in a hybrid setting. The visual means include both the *point-to-point mapping and GLC point-to-graph mappings* of the n-D data to 2-D data. The analytical means involve the adaptation of the *Shannon function* to obtain the worst case error estimate. It is illustrated below for classification tasks. In k-fold cross validation data are split into k equal-sized folds. Each fold is a validation/test set for evaluating classifiers learned on the remaining k-1 folds. The error rate is computed as the *average error* across the k tests and is considered as an *estimate of the error expectation.*

B. Kovalerchuk, *Visual Knowledge Discovery and Machine Learning*,
Intelligent Systems Reference Library 144,
https://doi.org/10.1007/978-3-319-73040-0_10

Four cross validation *schemes* are presented in Moreno-Torres et al. (2012), which are summarized below:

(1) Standard stratified cross validation (SCV) places an *equal* number of samples of each class on each partition to keep *the same class distributions* in all partitions.
(2) Distribution-balanced stratified cross validation (DB-SCV) keeps data distribution as *similar* as possible between the training and validation folds and maximizes the *diversity* on each fold to minimize the covariate shift.
(3) Distribution-optimally-balanced stratified cross-validation (DOB-SCV) is DB-SCV with the additional information used to choose in which fold to *place* each sample.
(4) Maximally-shifted stratified cross validation (MS-SCV) creates the folds that are as *different* as possible from each other. It tests the maximal influence partition-based *covariate shift* on the classifier performance by putting the *maximal shift* on each partition.

Here *covariate shift* means that the training and testing sets have different distributions (Shimodaira 2000), e.g., a unimodal distribution on the training set and a two-modal distribution on the testing/validation set.

This chapter provides (1) a justification for the use of the worst case estimates using the Shannon Function as a criterion, (2) hybrid visual and analytical ways to get such worst case estimates, and (3) illustrative case studies.

10.1.2 Challenges of k-Fold Cross Validation

The k-fold cross validation model error estimates vary depending on the way how data are split for cross validation and distributed in the splits. This leads to the difficulties judging how actually successful the learned model in predictions is.

The theorem proved in Bengio and Grandvalet (2004) had shown that there is *no* universal (valid under all distributions) unbiased estimator of the variance of k-fold cross validation.

Multiple attempts made to address k-fold problems by making additional *assumptions* and modifications to get better average estimates, e.g., (Dietterich 1998; Grandvalet and Bengio 2006). To the best of our knowledge much less was done to improve the worst-case estimates in both probabilistic and deterministic settings.

Estimates of the average error and its variance can be insufficient or even irrelevant to the supervised learning problem that is of user's interest. It is related to the Maximally-shifted stratified cross validation (MS-SCV) listed above as schema (4). It is found in extensive experiments on real data in Moreno-Torres et al. (2012) that:

- MS-SCV produces a *much worse accuracy* than all other partitioning strategies, and
- cross validation approaches that limit the partition-induced covariate shift (DOB-SCV, DB-SCV) are *more stable* when running a single experiment, and need a lower number of iterations to stabilize.

These results illustrate the problem. Limiting the covariate shift gives a more stable result on validation data. However, nobody can guaranty us that on new unseen data the covariance shift will be limited and limited in the same way. It is simply out of our control in many real world tasks. Therefore, the stable result under such limits can be biased showing a lower error rate than it can be on the real test data.

We address these challenges by supplementing limited covariate shift ("average" case) by the bounds for "worst" and "best" cases. This will balance the risk of using a given learning algorithm with "average" by providing information that is more complete. This is the ultimate goal of this chapter.

10.2 Method

10.2.1 Shannon Function

Below we formalize a way to evaluate the worst case as a compliment to k-fold estimates of the average error. It is done by adaptation of the minmax **Shannon function** (Shannon 1949) originally proposed for analysis of the complexity of switching circuits as Boolean functions. The Shannon function measures the *complexity of the most difficult function.* In particular, this function was applied to find an algorithm A_j that restores the worst (most complex) monotone Boolean function of n-variables for the smallest number of queries (Hansel 1966; Kovalerchuk et al. 1996).

Consider a labeled dataset D and a set of machine learning algorithm $\{A_j\}_{j \in J}$. Let $\{D_i\}_{i \in I}$, $I = \{1,2,\ldots,m\}$ be a set of splits of D to < Training data, Validation data > pairs. k-fold cross validation split is one of them. Each D_i is a pair of training and validation data, $D_i = (Tr_i, Val_i)$. $A_{jv}(D_i)$ is the **error rate** on validation data Val_i produced by A_j when A_j is trained on the training data Tr_i from D_i. The adaptation of the **Shannon function** $S(I, J)$ to supervised learning problem is defined as follows

$$S(I,J) = \min_{j \in J} \max_{i \in I} A_{jv}(D_i) \quad (10.1)$$

The algorithm A_b is called ***S* best algorithm** if

$$S(I,J) = \min_{i \in I} A_{bv}(D_i) \quad (10.2)$$

In other words, the S-best algorithm produces fewer errors on validation data on its worst k-fold splits among $\{D_i\}$ than other algorithms on their worst k-fold splits among $\{D_i\}$.

Let $\boldsymbol{D}_A = \{D_{i:}\ i \in I_A\}$ be a set of *all* passible k-fold splits for given k and data D, i.e., k-1 folds (bins) with the training data and one fold (bin) with the validation data. In contrast with the standard k-fold validation, here the validation sets for different D_i can *overlap*. Let $\boldsymbol{D}_T = \{D_{i:}\ i \in T\}$ is some set of splits.

Statement If $\boldsymbol{D}_T = \{D_i: i \in T\} \subseteq \boldsymbol{D}_A$ then $S(I_A, J) \leq S(I_T, J)$

This statement follows directly from definitions of these terms. For instance, if $S(I_T, J) = 0.2$ then adding more splits can give us a better split D_r in D_A such that $A_{jv}(D_r) < 0.2$ for some A_j.

In other words, for each $\boldsymbol{D}_T$ the value of $S(I_A, J)$ provides a **low bound** for $S(I_T, J)$. Similarly, for $\boldsymbol{D}_A$ the value of $S(I_T, J)$ provides an **upper bound** for $S(I_A, J)$. A standard k-fold split $\boldsymbol{D}_K = \{D_i: i \in K\}$ is one of $\boldsymbol{D}_T$. How close the bounds are to the actual worst case depends on the specific $\boldsymbol{D}_T$ and D_A. At least the average error rate for $\boldsymbol{D}_K$ can be computed quickly enough. Computing error rates for multiple $\boldsymbol{D}_K$ produced by random or non-random splits of data into folds will give several bounds.

Asymptotically this will lead to the actual Shannon worst case,

$$D_w = \arg(\min_{j \in J} \max_{i \in I} A_{jv}(D_i)) \tag{10.3}$$

Split D_w is called the *worst case split for S-best algorithm* A_b.

$$D_w = \arg(\max_{i \in I} A_{bv}(D_i)) \tag{10.4}$$

Informally, the worst case split is a split, which is most difficult for the S-best algorithm which produces fewer errors on validation data than other algorithms on their worst splits from $\{D_i\}$.

Split D_b is called the *best case split for S-best algorithm* A_b

$$D_h = \arg(\min_{i \in I} A_{jv}(D_i)) \tag{10.5}$$

Informally, the best case split is a split, which is easiest for the S-best algorithm, which produces fewer errors on the validation data than the other algorithms on their worst splits from $\{D_i\}$.

Split D_m is called the *median split for S-best algorithm* A_b,

$$D_h = \arg(\underset{i \in I}{\text{median}}(A_{jv}(D_i))) \tag{10.6}$$

Informally, the median split for the S-best algorithm produces the error rate that is close to the average error rate among $\{D_i\}$ for A_b algorithm.

Both the worst-case and best-case estimates provide the "bottom line" of the expected errors. As we mentioned above, for the tasks with a high cost of individual error, it is very important.

10.2.2 Interactive Hybrid Algorithm

The steps of **first part** of the **interactive hybrid algorithm for *S*-best algorithm** that is discovering data *structure* are as follows:

(S1) Visualize n-D data in 2-D;
(S2) Select border points of each class and color them in different colors;
(S3) Outline classes by constructing envelopes in the form of a convex or a non-convex hull;
(S4) Outline (a) overlap areas L for overlapped classes or (b) select closest areas C for separable classes;
(S5) Compute the size of the overlap areas L or areas C of the closest samples;
(S6) Set up ratio of training-validation data, |Tr|/|Val|, e.g. 90%:10% with (|Tr| + |Val|)/|Val| = k.

The steps of **second part** of the Interactive Hybrid algorithm for the **Worst case (IH-W) of *S*-best algorithm** are:

(W1) Form Val as areas L or C;
(W2) Adjust (increase or decrease) L or C to make $|L|$ = |Val|, or $|C|$ = |Val|;
(W3) Form training data Tr = D\Val and pair < Tr, Val >;
(W4) Apply each algorithm A_j to Tr to construct discrimination function F;
(W5) Apply F to Val to get error rate A_{jv}(Val);
(W6) Record A_{jv}(Val) and find max(A_{jv}(Val)), $j \in$J;
(W7) Repeat (W1)–(W6) to get values {max $A_{jv}(\mathrm{Val_i})$} $i \in I$ for a set of training-validation pairs $\{D_i\}$;
(W8) Find the Shannon worst case split, $\min_{i \in I} \max_{j \in J} (A_{jv}(\mathrm{Val_i}))$ and algorithm A_b that provides this split.

The interactive algorithm for the **best case (IH-B) of *S*-best algorithm** consists of using algorithm A_b from step W8 to get $\min_{i \in I} (A_{bv}(\mathrm{Val_i}))$.

The interactive algorithm for the **median case (IH-M) of *S*-best algorithm** consists of using algorithm A_b from step W8 to get median $_{i \in I}$ $(A_{bv}(\mathrm{Val}_i))$.

10.3 Case Studies

Case studies in this section show how the visual means support finding worst and best cases of splits with the use of the interactive hybrid algorithm.

10.3.1 Case Study 1: Linear SVM and LDA in 2-D on Modeled Data

Worst case Figure 10.1 shows two separable classes classified by linear SVM and simplified Linear Discriminant Analysis (LDA) algorithms. For the linear SVM for linearly separable classes we use its geometric interpretation (Bennett and Campbell 2000; Bennett and Bredensteiner 2000), which is based on the closest support vectors of the two classes. In this SVM the red line connects *closest support vectors from opposing classes*. In simplified LDA it connects *centers of training data of classes*.

The green lines that bisect these lines in the middle serve as SVM and LDA linear classifiers, respectively. Both classifiers are error-free on training data (blue and grey convex hulls), but not error free on validation data of the blue class (violet triangle on the left). This triangle illustrates the worst-case example of the splitting data to training and validation data in cross validation algorithm evaluation process.

In Fig. 10.1, the violet areas form the validation data (5% of the blue pentagon and 5% of the grey pentagon). The classification results of training data in Fig. 10.1 are the best cases for both linear SVM and simplified LDA on training data, because this pair D_i is error-free on training data (see green discrimination lines in Fig. 10.1). Both SVM and LDA are the winners *on training data* for this D_i.

Fig. 10.1 Examples of the worst case cross validation split of two separable classes classified by linear SVM and simplified LDA

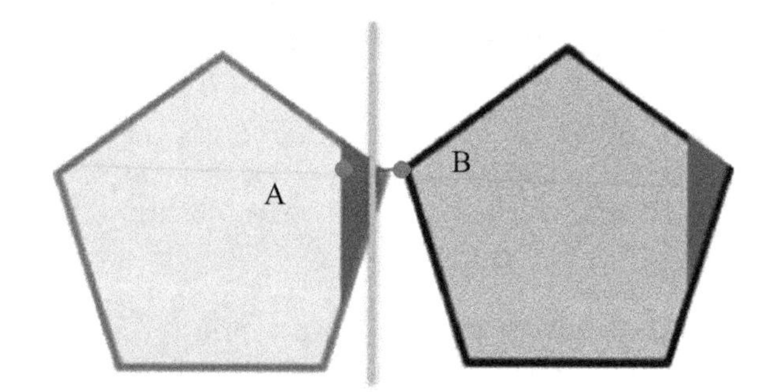

(a) Linear SVM: Narrow margin case

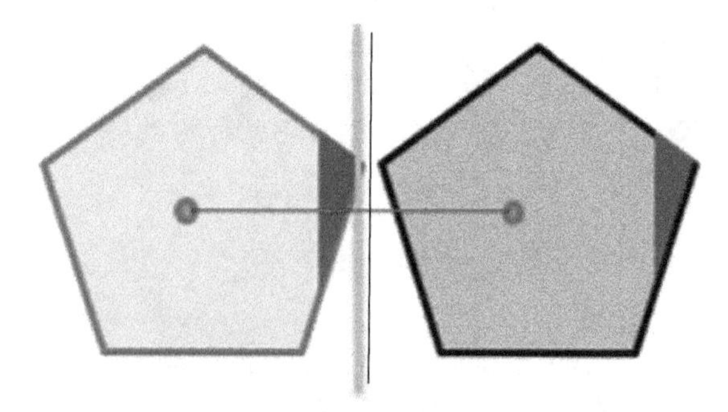

(b) Simplified LDA: narrow margin case

However, both are *erroneous on the validation data*, because a discrimination line exists which separates validation data with 100% accuracy. This line is the black middle line between two pentagons in Fig. 10.1b, but both algorithms did not find it. This black line was found by pure visual means. This shows that there are situations where visual classification can be more accurate than the classification by using known analytical machine learning methods. Thus, SVM and simplified LDA are not winners for the validation data in D_i and respectively are not *S*-best algorithms for this D_i.

Next, even if such worst D_i is included in 10-fold cross validation, the difference between average error estimates for two algorithms will likely be statistically insignificant if both algorithm equally accurate on the remaining nine training-validation pairs. This is a motivation for using the Shannon function and for search of the worst cases or at least *estimates the bounds of the worst cases.*

Why is it important to search for such rare worst training-validation pairs D_i? The ultimate goal of machine learning is *generalization* beyond the given data *D* to *unseen data.* The existence of worst training–validation pairs with large error indicates that the algorithm A_j *does not capture a generalization pattern* in some situations on given data *D*.

This increases the chances of misclassification on unseen data too. In the tasks with the *high cost of an individual error* (e.g., medicine), such situations must be traced and analyzed before use in real applications.

For instance, if a set of selected splits $\{D_i\}$ for the *S*-best algorithm is not error-free then we can treat the areas, where those errors occurred differently.

This treatment can include:

(1) refusal to classify data from those areas,
(2) use other machine learning algorithms,
(3) adding more data and retraining on extended data,
(4) cleaning existing data,
(5) modifying features,
(6) use other appropriate means, such as manual classification by experts.

The case study in this section uses 2-D artificial modeled 2-D data. For this analysis in real machine learning tasks, such 2-D data can be obtained from n-D data by point-to-point matching visualization algorithms such as PCA, MDS, SOM and others. Transformation of n-D to 2-D by these methods needs to be used cautiously due to lossy nature of these methods that was discussed in other chapters.

10.3.2 Case Study 2: GLC-AL and LDA on 9-D on Wisconsin Breast Cancer Data

The case study in this section is based on the graph representation of the n-D points in 2-D, not on a single 2-D point representation of an n-D point. For this study, *Wisconsin Breast Cancer Diagnostic (WBC) dataset* was used from ICI Machine Learning Repository (Lichman 2013) with 9 attributes for each record and the class label which was used for classification. These data have been explored in Sect. 7.3.1 in Chap. 7 with the accuracy over 95% on these data.

Figure 10.2 shows the screenshots, where these data are interactively visualized and classified with a linear classifier using GLC-AL algorithm defined in Sect. 7.2 in Chap. 7. In Fig. 10.2 the malignant cases are drawn in red and benign in blue. Figure 10.2 is based on Figs. 7.14 and 7.15 from Chap. 7.

In Fig. 10.2a, the GLC-AL linear classifier misclassified 31 samples with all of them from class 1 when all data (444 benign cases and 239 malignant samples) were used for training. The selected overlap area contains 38 samples (4.5%, with 28 samples from class 1, and 10 samples from class 2).

According to step W1 of the algorithm IH-W, we form the validation set Val as a set of samples in the overlap area L. We keep Val equal to L without adjustment, skipping the step W2. Next we use a shortcut for steps W2–W5, which allows us to get a bound for the error rate $A_{jv}(L)$, where A_{jv} is the GLC-L algorithm applied to $Val = L$ trained on Tr. The result of this shortcut is presented in Fig. 10.2b. It shows the overlapping cases

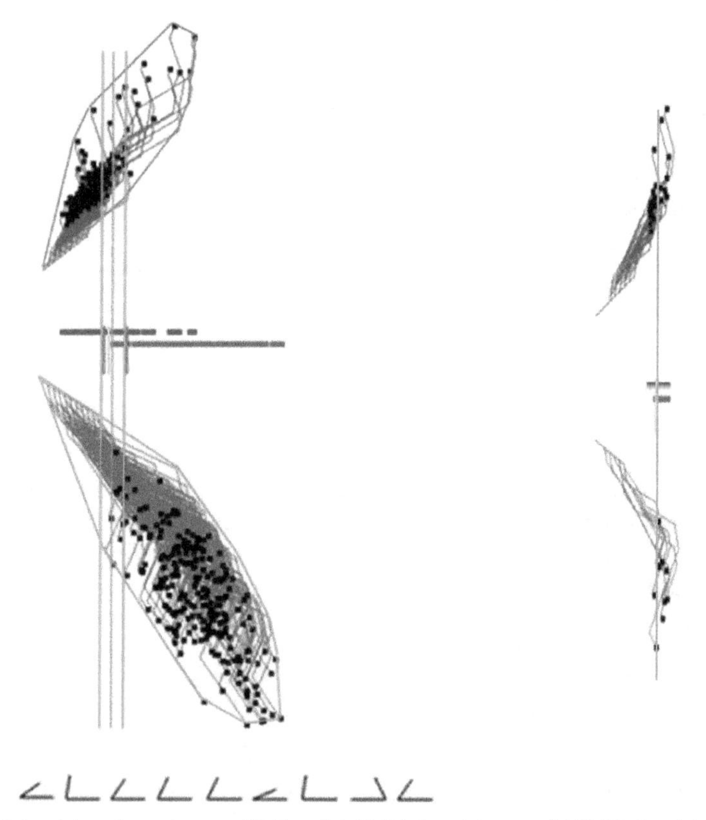

(a) Training data and worst-case validation data. Validation data are between green vertical lines where two classes overlap.

(b) Validation data from (a)

Fig. 10.2 Worst case cross validation example for 9-D Wisconsin Breast Cancer data of two classes (red and blue) in lossless GLC-L visualization

L, selected in Fig. 10.2a and the accuracy of classification of samples from L, when all of them and only them are used as training data. At the first glance, running GLC-L on L as training data, not validation data, contradicts steps W2–W5, which require running L as validation data. The trick is that, *training* GLC-L on L as training data, we expect to get a smaller error rate on L than *running the linear model* on L, constructed by GLC-L on training data Tr without any data from L in Tr.

In Fig. 10.2b, the accuracy is 73.68% (error rate 0.2632) with L as training data. The error rate 0.2632 is the upper bound for the error rate $A_{jv}(L)$, $A_{jv}(L) \leq 0.2632$. We cannot get a bound with the larger number of errors than 0.2632 for the algorithm GLC-L, if we continue to run GLC-L on the overlap area L for more epochs. It follows from the design of GLC-L. GLC-L keeps coefficients with the current lowest error rate. Having the error rate equal to 0.2632 GLC-AL will update it only by finding a smaller error rate, not a larger one.

This conclusion was made under assumption that we use L as Tr. Now we need to explore what will happen with the other splits when L is only a part of Tr, not equal to Tr. Can we get another error rate r for GLC-L, say $r = 0.3$, which is greater than 0.2632 for these other splits and, respectively, another upper bound for $A_{jv}(L)$? If such greater r exists our previous claim, that we cannot get more errors with GLC-L, will be wrong.

We cannot get such greater r for the same reason as above. The GLC-L design will not allow it. We already have a linear model in Fig. 10.2a that classified all samples from Tr = $D \backslash L$ with zero error rate, where D is the total given dataset. Thus, GLC-AL algorithm trained on Tr data that include L will only keep linear models that classify L better because for samples outside L GLC-AL already obtained models with zero error rate.

This shortcut can be applied for any GLC-L data. If such upper bound is a tolerable error rate then we can apply the coefficients found by GLC-AL on TR\L as training data for classification of new data. Thus steps W2–W5 of the algorithm IH-W for GLC-L can be simplified.

To compare the bound for GLC-L with the bounds for linear SVM and LDA steps, W4–W6 must be run for these algorithms. The algorithm with the smallest bound will be a candidate for the S-best algorithm on these data. In addition to this analytical option, an interactive option can be applied to the modified and simplified versions of linear SVM and LDA algorithms that work with 2-D GLC-L visual representations of n-D data. Both algorithms follow the steps used in case study 1 with two differences: (1) convex hull constructed by GLC-L algorithm are used, and (2) the overlap area is defined by the location of the last node of the graph (marked by black squares) between green lines. This way, to identify the overlap area, was used in Fig. 10.2.

Linear SVM in GLC-L visualization uses closest support vectors (SV) from two classes in GLC-L. For overlapping convex hulls of two classes we use the overlap area that is identified by a user interactively using two thresholds (see green lines in Fig. 10.2a). Two closest nodes of graphs from two different classes in the overlap area are called closest support vectors of these graphs. If the overlap area is empty (the case of linearly separable classes) then two closest nodes of the frames of two convex hulls are called closest support vectors of the frames. Having two closest support vectors *A* and *B*, we build a line that connects them and a line that bisects them in the middle and orthogonal to the first line. The closest nodes are defined by the distance between projections of the last points of the graphs for *A* and *B* to the horizontal line.

For the LDA we use for *A* the average point in the projection on the point of class 1 to the horizontal line and for point *B* we use the same in the class 2. Then the middle point *C* between *A* and B is used to construct the discrimination line. It is shown in Fig. 10.2a as a grey line.

What is important in the example in Fig. 10.2 is the abilities to build a visual classifiers (in this case for 9-D), and be able to compare error rates visually. It also allows chopping visually overlapping parts by setting up thresholds interactively and using these folds to construct validation data for the worst case.

10.4 Discussion and Conclusion

This chapter had shown a hybrid way to improve cross validation by using combined visual and analytical means. The main *benefit* of this hybrid approach is leveraging the abilities of the human visual system to *guide the discovery of* patterns in 2-D. This includes discovering splits of n-D data in 2-D visualization of these data. This approach creates an opportunity to avoid a blind computational search of worst splits among the exponential number of alternatives that can be the case in the pure computational approach. In essence, the visual approach brings *additional information* about the n-D data structure that the computational approach lacks. Adding such information from the visual channel can be viewed as a way to add more features and relations to the data, sometimes called privileged information (Vapnik and Vashist 2009), or prior domain knowledge (Mitchell 1997; Mitchell et al. 2005). The difference is that both privileged information and domain knowledge typically is not present in the original data. In contrast, the visual channel makes the hidden information already present in n-D data be readily available via the interactive process.

While this visual opportunity exists, it requires a relatively simple visualization for humans to be able to discover a pattern in them, i.e., within the abilities of the human visual channel. The ways to simplify the visual patterns in the General Line Coordinates are presented in Chap. 4. Such ways should be applied before and in concert with the interactive search for worst case splits in cross validation.

The focus on worst-case splits and adaptation of the Shannon function bring a new formal validation task. The main justification for the use of worst case estimates and Shannon Functions is three-fold:

(1) Existence of the tasks with a *high cost* of individual errors;
(2) Existence of the tasks with *high error rate* for the worst case splits;
(3) Abilities to limit using algorithm that are poor in worst cases.

In (1) and (2), the use of the average error rate can be too optimistic and risky where the worst-case estimate serves as warning, while (3) allows preventing risky decisions and predictions. We may have two algorithms A and B with the average error rates with a statistically insignificant difference, but A has much smaller worst-case error rate than B. This can be a reason to prefer A for the classification of new samples, because A was able to discover better difficult patterns than B showing stronger generalization ability. In addition, while error rate for A is better than for B in the worst case, in some worst folds it can be too big. The prediction on new unseen data in these folds can be blocked for both A and B.

The challenges for k-fold cross validation include: (1) selecting the number of folds k and running multiple k, (2) missing multiple splits that left untested, (3) large variance of error rates, (4) bias in estimated average errors and its variance, and (5) insufficiency or *irrelevance of* estimated average errors.

The hybrid approach allows dealing with these challenges as follows. First $k = 2$ is used to provide an upper bound of the worst error rate for all the other k for the given algorithm A. Then we increase k until the worst case bound will be below threshold T_{worst} selected by a user for the given task. This k is considered acceptable. On the other extreme, with $k = m$ (leave-one-out split), where m is the number of samples, we consider another threshold T_{best}, and decrease k until the best error rate will be still below T_{best}. Assume that we find k that satisfies both the T_{worst} and T_{best}. For instance, we can find that for $k = 8$ the worst error rate is bounded by 0.18 and the best error rate is bounded by the error rate 0.05, with average error rate as 0.12 with its variance ± 0.02. In other words, we have a wider interval [0.05, 0.18] than the average interval [0.10, 0.14].

The computational support of visual exploration and visual support of analytical computations are important parts in this hybrid approach to avoid brute force search. As examples in this chapter show, the visual approach allows a quick visual judgment that the error rate in one split is greater than in another one. A user can find visually a large overlap area of two classes and chop it to form several validation folds, e.g., getting 10-fold cross validation splits. This confirms our main statement that brute force search is not mandatory and is avoidable using an appropriate visualization.

The future studies are toward making hybrid interactions more efficient and natural in the computational and visual aspects but not limited by them. This includes adding speech recognitions to interactions allowing a user to give oral commands such as "decrease slightly the overlap area", "shift the overlap area to the right", "make an about 5% area on the top of the convex hull" and so on. This will require formalization of the linguistic variables involved in these commands in the spirit of the Computing with Words (CWW) approach (Kovalerchuk 2013). More complex commands such as "decrease *slightly* the overlap area, and shift the overlap area to be *close* to the envelope frame" will require more sophisticated uncertainty aggregation techniques (Kreinovich 2017).

References

Bengio, Y., Grandvalet, Y.: No unbiased estimator of the variance of k-fold cross-validation. J. Mach. Learn. Res. **5**, 1089–1105 (2004)

Bennett, K.P., Campbell, C.: Support vector machines: hype or hallelujah? ACM SIGKDD Explor. Newsl **2**(2), 1–13 (2000)

Bennett. K.P., Bredensteiner, E.J.: Duality and geometry in SVM classifiers. In: ICML 2000 Jun 29 (pp. 57–64)

Dietterich TG. Approximate statistical tests for comparing supervised classification learning algorithms. Neural Comput. **10**(7), 1895–1923 (1998)

Grandvalet, Y., Bengio, Y.: Hypothesis testing for cross-validation. Montreal Universite de Montreal, Operationnelle DdIeR. 2006 Aug 29; 1285

Hansel, G.: Sur le nombre des functions Bool´eenes monotones de n variables, C.R. Acad. Sci., Paris, **262**(20), 1088–1090 (1966)

Kovalerchuk, B.: Quest for rigorous combining probabilistic and fuzzy logic approaches for computing with words. In: Seising, R., Trillas, E., Moraga, C., Termini S. (eds.) On Fuzziness. A Homage to Lotfi A. Zadeh, Vol. 1, pp. 333–344, Berlin, New York: Springer (2013)

Kovalerchuk, B., Triantaphyllou, E., Despande, A., Vityaev, E.: Interactive Learning of Monotone Boolean Functions. Inf Sci **94**(1–4), 87–118 (1996)

Kreinovich, V. (ed.): Uncertainty Modeling, studies in computational intelligence 683. Springer (2017)

Lichman, M.: UCI machine learning repository (http://archive.ics.uci.edu/ml). Irvine, CA: University of California, School of Information and Computer Science (2013)

Mitchell, T.: Introduction to machine learning Machine learning. McGraw-Hill, Columbus (1997)

Mitchell, T.J., Chen, S.Y., Macredie, R.D.: Hypermedia learning and prior knowledge: domain expertise versus system expertise. J. Comput. Assist. Learn. **21**(1), 53–64 (2005)

Moreno-Torres, J.G., Sáez, J.A., Herrera, F.: Study on the impact of partition-induced dataset shift on k-fold cross validation. IEEE Trans Neural Netw Learn Syst **23**(8), 1304–1312 (2012)

Shannon, C.E.: The synthesis of two-terminal switching circuits. Bell Syst Tech J **28**, 59–98 (1949)

Shimodaira, H.: Improving predictive inference under Covariate Shift by Weighting the Log-likelihood Function. J Stat Plan Infer **90**(2), 227–244 (2000)

Vapnik, V., Vashist, A.: A new learning paradigm: learning using privileged information. Neural Netw. **22**(5), 544–57 (2009)

Chapter 11
Pareto Front and General Line Coordinates

What a good thing Adam had when he said a good thing,
he knew nobody had said it before.
Mark Twain

11.1 Introduction

Optimizing simultaneously several likely *conflicting objectives* is the goal of multi-objective optimization. Mathematically it is an *ill-posed problem.* Formulating it as finding of all *non-dominated solutions* is a known way to make it a mathematically correct problem.

The concept of **Pareto front** (PF) that formalizes the idea of *all non-dominated solutions* plays an important role in *multi-objective optimization* (MOO) problems and related domains (Ehrgott 2006; Ehrgott and Gandibleux 2014). MOO problems with more than three objectives are called *many-objective optimization* problems (Chand and Wagner 2015). The Pareto Front is a mathematically correct solution of multi-objective optimization problems with several conflicting objectives. However, it is only a partial solution for many real-world situations, where only few of the PF alternatives can be implemented, due to resource limitations and other reasons. Commonly Pareto Front is narrowed by linear aggregation of contradictory criteria, where the challenge is assigning weights to criteria. Lossless visualization of multi-dimensional data of the Pareto Front is promising way to assist in interactive selecting appropriate weighs. This chapter shows a way to accomplish this with GLC-L visualization method defined Chap. 7. It also shows a way to visualize the approximation set for the Pareto Front with Collocated Paired Coordinates, defined in Chap. 2 in comparison with Parallel Coordinates to assists in finding "best" Pareto points.

Below we provide basic definitions used in this chapter.

Definition Two n-D points **x** and **y** are *incomparable* if indexes i and j exist such that $x_i > y_i$ and $x_j < y_j$. Example: **x** = (1, 2, 3, 4) and **y** = (2, 1, 3, 4), $x_2 > y_2$ and $x_1 < y_1$.

B. Kovalerchuk, *Visual Knowledge Discovery and Machine Learning*,
Intelligent Systems Reference Library 144,
https://doi.org/10.1007/978-3-319-73040-0_11

Definition A given set of n-D points **P** is a *non-dominated* set if every n-D point **x** in **P** is incomparable with any other n-D point **y** in **P**.

Definition A pair <**F**, **C**> is called a *multi-objective problem*, if **F** is a set of objective functions $\{F_i\}$ and **C** is a set of constrains on a set of objects $\{\alpha\}$.

Definition An n-D point **a** = $(F_1(\alpha), F_2(\alpha),\ldots,F_n(\alpha))$ is called a *feasible* n-D *point* (*solution*) of a multi-objective problem <**F**, **C**> if $\forall\ C_i \in \mathbf{C}\ C_i(\alpha) =$ True.

Definition A set of all non-dominated feasible n-D points of the *multi-objective problem* <**F**, **C**> is called *Pareto Front* (*PF*) for <**F**, **C**>.

In contrast with a generic set of non-dominated n-D points, MOO assumes a specific problem with a given pair <**F**, **C**>.

Below in this section, we follow the review in (Ibrahim et al. 2016) to summarize the role, main challenges, requirements and opportunities of *PF visualization.*

The PF allows and effective *interactive optimization.* Visualization of true Pareto front is difficult because

- It should show the *location, range, shape,* and *distribution* of obtained *non-dominated solutions.*
- It should preserve the *Pareto dominance relation* and relative closeness to reference points in the visual representation.
- Non-dominated solutions produced by MOO algorithms may be only approximations of the true PF (called *approximation sets*).
- Visualization of obtained non-dominated solutions must reveal its *relation* to the true PF.
- Scatter plots visualize only 2-D and 3-D PFs and their approximation sets, more advanced approaches are needed for four or more objectives.
- Existing visualization tools (e.g., parallel coordinates) fail to show the *shape* of the Pareto front.

Thus, visualization of PF consists of two problems:

- Visualization of the *true PF* and
- Visualization of an *approximation set* of the true PF with its likely relationship to the true PF.

There is no single formal definition of the approximation set beyond requiring that it is a distinct set of n-D objective points that are non-dominated relative to each other. These sets can be subsets, supersets of PF, or other sets. The quality of an approximation set is measures by such characteristics as convergence, spread, and distribution of objective vectors.

Visualization of the approximation set allows to:

- estimate the location, range, and shape of PF,
- assess conflicts and trade-offs between objectives,
- select preferred solutions,
- monitor the progress or convergence of an optimization run, and

- assess the relative performance of different MOO algorithms better than formal measures due to difficulties to capture in a single measure all variety of aspects of PF.

The major *requirement* of the PF and its approximation set visualization is preserving the *Pareto dominance* between n-D points, i.e., making it evident from visualization. Other requirements are:

- maintaining shape, range, and distribution of PF points,
- being stable under addition or removal of points within the range of the approximation set,
- handling a large number of PF points of large dimensions and multiple approximation sets for comparison, and
- being simple to understand and use.

For two or three objectives, we have scatter plots that satisfy most of these requirements. This situation is similar to what we have in general for 2-D and 3-D data visualization (see Chap. 10). All challenges are coming at the higher dimensions.

11.2 Pareto Front with GLC-L

While PF provides a mathematically correct solution of MOO problems, it is only a partial solution for many real world situations, where only one or few of the PF alternatives can be actually implemented as a solution, due to multiple reasons such resource limitations. For instance, the aircraft manufacturer cannot build airplanes for every alternative in PF. Thus, the next task is *narrowing* PF. The common way of doing this is linear aggregation of multiple criteria that is going back to (Gass and Saaty 1955). The major challenge in linear aggregation is assigning weights to criteria. Visualization is natural way to assist in interactive selecting appropriate weighs.

Below we show a way to accomplish this with GLC-L that was defined in Chap. 7. An analyst needs to see if weights in a linear function F are consistent with analyst's preferences of alternatives. Figure 11.1 shows two 4-D objective alternatives $\mathbf{a} = (1,1,1,1)$ and $\mathbf{b} = (1.2,0.5,1.4,0.7)$ visualized in GLC-L coordinates X_1–X_4 with angles (Q_1, Q_2, Q_3, Q_4). In GLC-L all vectors a_i and b_i shifted to be connected one after another and the end of last vector projected to the black line. X_1 is directed to the left due to negative coefficient k_1. Coordinates for negative ki are always directed to the left.

Projections of **a** and **b** show that weighted **a** is greater than weighted **b**. However, for the analyst **b** is better than **a,** this contradiction is evident in GLC-L. GLC-L shows the contribution of each weights and criteria and allows adjusting them interactively by changing angels as shown in Fig. 11.2 to make weights consistent with preference of **b** over **a**, **b**, **b** > **a**

Assume that an analyst has a matrix of preferences P on the alternatives in PF. If **a** > **b** in P, but $F(\mathbf{a}) < F(\mathbf{b})$ in GLC-L then **a** and **b** and the endpoints of projections of **a** and **b** to U axis are colored red. Then in line with Fig. 11.1 the number of red lines and dots (for endpoints) can indicate the level of inconsistency between F and P.

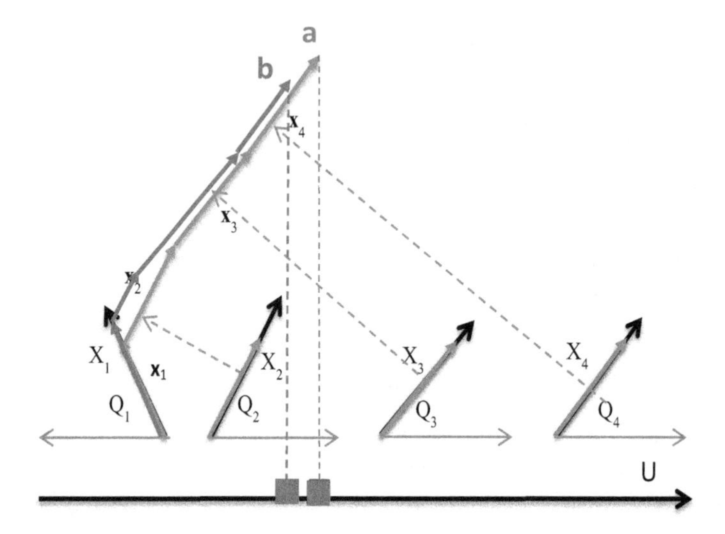

Fig. 11.1 4-D points **a** = (1, 1, 1, 1) and **b** = (1.2, 0.5, 1.4, 0.7) in GLC-L coordinates X_1–X_4 with angles (Q_1, Q_2, Q_3, Q_4) with vectors $\mathbf{x}_i$ shifted to be connected one after another and the end of last vector projected to the black line. X_1 is directed to the left due to negative k_1. Coordinates for negative k_i are always directed to the left

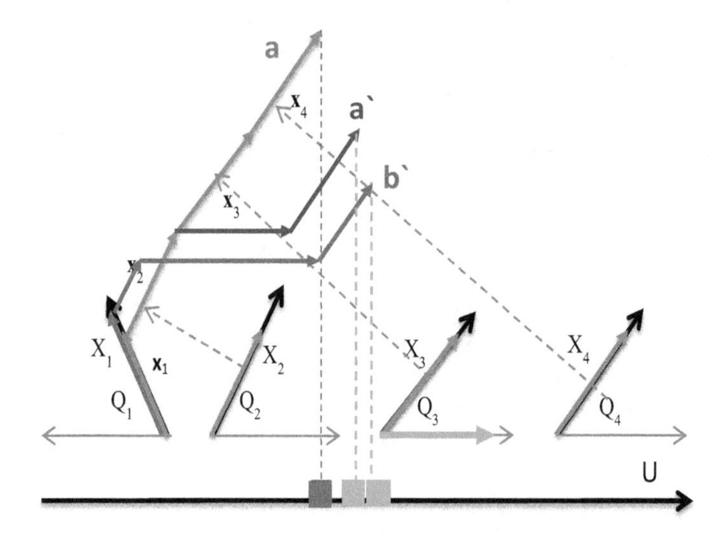

Fig. 11.2 Adjusting coefficients of the weighted sum of criteria to meet expert's preferences

Next, the analyst may have a full order of all alternatives in addition to preferences of pairs of alternatives. While the order of alternatives is transitive the preferences on pairs may not. In addition, the order can be with the strength of the preference on some numeric scale. GLC-L can show consistency of F with such order and preferences. Figures 11.1 and 11.2 illustrate this too. The distance between projections indicates this. If the user expects the small difference, but it is large according to F as in Fig. 11.1, then the analyst can make it smaller as in Fig. 11.2. In Fig. 11.2 it is not only smaller but with a different sign (negative).

Another goal is of using PF is decreasing the occlusion and clutter by visualizing only PF of a dataset for the tasks such observing a set of students and selecting one representative student for each category such as "Best", "OK", "not OK". Only PF students in each category will be visualized.

The comparison of GLC visualization of PF with Parallel Coordinates is provided in the next section. The comparison with lossy methods such as RadVis for Iris data is shown in Sect. 5.7.4 in Chap. 5. While those data are not directly about PF, they are representative for the comparison situation.

PF alternatives can be ordered by the weighted Euclidean distances to the "ideal" alternative that is not a part of PF, but can be a majorant of all alternatives in PF. The assignment of weights interactively with GLC-L is illustrated in Fig. 7.5 in Chap. 7. The distance-based idea is explained below. Let **a** be such ideal alternative (prototype) expressed as an n-D point, where n is the number of multi-objective criteria and **y** be an n-D point from PF for these n criteria. Then we compute a squared weighted Euclidian distances

$$d_{\mathbf{k}}(\mathbf{a}, \mathbf{y}) = k_1(a_1 - y_1)^2 + k_2(a_2 - y_2)^2 + \cdots + k_n(a_n - y_n)^2$$

between **a** and **x**. To visualize it in GLC-L we rewrite $d_{\mathbf{k}}(\mathbf{a}, \mathbf{y})$ using another notation, where

$$x_1 = (a_1 - y_1)^2, x_2 = (a_2 - y_2)^2, \ldots, x_n = (a_n - y_n)^2$$

Thus, in this notation, $d_{\mathbf{k}}(\mathbf{a}, \mathbf{y}) = k_1x_1 + k_2x_2 + \cdots + k_nx_n$. This is exactly GLC-L linear from that was visualized in Chap. 7. Therefore, we can adjust angles in this visualization in the same way as it is done in Fig. 7.5 in Chap. 7. This idea is in line with generalization of GLC-L to non-linear relations presented in Chap. 7.

Example 1 We illustrate this PF approach in comparing it with Grade Point Average (GPA) based practice for selecting students to be admitted to the Computer Science major. Consider grades (3, 2.5, 4, 2, 3) and (4, 4, 2, 4, 3) of two students S_1 and S_2, respectively, in five classes Cl_1–Cl_5. Their GPA in these classes are 2.9 and 3.4, respectively. Based on these numbers, the second student will be admitted to the Computer Science major more likely than the first one. However, these classes have different importance for admission of these students to the Computer Science major program. The experts may say that Cl_3 is the most important and rather the second student must be admitted. Thus, assume that we have expert's preference (3, 2.5, 4, 2, 3) > (4, 4, 2, 4, 3). To set up weights of the classes to the make weighted GPA consistent with expert's opinion, we experiment with different weights w_1–w_5 and find (w_1, w_2, w_3, w_4, w_5) = (0.5, 0.6, 1.0, 0.2, 0.3).

Applying these weights gives us weighted GPA, $\mathrm{WGPA}_1 = 1.66$ and $\mathrm{WGPA}_2 = 1.62$ that reverse GPA preferences that is matched with expert's preference. These experimenting can be done with GLC-L interactively as shown in Fig. 11.2. Multiple different weights can satisfy this requirement. GLC-L has an advantage of visualizing the alternatives and abilities of adjusting and selecting weighs to meet better expert's opinion.

Example 2 This example uses actual grades of over 50 students in 6 courses to be admitted to the Computer Science major program. First, the Pareto front was found in 6-D. Next the expert preliminary assigned weights to these 6 courses. Then a group of four students in PF was selected to test these weights on consistency to the expert opinion.

Figure 11.3a visualizes in GLC-L these four students as lines along with the green line for the "ideal" student with all 6 A grades. Figure 11.4 zooms the ends of these lines and their projections. It uses weights that the expert preliminary assign. This visualization shows that these original weights are not consistent with expert's ordering of students. Interactive adjustment of coefficients made the weighted sum consistent with expert order of the students: $St_1 > St_5 > St_3 > St_2 > St_4$ to be admitted to the major.

Figures 11.3b, c visualize these five students using other coefficients that provide weighed sums that are consistent with the expert's order of students. The coefficients used in Fig. 11.3c are equal to 0 in three courses that effectively provided dimension reduction from 6-D to 3-D. The expert was able to justify it by noting that remaining courses are more advanced and can be sufficiently representative for the admission.

11.3 Pareto Front and Its Approximations with CPC

This section demonstrates visualization of PF for 4-D data in Collocated Paired Coordinates (CPC) defined in Chap. 2. Figure 11.5 shows an example of six 4-D points in CPC. Three orange arrows that encode 4-D points **a–c** represent Pareto Front. Other three 4-D points **x, y, z** show as blue arrows are dominated by at least one of points from PF.

Dotted blue arrows show dominance directions between 4-D points. Point **z** has two dominance arrows to point **a** that link node (a_1, a_2) with node (z_1, z_2) and node (a_3, a_4) with node $(z_3 \cdot z_4)$. The direction of all dominance arrows is to up and right. Figure 11.6 shows data from Fig. 11.5 in Parallel Coordinates for comparison.

Figure 11.7 shows two examples of approximation of PF. The first approximation consists of 4-D points **r** and **m** shown as green and grey arrows, respectively. Point **r** is a majorant for PF and point **m** is a minorant for PF, i.e., all PF points are less than **r** and greater than **m**.

These dominance arrows indicate that (a_1, a_2) dominates (z_1, z_2) and (a_3, a_4) dominates $(z_3 \cdot z_4)$. In this way, we can visually establish that **z** is not in the PF. If only one dominance arrow or none of them connects two 4-D points then these 4-D points are incomparable and therefore are candidates to PF. The second approximation consists of 4-D points **r** and **q.** The point **q** shown as a black arrow is a minorant of all six 4-D points from Fig. 11.7. Figure 11.8 shows data from

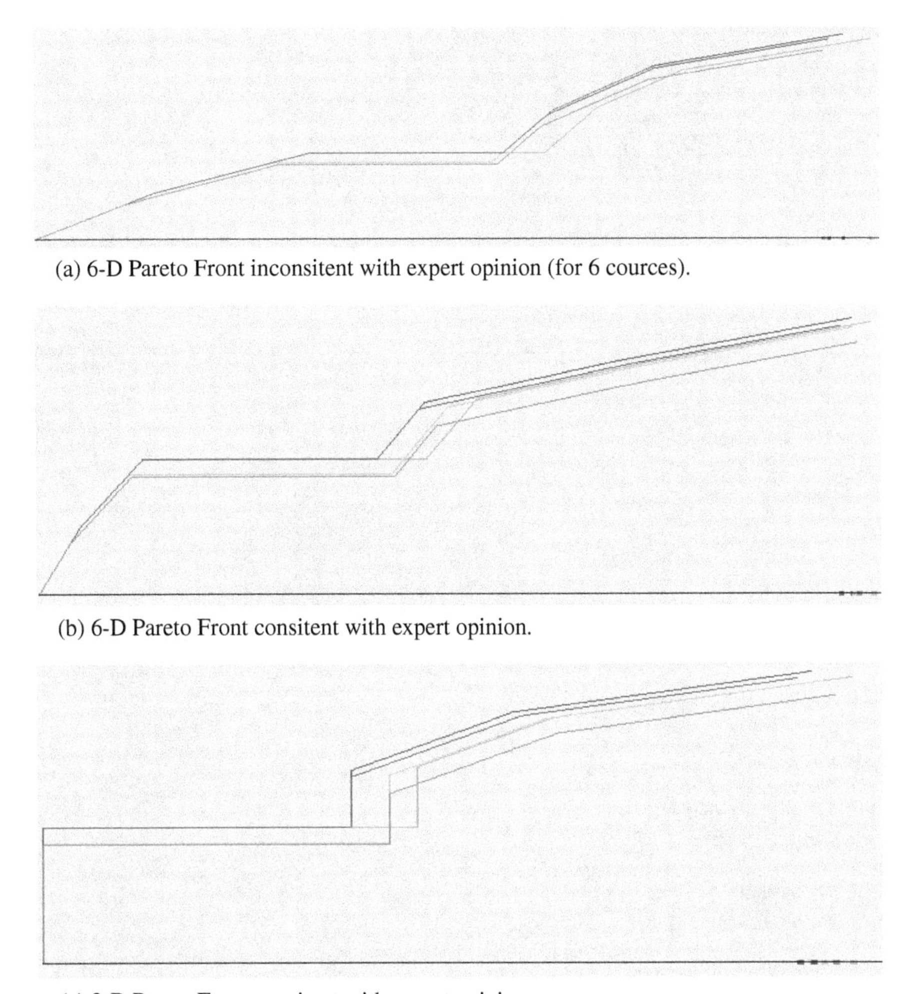

(a) 6-D Pareto Front inconsitent with expert opinion (for 6 cources).

(b) 6-D Pareto Front consitent with expert opinion.

(c) 3-D Pareto Front consitent with expert opinion,

Fig. 11.3 6-D and 3-D Pareto Fronts in GLC-L

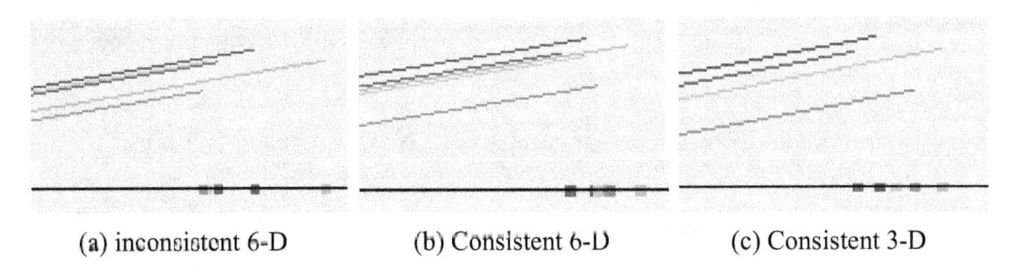

(a) inconsistent 6-D (b) Consistent 6-D (c) Consistent 3-D

Fig. 11.4 Zoomed ends of GLC-L graphs and projections from Fig. 11.3

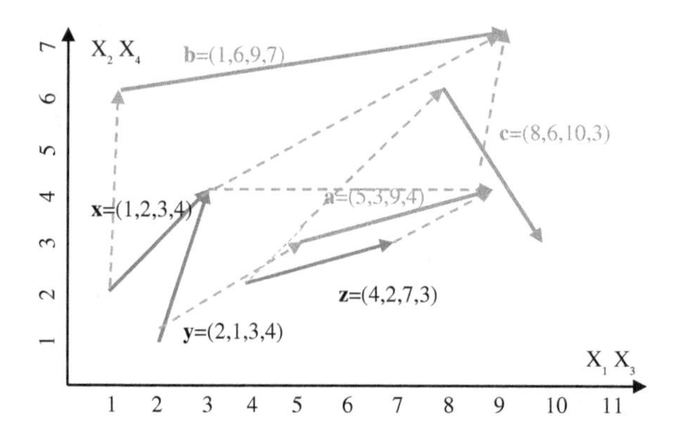

Fig. 11.5 Six 4-D points as arrows in CPC with Pareto Front (orange arrows). Dotted blue arrows show dominance directions between 4-D points (arrows)

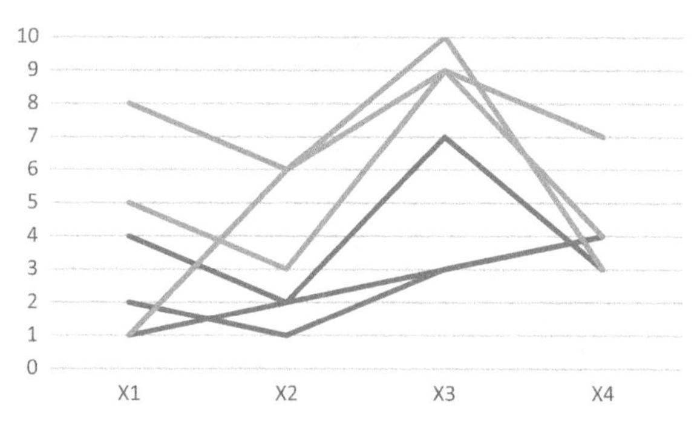

Fig. 11.6 Data from Fig. 11.5 in Parallel Coordinates

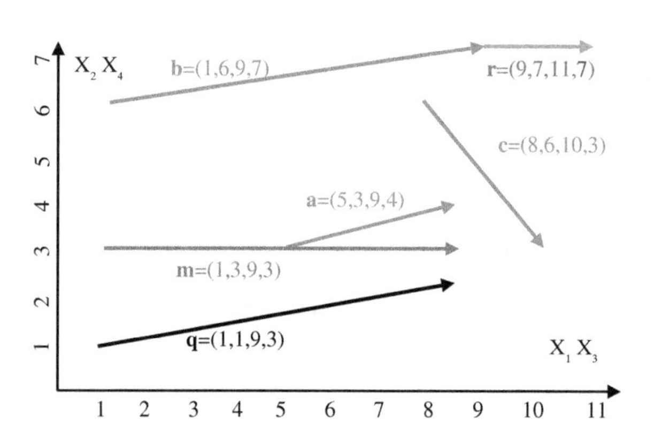

Fig. 11.7 Approximation of Pareto set (green arrow **r** and grey arrow **m**). Point **r** is a majorant and point **m** is a minorant for PF. Point **q** is a minorant of all six 4-D points from Fig. 11.5 in Collocated Paired Coordinates

Fig. 11.7 in Parallel Coordinates for comparison. Next, we consider a linear PF that is defined below.

Definition Pareto Front is a *linear Pareto Front* for a set of n-D points $\{\mathbf{p}_i\}_{i=1:\,m}$ if any affine sum **s** of $\{\mathbf{p}_i\}$ is in PF, i.e., $\mathbf{s} = a_1\mathbf{p}_1+a_2\mathbf{p}_2+\cdots+a_m\mathbf{p}_m$ for any $\{a_i\}_{i=1:\,m}$ such that $a_1+a_2+\cdots+a_m = 1$ and $\forall i\ a_i \in [0, 1]$.

A set of these n-D points $\{\mathbf{p}_i\}_{i=1:\,m}$ is called the *basis*. As an example, we construct a linear Pareto Front for four 4-D points $\mathbf{p}_1$–$\mathbf{p}_4$ shown in Table 11.1.

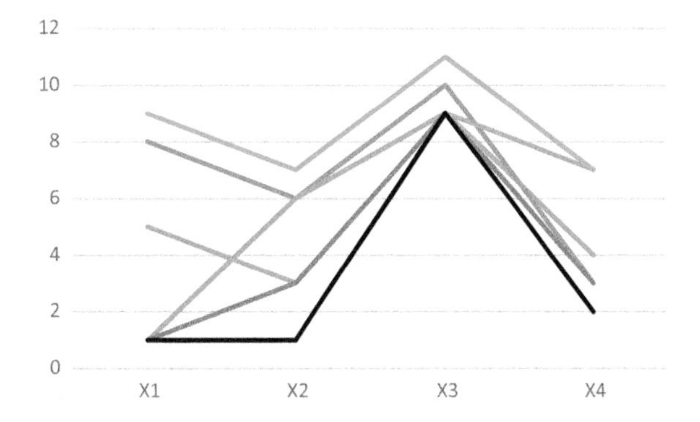

Fig. 11.8 Data from Fig. 11.7 in Parallel Coordinates

Table 11.1 Four 4-D basis points for linear affine PF

Point	X_1	X_2	X_3	X_4
$\mathbf{p}_1$	0	0	0	1
$\mathbf{p}_2$	0	0	1	0
$\mathbf{p}_3$	0	1	0	0
$\mathbf{p}_4$	1	0	0	0

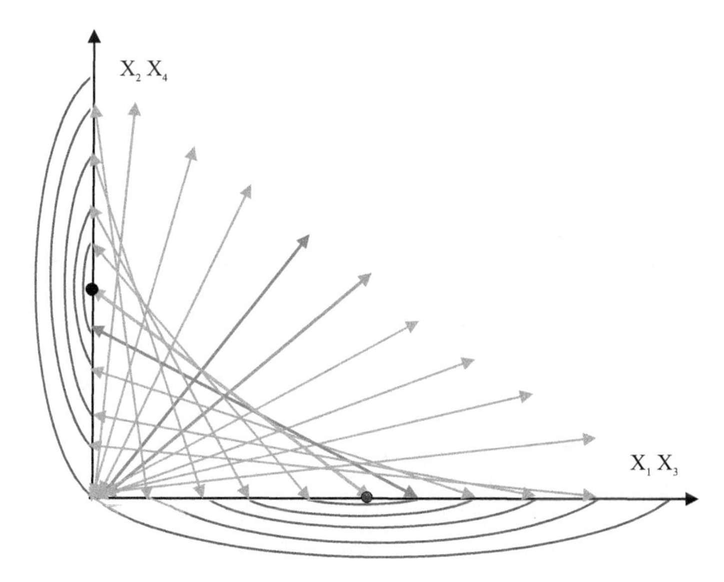

Fig. 11.9 Subset of a linear Pareto Front for four 4-D points $\mathbf{p}_1$–$\mathbf{p}_4$ visualized in Collocated Paired Coordinates. Each curve represents two 4-D points similarly to straight lines

For illustration of this linear PF we use values of coefficients a_i from [0, 1] with step 0.1, i.e., 0, 0.1, 0.2,…,0.9,1 and any combinations with only two nonzero coefficients $a_i > 0$, $a_j > 0$, e.g., $0.3\mathbf{p}_2+0.7\mathbf{p}_4$, $0.1\mathbf{p}_1+0.9\mathbf{p}_3$. This subset of the linear Pareto Front is shown in Fig. 11.9. It is one of the possible approximations of the

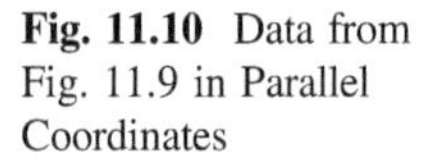

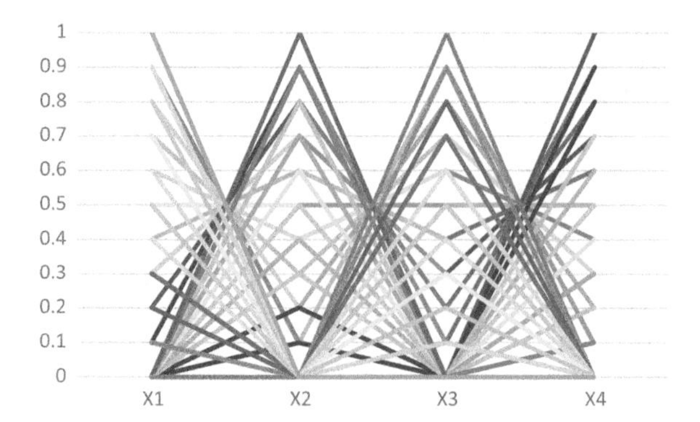

Fig. 11.10 Data from Fig. 11.9 in Parallel Coordinates

full linear PF for 4-D points $\mathbf{p}_1$–$\mathbf{p}_4$. Two black dots on X and Y coordinates represent two 4-D points (0.5, 0, 0.5, 0) and (0, 0.5, 0, 0.5).The orange two-sided arrow represents other two 4-D points (0, 0, 0.5, 0.5) and (0.5, 0.5, 0, 0). The green two-sided arrow represents other two 4-D points (0.5, 0, 0, 0.5) and (0, 0.5, 0.5, 0).

Figure 11.10 shows the same data in Parallel Coordinates as in Fig. 11.9. Figure 11.10 is much more cluttered than Fig. 11.9, because it used three lines per each 4-D point, versus only one line in CPC in Fig. 11.9.

In CPC representation of PF, similar n-D points are located next to each other, which allows an analyst to interactively exploring similar options to select the "best" one. The bold blue arrow going up represents (0.4, 0.6, 0, 0) and the orange arrow that is also going up next to it represents (0.5, 0.6, 0, 0). In the same way, the bold blue arrow next to the green arrow, represent a 4-D point that is similar to the green one.

This CPC visualization of PF can be provided for higher dimensions than 4-D and for other types of PF not only linear explored above such as hyper-sphere and hyper-cylinder (hyper-tube). Also similarly, to visualization in CPC we can visualize PF in other paired coordinates defined in Chap. 2 such as SPC. The advantage of Parametrized SPC is in abilities representing any given n-D point as a single 2-D point losslessly as shown in Sect. 2.2 in Chap. 2. This ability combined with visualization of the similar points as short graphs next to that 2-D point provides an intuitive way to analyze similar n-D points in PF.

Another opportunity with CPC for PF is in abilities to generate multiple approximate sets of Pareto Front and to split CPC space to quadrants and approximating PF in quadrant including finding and visualizing local majorants and minorants in each quadrant.

References

Chand, S., Wagner, M.: Evolutionary many-objective optimization: a quick-start guide. Surv. Oper. Res. Manage. Sci. **20**(2), 35–42 (2015)

Ehrgott, M.: Multicriteria optimization. Springer Science & Business Media, Germany (2006)
Ehrgott, M., Gandibleux, X.: Multi-objective combinatorial optimisation: Concepts, exact algorithms and metaheuristics. In: Al-Mezel, S.A.R., Al-Solamy, F.R. M., Ansari, Q.H. (eds.) Fixed Point Theory, Variational Analysis, and Optimization, pp. 307–341. CRC Press (2014)
Gass, S., Saaty, T.: Parametric objective function Part II. Oper. Res. **3**, 316–319 (1955)
Ibrahim A, Rahnamayan S, Martin MV, Deb K.: 3D-RadVis: visualization of Pareto front in many-objective optimization. In: Evolutionary Computation (CEC), 2016 IEEE Congress on 2016 Jul 24 (pp. 736–745). IEEE

Chapter 12
Toward Virtual Data Scientist and Super-Intelligence with Visual Means

All generalizations are false, including this one.
Mark Twain

The Big data challenge includes dealing with a big number of, heterogeneous and multidimensional, datasets, of *all possible sizes,* not only with the data of big size. As a result, a huge number of Machine Learning (ML) tasks, which must be solved, dramatically exceeds the number of the data scientists, who can solve these tasks. Next, many ML tasks require the critical input, from the subject matter experts (SME), and end users/decision makers, who are not ML experts. A set of tools, which we call a "virtual data scientist" is needed to assist the SMEs, and end users to construct the ML models, for their tasks, to meet this Big data challenge, with a minimal contribution from data scientists. This chapter describes our vision of such a "virtual data scientist", based on the visual approach of the General Line Coordinates.

12.1 Introduction

This chapter considers a problem of *automated creation of empirical ML models* of the real, complex processes from the data with the *dominant* role of SMEs, who are *not data scientists,* with the minimal involvement of Data Scientists (DSs). We approach this problem from the Visual Knowledge Discovery viewpoint. This goal is within the DARPA recent vision of Machine Learning (DARPA 2016).

Tasks The Machine Learning modeling tasks, which need the active involvement of SMEs include (DARPA 2016):

B. Kovalerchuk, *Visual Knowledge Discovery and Machine Learning*,
Intelligent Systems Reference Library 144,
https://doi.org/10.1007/978-3-319-73040-0_12

(1) **Formally defining the modeling problems** (defining the objectives, attributes/features, relations, evaluation metrics, labeling data, etc.),
(2) **Constructing the ML models for the formally defined problems** (selecting a class of the ML methods, constricting the ML model within a selected class of methods, by optimizing model parameters, using the input training data),
(3) **Curating the automatically constructed models** (providing the explanatory data and model visualization, finding spurious correlations, evaluating the predictive prospects on new data, etc.).

This list is not complete. Other tasks include the feature subset selection, dimensionality reduction, feature extraction, eliminating spurious attributes; dealing with noisy visually overlapping data, solving the clustering tasks, tasks with the high dimensional data, from heterogeneous sources, etc.

This chapter considers the difficulties, in solving these tasks, and then focuses on the SMEs involvement, for model construction (task 2 above), as one of the most difficult ones for the SME. Typically, the SME *cannot select an initial class of ML methods,* such as the Support Vector Machine (SVM), Bayes, Artificial Neural Networks (ANN), k-Nearest Neighbors (k-NN) or others, not being an ML expert. Moreover, for Data Scientists it is also a significant challenge. One of challenges is that humans *cannot see the n-D data, with a naked eye*. Thus, at first glance, it seems unrealistic to expect selecting a class of ML methods, and building the ML models with them, by a non ML expert. This chapter shows that the GLC approach is a promising way, for doing this.

Below, we first present the typical deficiencies of the ML projects, and outline the ways for resolving them. Then, we outline the vision of constructing the n-D ML models, with the SME, by generalizing the success in 2-D. After that, we present the visual approach, for constructing the ML classification models, with the SME. A visual approach to the curation of the ML models, and the preattentive n-D data visualization, for the classification and clustering, with the outlining of the future research, conclude this chapter.

12.2 Deficiencies

Typical deficiencies D1–D5, of the ML project, are listed below, with the outline and a discussion of the ways to resolve them, including the role of the SME, and the visual means.

D1. Questionable Input (*input training data deficiency*): *incomplete, noisy biased, and redundant data.* It commonly leads to the overfitting, irrelevant correlations, and the wrong predictions.

Adding the relevant data, removing the redundant data, denoising the data, and dealing with biases are the time consuming tasks, and often are out of competence, of the data scientists, who build the ML models. Visual means can assist the SME, and the data scientists in doing this.

D2. Inaccurate Model (*with sufficient/representative input data*). The technologies, to improve the models, can potentially resolve this deficiency.

The known solutions are *incremental learning* ML models, with additional data, and *discovering the limited subareas,* where the models that are built deficient, but the data can still be useful. Visual means can assist the SME, and the data scientists, in the incremental learning, and finding these subareas.

D3. Unexplained Model *the lack of explanatory power, of the accurate—enough, ML models.* These models are *black box* models, for the SME, who can refuse to use the black box models. The technologies, to convert the black box models, into models, which make sense to the SME, can resolve this deficiency. Extracting the logical rules, from the neural networks is one of these technologies, which can be also be visualized for the SMEs, to understand the underlying learned models. Visualization, of such logical rules, and other explainable models, can assist the SME, in understanding these models.

D4. Inconsistent Model (*the explanatory ML model, derived from the data, but rejected by the SME*). In these cases, the explanation is wrong for the SME, even when the model is accurate enough, on the given data.

For instance, an SME can claim, that the high accuracy is the result of over-fitting, by limited training data, which are not representative for the task (Kovalerchuk et al. 2000). Another claim can be that the model is based on spurious variables, overgeneralizes, or undergeneralizes the training data, and will lack the predictive capabilities, on the new data, due to these reasons.

The ways, to resolve this deficiency, are the same as for D2 and D3, but with a better explanation, and focus. The visual means can help, in this model rebuilding process, including the faster discovery of inconsistencies.

D5. Lack of Skills The need for tight collaboration between the SMEs and the data scientists, often is mandatory in machine learning modeling. While the obvious approach is mutual knowledge acquisition, from the subject domain, and ML, it is often time consuming, and impractical. Another approach is developing the tools (including visual tools), to enable an SME to build the ML models, with the minimal participation of the data scientists. This is our interest in this chapter.

The deficiencies D1–D5 can be found, in all tasks 1–3. Questionable input (D1), and lack of skills (D5), are related to defining the problems (task 1). Inaccurate unexplained, or inconsistent models (D2–D4), and lack of skills (D5) are related both to constructing and curating models (tasks 2 and 3).

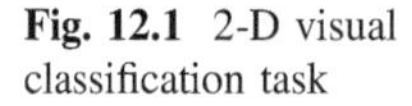
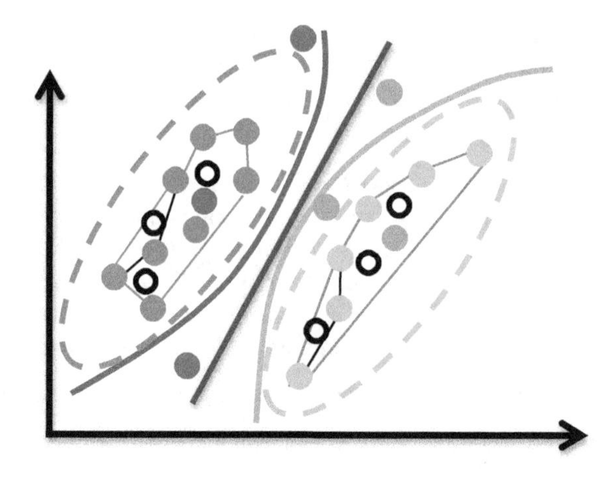

Fig. 12.1 2-D visual classification task

12.3 Visual n-D ML Models: Inspiration from Success in 2-D

The major challenge, in all three tasks, is the high dimensionality of data in ML. The SMEs can build the models in 2-D, with the two variables, visually and with minimal ML knowledge, as we later illustrate, in Fig. 12.1.

Such models include as linear regression, linear, and non-linear discrimination functions. In contrast, in n-D, it is impossible to see the n-D data, with a naked eye. As a result, the SME and the data scientists cannot build the n-D ML models visually, without the special visual means ("n-D glasses").

Levels of Generalization in the Construction of the ML Models Different ML analytical methods favor different levels of generalizations, of the training data, in constructing the ML models. This level can be over-generalization, or under-generalization of the given training data. It is difficult to select, and justify the right level without the SME. It is especially difficult for the high-dimensional data.

Figure 12.1 illustrates the fundamental difficulty of over-generalization, or under-generalization of the training data in developing the data classification models. It shows the *five levels of generalization* of the training data of the two classes (blue and green dots, respectively) from the widest to narrowest generalizations:

1. The *widest* generalization is given by a *brown straight line*. Here every point on the right is classified, into the green class, and every point on the left is classified, into the blue class.
2. The second level of generalization of training data is presented by the *blue and green curves*, i.e., only the points, which are on the right of the green curve are classified into the green class and, respectively, only points, on the left of the blue curve, are classified into the blue class. The points, between these two curves, are not classified at all.

3. The third level of generalization, of the training data, is presented by the *blue and green ovals*, i.e., only the points, which are within the ovals are classified into these classes.
4. The forth level of generalization is presented by the blue and green *convex hulls,* around these training data. It presents a more conservative generalization, than the levels 1–3. Only the points, which are within these convex hulls, are classified into these classes. The system will refuse to classify the points, which are outside of these convex hulls.
5. The fifth level of generalization is presented by the *non-convex hulls,* around the blue and green dots, which is a more conservative generalization, than 1–4. The black lines, in Fig. 12.1, show the deviations of these hulls, from the convex hulls.

In Fig. 12.1, all the testing data are dots, with black outlines. All of them are within the blue and green convex hulls, and the two of them are outside, of the non-convex hulls. Thus, levels 1–4 of the generalization classify all of them, and the non-convex hulls generalization refuses to classify the two of them.

Now, assume that the new data, to be classified, are the red and orange dots, in Fig. 12.1. All five generalizations will consistently classify the red, and the orange dots, which are in the middle of respective ovals. However, these generalizations will be inconsistent, for the four other new points. For instance, the red dot, on the top, will be classified into the blue class at levels of generalizations 1 and 2, and will be refused to be classified at the levels 3–5.

Classification Confidence and Visualization The solution, for this situation, proposed in literature, is the provision of a *classification confidence measure,* for each prediction. First, not all the ML methods provide such measures. If measures are provided, they are rarely comparable across different ML methods, but are ML method specific. For instance, for the brown straight line it can be how large is Euclidean distance from this line to the point, but for the ovals (ellipses) it can be how large is Mahalanobis distance from the center of oval to the point. Next, both of the distances can be irrelevant, to the user task and goal, because they are *not derived, from the user task* at hand, but are *externally imposed, by the respective ML methods*.

To deal with this challenge, we propose, as a part of the virtual data scientist vision:

- *visualizing multiple levels of generalizations,* produced by the n-D ML models, for the SME, such as shown in the Fig. 12.1, and
- accompanying it by the, formally computed, *confidence measures,* provided by the different ML methods.

This will allow the SMEs to make a better-informed decision, on selecting the level of generalization. While this approach works for 2-D, we need the visualization means to produce the figures, like Fig. 12.1, *for the n-D data.*

Multiple *data dimensionality reduction methods* provide the opportunity, to present the different levels of generalization, of the n-D data in 2-D. These methods include Principal Component Analysis (PCA), Multidimensional Scaling (MDS), manifolds, Self-Organized maps (SOM), and others. As was discussed in Chap. 1, while these methods are very valuable, not all of the n-D data can be represented in this way.

The first two PCA components may not capture the relations between the n-D points, accurately enough. MDS, SOM, and manifolds may not represent the distances between the n-D data points accurately in 2-D, for some n-D datasets (Duch et al. 2000).

In other words, these methods can be *inadequate,* in representing some of the n-D data in 2-D. As a result, the constructed ML models can be inaccurate, and/or uninterpretable by the SMEs. The combination of the GLC approach, with such lossy methods, allows minimizing the impact of the information loss.

12.4 Visual n-D ML Models at Different Generalization Levels

Convex Hull Generalization in Model Construction The visual approach, to the classification model construction, is illustrated below with the data, from Table 12.1. In this table, the class 1 is represented, by its 16 four-dimensional border points, i.e., all 4-D points of class 1 are within a convex hull H_1, formed by these 4-D points. Similarly, the class 2 is represented by its border 4-D points, which form a convex hull H_2. Respectively, the formal solution for this classification task is:

Table 12.1 4-Data for 2 classed (16 4-D points for each class)

#	Class 1				#	Class 2			
a_1	**1**	**5**	**2**	**4**	b_1	**2**	**4**	**3**	**3**
a_2	1	5	3	4	b_2	2	4	4	3
a_3	1	5	4	6	b_3	2	4	5	5
a_4	1	5	5	6	b_4	2	4	6	5
a_5	2	5	2	4	b_5	3	4	3	3
a_6	**2**	**5**	**3**	**4**	b_6	**3**	**4**	**4**	**3**
a_7	2	5	4	6	b_7	3	4	5	5
a_8	2	5	5	6	b_8	3	4	6	5
a_9	3	7	2	4	b_9	4	6	3	3
a_{10}	3	7	3	4	b_{10}	4	6	4	3
a_{11}	**3**	**7**	**4**	**6**	b_{11}	**4**	**6**	**5**	**5**
a_{12}	3	7	5	6	b_{12}	4	6	6	5
a_{13}	4	7	2	4	b_{13}	5	6	3	3
a_{14}	4	7	3	4	b_{14}	5	6	4	3
a_{15}	4	7	4	6	b_{15}	5	6	5	5
a_{16}	**4**	**7**	**5**	**6**	b_{16}	**5**	**6**	**6**	**5**

$$\text{If } \mathbf{x} \text{ inside } H_1 \text{ then } \mathbf{x} \in \text{ class } 1,$$
$$\text{If } \mathbf{x} \text{ inside } H_2 \text{ then } \mathbf{x} \in \text{ class } 2.$$

Below we show, that one can find this solution visually, in 2-D, while the data are in 4-D using the Parametrized Shifted Paired Coordinates (PSPC), defined in Chap. 3. This will allow the SMEs, who are not ML experts, to operate with these 4-D data, in a familiar 2-D space.

The 4-D point (3, 5, 4, 4) is used, as a parameterization parameter of the PSPCs, which are described below. Figure 12.2 shows this point, in those shifted coordinates, in red. Note, that it is shown, as a 2-D point losslessly, i.e., all of its four dimensions can be restored from this 2-D point. It is done first, by projecting it, in the Fig. 12.2, to X_1 and X_2 coordinates, and getting (3, 5), then by projecting it to X_3 and X_4 coordinates, and getting (4, 4).

Only those n-D points that have the same overlaying projection lines for all odd coordinates and the same overlaying projection lines for all even coordinates will be represented as single 2-D points. This is the case, for points (1, 5, 2, 4), and (2, 5, 3, 4), shown in blue and green, in Fig. 12.2. Other n-D points will have the n/2 2-D points, for the even n, and $(n + 1)/2$, for the odd n. For the odd n, the last coordinate is duplicated, to get the even number of coordinates.

Other points are represented by graphs, which consist of the two points connected by an arrow. This is illustrated in Fig. 12.2 for the point (1, 5, 3, 4). The first 2-D point (1, 5) in (X_1, X_2) is shown in blue, and the second point (3, 4) in (X_3, X_4) is shown in green. Figure 12.3 shows the 4-D points, from Table 12.1, which have lossless representation, as single 2-D points, in this PSPC system.

This is an important advantage of the PSPCs. It allows representing losslessly, not only the 4-D points, as the 2-D points, but also the points of higher dimensions. For instance, for the 6-D parameterization point (3, 5, 4, 4, 6, 1), coordinates X_1, X_2 are shifted to the left by 3 and 5 positions, respectively; the coordinates X_3, X_4 are shifted to the left by 4 positions each; and coordinates X_5, X_6 are shifted to the left

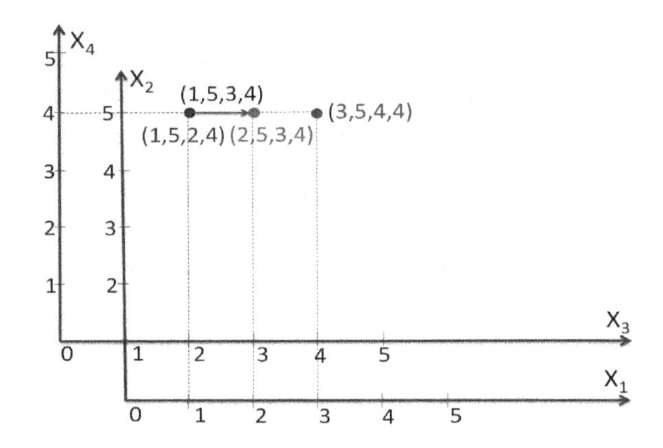

Fig. 12.2 Shifted 4-D coordinates, based on the parameterization of the 4-D point (3, 5, 4, 4)

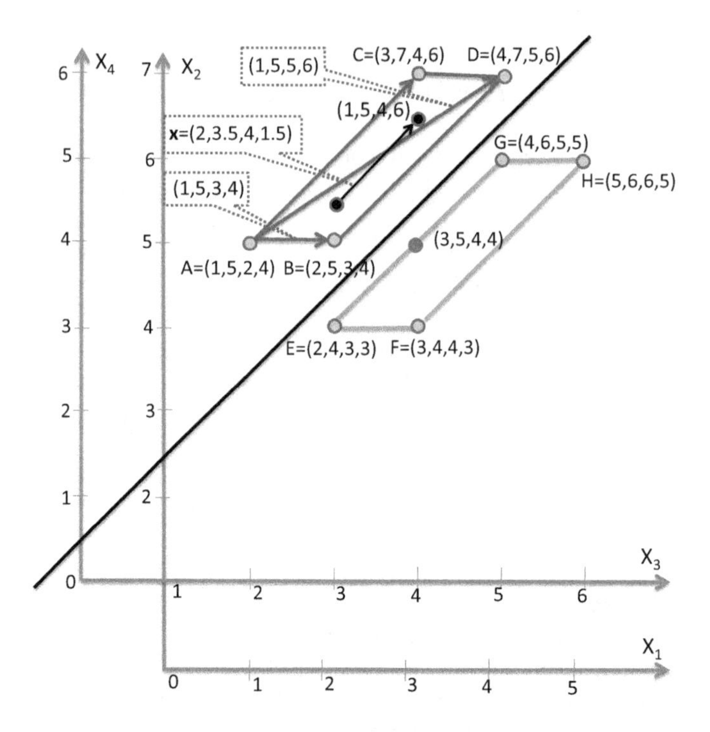

Fig. 12.3 4-D points, which have the lossless representation, as single 2-D points, and the selected another 4-D point in PSCs, based on the parameterization point (3, 5, 4, 4), shown in red

by 6 and 1 positions, respectively. Bold points, in Table 12.1, are corners of two 4-D hyper-rectangles.

Each of these 4-D hyper-rectangles is mapped one-to-one to 2-D quadrilaterals (quads for short) with corners (A, B, C, D) and (E, F, G, H), respectively, shown in Fig. 12.3. Thus, for accurate classification of all points from classes 1 and 2 visualization in Fig. 12.2 is *sufficient*. For this, a user simply draws a given 4-D point x = (x_1, x_2, x_3, x_4), e.g., (2, 3.5, 4, 1.5) in the coordinates X_1−X_4, in Fig. 12.3 as the two 2-D points (x_1, x_2) = (2, 3.5) and (x_3, x_4) = (4, 1.5), connected by an black arrow. As was shown, it can be a single 2-D point, if these two 2-D points are in the same location. Then a user visually checks, whether this arrow is within one of the rectangles. This is the case for $\mathbf{x} = (x_1, x_2, x_3, x_4) = (2, 3.5, 4, 1.5)$, which is shown as a black arrow in Fig. 12.2, in the rectangle with the corners (A, B, C, D). Thus (2, 3.5, 4, 1.5) belongs to class 1. This *100% accurate visual test* leads to the *100% accurate analytical discrimination rule* for any $\mathbf{x} = (x_1, x_2, x_3, x_4)$, which tests that the point **x** belongs to the respective class.

Hyper-plane Generalization in Model Construction Next, we consider a typical, supervised learning *classification task,* where *only the training data* are given, such as in Table 12.1. The information, that the 4-D points in Table 12.1 form the two convex hulls, around these two sets of points, is not provided; i.e., no guidance is provided, and no limiting overgeneralizations of classes, by convex hulls, are provided. Without such limitations, a black diagonal line *F*, in Fig. 12.3 is acceptable. It discriminates, these two datasets, with 100% accuracy. A user can draw it on the screen, by using a mouse function. Next, its equations can be easily be *extracted visually* or *automatically* as $x_2 = x_1 + 2.5$, in the coordinates (X_1, X_2), and as $x_4 = x_3 + 0.5$, in the coordinates (X_3, X_4).

Let's illustrate the use of this line *F*, for the *visual decision* for the point (1, 5, 3, 4), which is shown, in Fig. 12.3, as a horizontal arrow from the point (1, 5) to the point (3, 4). The class 1 is above the line *F*, and this arrow is also above it, therefore (1, 5, 3, 4) is in class 1. This discrimination is done, *completely visually,* in 2-D, for a 4-D point, similarly to the visual solution, for the previous task. For *analytical discrimination* of any 4-D point, one needs to test both of the linear inequalities: $x_2 > x_1 + 2.5$, and $x_4 > x_3 + 0.5$. For the 4-D point (1, 5, 3, 4), both are obviously true, therefore it belongs to class 1. This discrimination rule is derived directly, from the visual representation. Formally for $\mathbf{x} = (x_1, x_2, x_3, x_4)$ the discrimination rule is:

$$\begin{gathered}\text{If } x_2 > (x_1 + 2.5)\ \&\ (x_4 > x_3 + 0.5),\ \text{ then } x \in \text{Class 1, else}\\ \{\text{If } x_2 < (x_1 + 0.5)\ \&\ (x_4 < x_3 + 0.5),\ \text{ then } x \in \text{Class 2,}\\ \text{else } (x \notin \text{Class 1})\ \&\ (x \notin \text{Class 2})\}.\end{gathered}$$

After such a rule is extracted, it can be used in the same way as any other rule, built, by using the analytical ML methods, to compute the class prediction for the new n-D points. Note, that while the visual solution is simple, this analytical rule is more complex, than a single linear discrimination function. The rule can be simplified by changing the coordinates, as shown, in Fig. 12.4, with two new non-orthogonal coordinate systems (X'_1, X'_2), and (X'_3, X'_4). Next, the coordinates of the points, from Table 12.1, are recomputed into these coordinates. In these new coordinates, a simpler discrimination rule for $\boldsymbol{x}' = (x'_1, x'_2, \ldots, x'_n)$ is equivalent to a decision tree, in these coordinates:

$$\begin{gathered}\text{If } (x'_2 > 2.5)\ \&\ (x'_4 < 0.5) \text{ then } \boldsymbol{x}' \in \text{Class 1, else}\\ \{\text{If } (x'_2 < 2.5)\ \&\ (x'_4 > 0.5) \text{ then } \boldsymbol{x}' \in \text{Class 2, else } (\text{x} \notin \text{Class 1})\ \&\ (\text{x} \notin \text{Class 2})\}.\end{gathered}$$

Both the 4-D, convex hull and hyper-plane based, generalizations are visualized in 2-D, in Figs. 12.3 and 12.4. These figures play the same role, for these data, as Fig. 12.1 for the 2-D data, used as inspiration for such visual exploration, of alternative levels of generalization, for 4-D data.

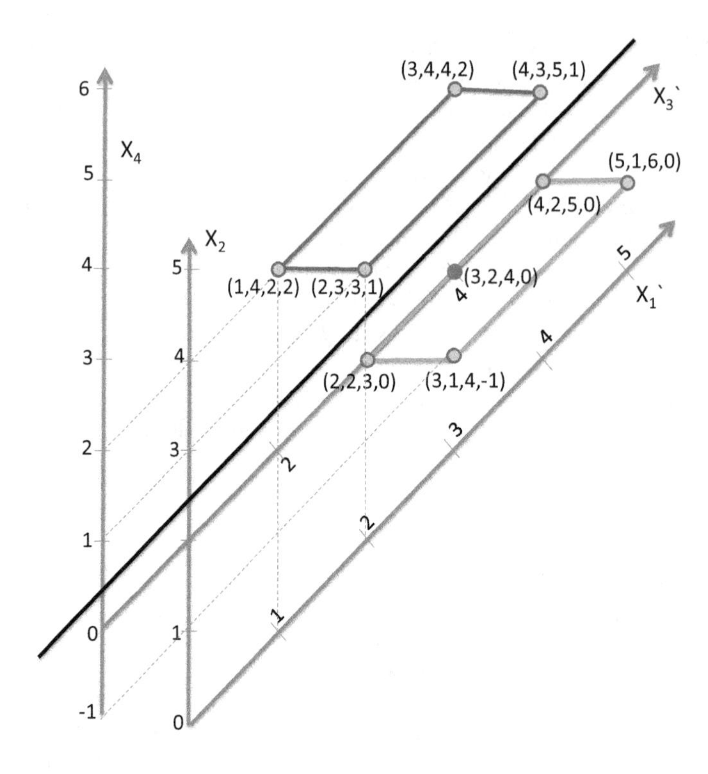

Fig. 12.4 Transformed solution, with non-orthogonal shifted coordinates

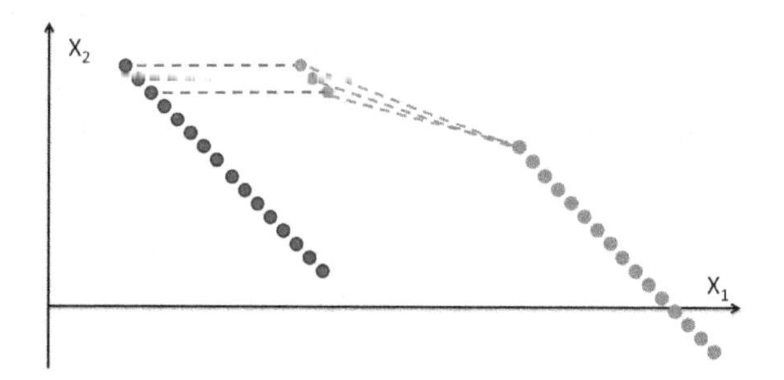

Fig. 12.5 Example of machine learning task with the nearest neighbors solution

12.5 Visual Defining and Curating ML Models

Questionable Input (deficiency D1) The deficiencies, of the training data, often are the main reason for the failure of the ML projects. The fundamental assumption of the training data is: that these data are *representative for new unknown data,* for which the class must be predicted. It means that, for instance, grey points, in

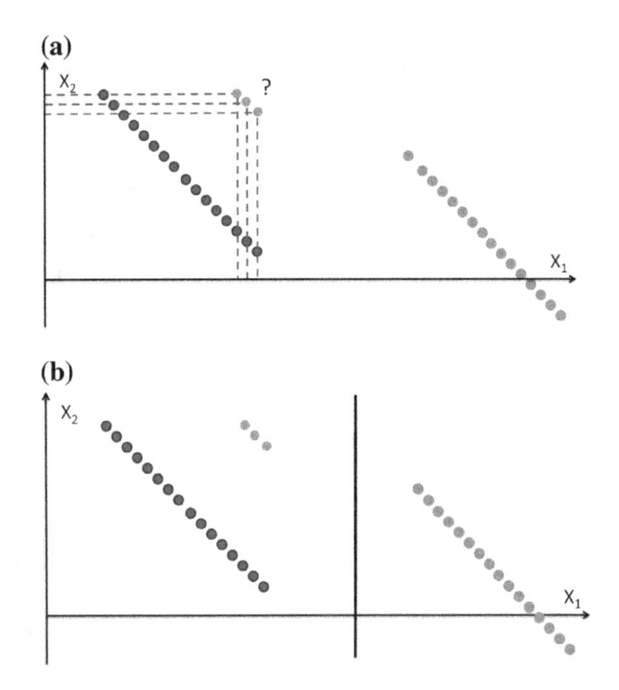

Fig. 12.6 Example of the machine learning task with the orthogonal projections, solution and projection to X_1

Fig. 12.5, will not come as the new data, when the training data of the two classes are limited by the blue and yellow points in this figure.

If these grey data come, then retraining will be needed, with the added training labeled data, which are close to these grey points. The problem is that recognizing the need, to retrain the system, when the grey data will come is difficult, because an ML method may have no *internal mechanism, to trigger the retraining process*. For instance, k-NN can classify the grey points as belonging to the blue class, due to the shorter distances to that class (see Fig. 12.5). Similarly the projection method, in Fig. 12.6a, will classify these points into the blue class too, because both projections of the gray points are within the projections of the blue points into coordinates X_1 and X_2.

The same result gives the *linear, discriminant-function, algorithm,* with a vertical line (see Fig. 12.6b).

In contrast, the *linear discriminant function method,* with a tilted line in the middle, between the green and yellow points, puts the gray points into the yellow class (see Fig. 12.7). Note that the projection method, with the original coordinates in Fig. 12.6a, classifies the grey points as belonging to the blue class. The same projection method, in the rotated coordinates, shown in Figs. 12.7 and 12.8, classifies them into the yellow class.

The deficiencies of the training data, for the grey points, along with the deficiencies of the ML generalization models, for these grey points, led to the presented

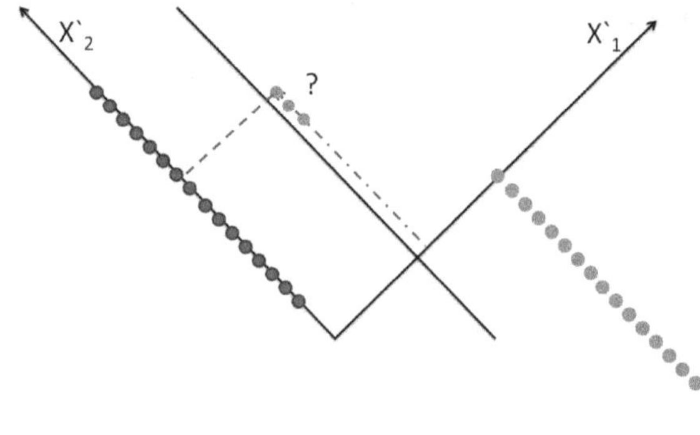

Fig. 12.7 Example of the ML task, with the tilted projection solution

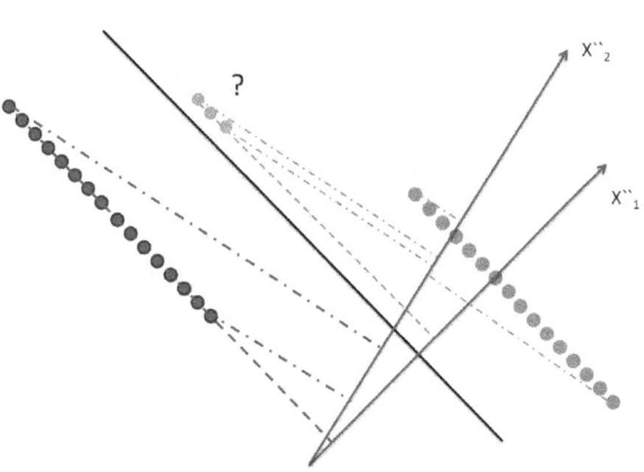

Fig. 12.8 Example of the machine learning task, with the skewed coordinates

predictive inconsistencies. Visualizations, in Figs. 12.5, 12.6, 12.7 and 12.8, made both deficiencies available, for the analysis by the SME.

As we see from this example, the visual approach is, in essence, the same as the presented one, for model construction, in Sect. 12.4. Here, we visualized the training, and the new (grey) data, along with the classification models. Figures 12.5, 12.6 and 12.7 visualize the 2-D data and the 2-D classification models. Such a type of visualization can be made, to assist the SME in dealing with the n-D data, and models, using multiple GLCs, as we have seen in Sect. 12.4, and other chapters.

Commonly **the generalization principle**s, in ML methods, are *external* for the given training data, and task. To recognize and resolve these inconsistencies additional information is needed. SME is a natural source of it. The proposed visualization approach allows the SME to see these inconsistencies in 2-D, decide about the additional training data, and/or ML models to use.

We envision the following process:

- The SME observes the new n-D data, to be classified, which are visualized in 2-D in CPCs, PSPCs, or other appropriate GLC coordinates,
- gets classification from the models from the different ML methods and

- triggers the retraining based on the visual judgment, that the training data are not representative, for the new data.

In this process, observing data in figures like Figs. 12.1, 12.3 and 12.4 SME can *request retraining* for points in the areas outside of convex hulls and keep classifications provided by automatic ML methods within convex hulls.

12.6 Summary on the Virtual Data Scientist from the Visual Perspective

The Sects. 12.1–12.5 above described a vision of a "virtual data scientist" assisting the SMEs, and the end users in building the empirical ML classification models, using the different GLCs. The feasibility of this new approach to construct and cure ML models was demonstrated with real world and simulated data not only in this chapter, but also in several other chapters. Table 12.2 summarizes the visual representations of real data and ML models from different chapters of this book that are relevant to the vision of the future virtual data scientist presented in this chapter.

These representations include the interactive visual classification, clustering, and dimension reduction with multiple GLCs on the real-world data.

Future research is on elaborating this vision of the virtual data scientist, and ways to implement it, for increasingly complex ML problems, from multiple domains, including the large and highly overlapped, and imbalanced classes.

12.7 Super Intelligence for High-Dimensional Data

The visual discovery approaches, in the n-D data, create an exciting opportunity for progress in the *super-intelligence studies*. While significant progress in Artificial Intelligence, Computational Intelligence, and Machine Leaning, improved the human abilities to discover the patterns, in the n-D data, the direct human cognitive abilities to do this, with a naked eye, are extremely limited, to the relatively small 2-D and 3-D datasets.

Lifting this human cognitive limitation is in a drastic contrast, with the opposite goal of reaching the human-level machine intelligence for the human abilities, which is the goal of other aspects of Artificial Intelligence, Computational Intelligence, and Cognitive Science. This opposite goal is deciphering the brain's existing cognitive abilities, and mimicking human intelligence, which uses the naked eye very successfully, to recognize and discover visual patterns, e.g., faces and facial expressions, in our physical 3-D world. Thus we need both:

- the deciphering of the brain, and

Table 12.2 Sections of chapters that are relevant to the vision of the future virtual data scientist

#	Chapter, Section, Figures	Description
1	Chap. 4, Sect. 4.6. Figures 4.25, 4.26, 4.27, 4.28, 4.29 and 4.30	Case study 4.5. User knowledge modeling dataset with SPC
1	Chap. 5, Sect. 5.1. Figures 5.1, 5.2 and 5.3	Case study 1: Glass processing with CPCs, APCs and SPCs
2	Chap. 5, Sect. 5.4. Figures 5.9, 5.10, 5.11 and 5.12	Case study 4: Challenger USA space shuttle disaster with PCs and CPCs
3	Chap. 5, Sect. 5.5. Figures 5.13 and 5.14	Case study 5: Visual n-D feature extraction from blood transfusion data with PSPCs
4	Chap. 5, Sect. 5.7. Figures 5.16, 5.17, 5.18, 5.19 and 5.20	Case study 7: Iris data classification in two-layer visual representation with CPCs
5	Chap. 5, Sect. 5.8. Figures 5.25, 5.26, 5.27, 5.28, 5.29, 5.30, 5.31 and 5.32.	Case study 8: Iris data with PWCs
6	Chap. 5, Sect. 5.9. Figures 5.33 and 5.34	Case study 9: Car evaluation data with CPCs, SPCs, APCs, and PCs
7	Chap. 5, Sect. 5.10. Figures 5.35, 5.36, 5.37 and 5.38	Case study 10: Car data with CPCs, APCs, SPCs, and PCs
8	Chap. 5, Sect. 5.12. Figures 5.40, 5.41 and 5.42	Case study 12: Seeds dataset with in-line coordinates and shifted parallel coordinates
	Chap. 5, Sect. 5.13. Figures 5.43, 5.44, 5.45, 5.46 and 5.47	Case study 13: Letter recognition dataset with SPC
9	Chap. 6, Sects. 6.2–6.4. Figures 6.3, 6.4, 6.5, 6.6, 6.7, 6.8, 6.9, 6.10, 6.11, 6.12, 6.13, 6.14, 6.15, 6.16, 6.17 and 6.19	Experiments 1–4: CPC Stars versus traditional stars for 192-D data; stars versus parallel coordinates for 48-D, 72-D and 96-D data; stars and CPC stars versus PC for 160-D data; CPC stars, stars and PC for feature extraction in 14-D and 170-D
10	Chap. 7, Sect. 7.3.1 Figures 7.7, 7.8, 7.9, 7.10, 7.11, 7.12, 7.13, 7.14, 7.15 and 7.16	Case study 1: Wisconsin Breast Cancer Diagnostic (WBC)
11	Chap. 7, Sect. 7.3.2 Figures 7.17, 7.18, 7.19, 7.20, 7.21, 7.22, 7.23, 7.24, 7.25 and 7.26	Case study 2: Parkinson's data set classification
	Chap. 7, Sects. 7.3.3 and 7.3.4. Figures 7.27, 7.28, 7.29, 7.30, 7.31, 7.32, 7.34, 7.35, 7.36, 7.37, 7.38, 7.39, 7.40, 7.41, 7.42, 7.43, 7.44 and 7.45	Case studies 3 and 4: Subsets of MNIST database of digits
	Chap. 7, Sect. 7.3.5 Figures 7.46, 7.47, 7.48 and 7.49	Case study 5: S&P 500 data that include the time of the Brexit vote
	Chap. 9, Sects. 9.5–9.7 Figures 9.1, 9.2 and 9.3	Visual text mining: Discovery of incongruity in humor modeling

- the enhancing of the brain to be able to deal with abstract high-dimensional data, as it is done with 2-D and 3-D data.

Compare this situation with building a machine that will fly as a bird. It is difficult to decipher the mechanism of bird flying. The history of aviation had shown, that direct attempts, to mimic it, failed many times. Next, the machine that intends only to mimic a flying bird will be limited. It will not fly to the Moon and the Planets. For flying that far a machine with *"super-bird" flying capabilities* is needed.

Similarly deciphering the brain's ability, to work visually with 2-D data, will hardly give us a way to build a *super-intelligence,* to deal visually with the large and *abstract n-D data*. This is a separate, and very challenging task. Evolution has developed our brain in a particular form, to adapt to a particular physical 3-D environment, which did not include the abstract high-dimensional data (n-D data) to be analyzed, until the very recent *Big data era.*

This separate task requires the ideas, beyond what is on the surface when the humans solve their typical cognitive tasks in 2-D and 3-D. In the same way, exploring how a bird is flying hardly will help in building a rocket to fly to the Moon, which requires discovering the more general flying principles. Similarly, dealing with Big n-D requires discovering the more general cognitive principles, than we use for the 2-D and 3-D data.

Is it always more difficult to discover the more general principles, than the more specific ones? The history of the science tells us, that it is not always the case. The modern flight theory, which includes the propulsion theory, and aerodynamics explains not only bird flight, but also rocket, and aircraft flights. However, this more general theory does not tell us anything, about the physiology of bird flight, at the level of muscles, and the bird brain control of the flight. Thus, higher generality does not mean the abilities to explain all aspects of the bird flight. However, it can help to discover, and understand the mechanism of other related activities. For instance, the propulsion theory allows the understanding of an octopus motion. In our case, it is discovering cognitive principles, to deal with the n-D data.

This brings us to the important point, that for understanding some fundamental brain cognitive principles, it is not necessary to study the brain itself first. Respectively, to build such a more general theory, we can work on the task that the brain does not support well, which is dealing with n-D abstract data. The goal is to understand and enhance the brain's capability, to deal with such n-D data. It includes experiments, with the same n-D data, where a human may, or may not, recognize the pattern, depending on the 2-D lossless representation, of these n-D data. These experiments can tell us about the human abstract pattern recognition abilities, providing the data to build a cognitive pattern recognition/discrimination model.

After a discrimination model is built, the next question is: "What is the mental process in the brain, behind this ability or inability?" The common approach in such tasks is: collecting, and analyzing the functional MRI data, when the subjects solve the task. In (Murray et al. 2002) functional MRI was used to measure the activity in

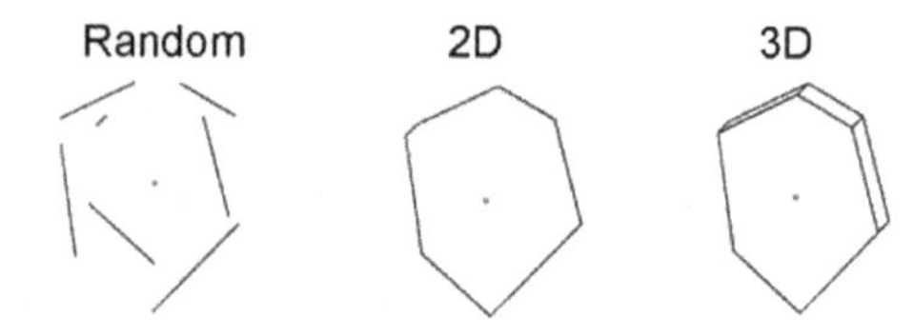

Fig. 12.9 Examples of different stimulus conditions (Murray et al. 2002)

a higher object processing area, the lateral occipital complex, and in the primary visual cortex, in response to the visual elements, which were either grouped into objects, or randomly arranged. These authors observed the significant activity increases, in the lateral occipital complex, and the concurrent reductions of activity, in the primary visual cortex, when the elements formed the coherent shapes.

Based on this observation, they suggested that the activity in the early visual areas is reduced because of grouping processes performed in the higher areas. These findings were used as an evidence for the brain predictive coding models of vision (Mumford 1992; Rao and Ballard 1999), which postulate that inferences of high-level areas are subtracted, from incoming sensory information, in lower areas, through cortical feedback. Note, that this study was conducted, for 2-D and 3-D shapes, such as those shown in Fig. 12.9, without any relation to the higher-n n-D data.

The predictive coding models of vision represent one side, of the two fundamental alternatives: local and distributed representation models/hypotheses, for the brain to be biologically adequate, representations for observed high-level structures, and cognitively adequate models. There are several, distributed representation, cognitive models with the bottom-up, and top-down signals (Carpenter and Grossberg 2016) including the dynamic logic model, which we advocate (Kovalerchuk et al. 2012), because of its ability to overcome the combinatorial complexity. On the other hand, while the current deep learning large Neural Networks may not be biologically adequate, their applied results are impressive.

The concept of the lossless reversible visualization, of n-D data, can be viewed, as a cognitive enhancer, for discovering the n-D data patterns. It simplifies the representation of the n-D data in 2-D, for the better perceptual and cognitive abilities, for the visual pattern discovery. Figure 12.10 summarizes the vision of the Virtual Data Scientist, and the Visual Super Intelligence.

The **future studies** are two-fold:

- enhancement of the methods for lossless representation, and the knowledge discovery, of the n-D data, in 2-D,
- clarification of the brain cognitive processes, associated with analysis of the abstract n-D data.

For the second issue future studies include gaze analysis: when humans analyze visual representations of abstract n-D data and discover n-D patterns. While the eyes provide the initial input of such visual information, visual perception, and cognition deeply involve the brain. Therefore, the gaze analysis will help, to look deeper into this complex process. Combining the eye-tracking methodology, the

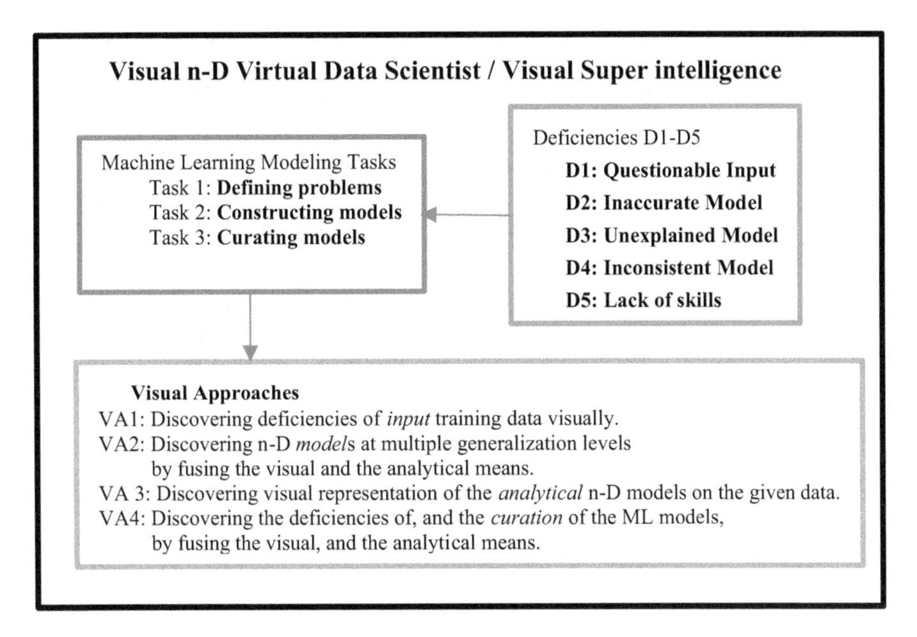

Fig. 12.10 The vision of the, visual n-D, virtual data scientist, and the visual super intelligence

mathematical models from different fields, and the behavioral information, which emerges in the analysis of n-D data, will be a source of new knowledge of the cognitive processes. This will include the future experiments, which compare observers' performance, in discovering the n-D data patterns, by analyzing the 2-D graphs as a function of their fixations, and the simulations by the computations of these fixations.

These future studies will also help: (a) to reveal the individual variability among the people, in their perceptual and cognitive abilities, for recognizing the abstract forms, and (b) to understand the visual and cognitive perception along with improving the accuracy, increasing efficiency, and decreasing the cost of the n-D data analysis.

References

Carpenter, G.A., Grossberg, S.: Adaptive resonance theory. In: Sammut, C., Webb, G. (eds.) Encyclopedia of Machine Learning and Data Mining, pp. 6–1. Berlin: Verlag (2016). https://doi.org/10.1007/978-1-4899-7502-7

DARPA, Data Driven Discovery of Models (D3M), 2016, https://www.fbo.gov/utils/view?id=68645e610e1e1ed5544e990a0c7dd91a

Duch, W., Adamczak R., Grąbczewski K., Grudziński K., Jankowski N., Naud A.: Extraction of Knowledge from Data Using Computational Intelligence Methods. Copernicus University: Toruń, Poland (2000). https://www.fizyka.umk.pl/~duch/ref/kdd-tut/Antoine/mds.htm

Kovalerchuk, B., Vityaev, E., Ruiz, J.: Consistent knowledge discovery in medical diagnosis. IEEE Eng. Med. Biol. **19**(4), 26–37 (2000)

Kovalerchuk, B., Perlovsky, L., Wheeler, G.: Modeling of Phenomena and Dynamic Logic of Phenomena. J. Appl. Non-classical Logics **22**(1), 51–82 (2012)

Mumford, D.: On the computational architecture of the neocortex. II. The role of cortico-cortical loops. Biol. Cybern. **66**, 241–251 (1992)

Murray S., Kersten D., Olshausen B., Schrater P., Woods D.: Shape perception reduces activity in human primary visual cortex. PNAS, **99**(23), 15164–15169 (2002). www.pnas.orgycgiydoiy10.1073ypnas.192579399

Rao, R.P., Ballard, D.H.: Predictive coding in the visual cortex: a functional interpretation of some extra-classical receptive-field effects. Nat. Neuroscience. **2**, 79–87 (1999)

Chapter 13
Comparison and Fusion of Methods and Future Research

Science never solves a problem without creating ten more.
George Bernard Shaw

In this chapter, we first compare GLCs with other visualization methods that were not analyzed in the previous chapters yet. Then we summarize some comparisons that were presented in other chapters. Next, the *hybrid approach* that fuses GLC with other methods is summarized along with the outline of the future research.

13.1 Comparison of GLC with Chernoff Faces and Time Wheels

Chernoff faces Table 13.1 presents the comparisons of Paired Coordinates with Chernoff faces, Parallel and Radial/Star coordinates. C*hernoff faces* are multi-part glyphs in the shape of a human face and the individual parts, such as eyes, ears, mouth and nose represent the data variables by their shape, size and orientation (Chernoff 1973).

The use of the Chernoff's faces is based on the human ability to easily recognize faces and small changes in them. However, faces are not necessarily superior to other multivariate techniques (Morris et al. 1999), In general, as it was noticed in the literature (Spence 2001), icons have advantages over other representations in the case of the semantic relation between the icons and the task.

The arbitrary match of the face features with the attributes of the n-D point has no such semantic match. The features of the faces such as the curvature of the mouth, the eye size and the density of the eyebrow are of different importance for our interpretations of the whole face (De Soete 1986), and an arbitrary match will lead to a very different conclusion about the n-D points based on the facial metaphor.

Table 13.1 shows the advantages of Line Coordinates such as Collocated, Parallel and Radial/Star Coordinates over Chernoff faces. There are multiple modifications of Parallel Coordinates methods that intend to improve them. Most of these improvements are also applicable to the Collocated Coordinates.

B. Kovalerchuk, *Visual Knowledge Discovery and Machine Learning*,
Intelligent Systems Reference Library 144,
https://doi.org/10.1007/978-3-319-73040-0_13

Table 13.1 Comparison of Chernoff faces, Parallel, Star and Collocated Paired Coordinates

Characteristics	Chernoff faces	Parallel and radial coordinates	Collocated paired coordinates
Visuals allow reading values of variables	No. Schroeder (2005)	Yes	Yes
Easy to quantify the differences	No. Schroeder (2005)	Yes	Yes
Pre-attentive perception (not serial)	No. Schroeder (2005)	Yes for simple n-D points such as with all equal attributes	Yes for multiple n-D points, see Chap. 4
Easy to use	No. Schroeder (2005)	Yes for multiple n-D points	Yes for multiple n-D points, e.g., see Sect. 5.6 in Chap. 5
Preserves dimension (no reduction)	Up to 18 variables Schroeder (2005)	Yes	Yes
Represent the object as a whole	Yes	Yes	Yes
Intuitive	Yes. Schroeder (2005)	Less intuitive	Less intuitive
Familiar metaphor	Yes	Less familiar	Less familiar
Order of variables im-pacts visualization	Yes. Chuah and Eick (1998)	Yes, but less significantly, visuals are more uniform	Yes, but less significantly, visuals are more uniform
Redundant visuals	Yes. Faces' symmetry doubles visuals	No	No
Useful for trend study	Yes. Morris et al. (1999)	Yes	Yes
Useful for decision making	Yes, but less than for trends. Morris et al. (1999)	Yes	Yes
Free from multiple crossing lines	Yes	No. Difficult to spot structure in large dimensions. (Blaas et al. 2008), Candan et al. (2012), Fua et al. (1999)	No, but less than for Parallel Coordinates, because two times fewer lines are used
Easily to spot correlations	No	Yes. Fua et al. (1999)	Yes
Feasible interactive exploration for Big Data (>10^7 n-D points)	No	No, visual *clutter* and *slow* performance. Blaas et al. (2008)	No, but less than for Parallel Coordinates, because two times fewer lines are used
Easy to keep 2-D, 3-D *spatial context* of data	No	*Spatial context* of 3-D data is usually lost. Blaas et al. (2008)	Yes

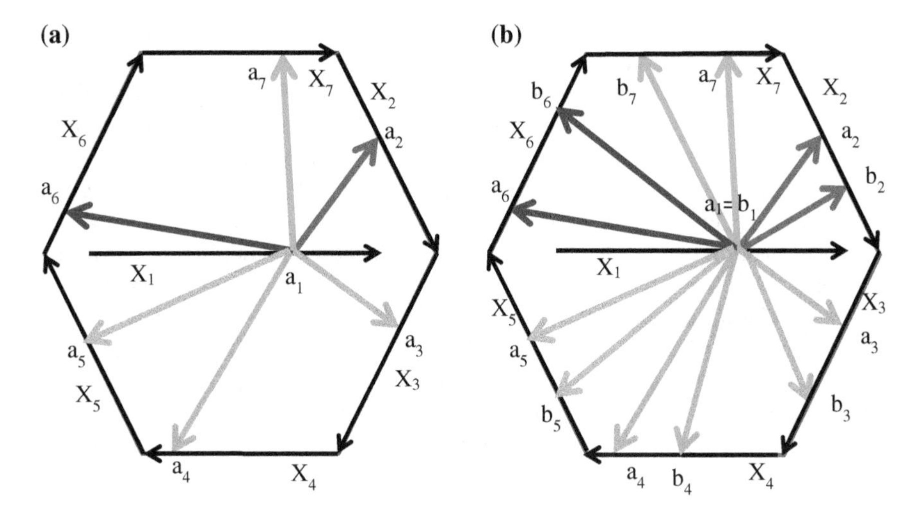

Fig. 13.1 **a** 7-D point **a** = (7, 4, 4, 9, 6, 2, 8) in the TimeWheel. **b** 7-D points **a** = (6, 4, 4, 9, 6, 2, 8) and **b** = (6, 7, 8, 6, 3, 8, 3) in the TimeWheel

TimeWheel Figure 13.1 shows the TimeWheel visual representation (Tominski et al. 2004). In this representation n-1 coordinates are located on the sides of the n-Gon and one coordinate is located horizontally between opposite nodes of the n-Gon. Typically this coordinate is time. Consider a set of 7-D points {**x**} = {(x_1, x_2,…,x_6, x_7)} where x_1 is a timestamp, x_2 is respiration rate, x_3 is heart rate, x_4–x_7 are other medical characteristics. For each **x** a brown line links x_1 and x_2 values of **x**. Similarly a green line links x_1 and x_3 values of **x.** Other pairwise links (x_1, x_i) are shown by other colored lines. This set of colored lines losslessly represents a 7-D point. See Fig. 13.1a for a 7D point **a** = (7, 4, 4, 9, 6, 2, 8) and jointly with a 7-D point **b** = (6, 7, 8, 6, 3, 8, 3) in Fig. 13.1b.

At the first glance, the TimeWheel is similar to our n-Gon representation shown in Figs. 2.4 and 2.6 in Chap. 2, because coordinates are located on the sides of the n-Gon. The first technical difference is in the location of x_7 in the middle of the n-Gon. The second one is that the TimeWheel is a **lossless** visual representation of an n-D point only if we have all n-D points {**x**} with **different** x_1 values, e.g., different timestamps. Otherwise, if two n-D points **a** and **b** have equal timestamps $a_7 = b_7$, we will not be able to restore the other a_i and b_i because two lines will start from the same point $a_7 = b_7$ for each coordinate (see Fig. 13.1b).

13.2 Comparison of GLC with Stick Figures

A Stick Figure (SF) of n lines ("sticks"), connected with different angles, encodes $2n$ attributes (Pickett and Grinstein 1988). It is done by encoding each pair of attributes (X_i, X_{i+1}) by the length and the angle of the stick (see Fig. 13.2a). A stick figure can look like a human body skeleton, which is a familiar metaphor. Other forms of SFs may not have this familiar metaphor. SFs are useful when figures are shown side-by-side. Otherwise, the occlusion severely limits the discovery of the patterns visually. While also suffer from occlusion, many GLCs including CPCs and SPCs allow the discovery of patterns when multiple n-D points are drawn in the same coordinates in a single display, as case studies in this book show. As any glyph approach, stick figures can be combined with Cartesian Coordinates. In (Grinstein et al. 1989) income and age are used to identify the locations of multiple small SFs that create a "texture".

SFs are similar conceptually to Chernoff Faces (CFs) that have been compared with GLC above in Table 13.1. A significant part of this comparison is applicable for comparison of SF and GLC. The major difference of paired GLCs from CFs and SFs is mapping data attributes to visual features. In paired GLCs, two attributes are encoded by a single 2-D point (a *node* of the graph). In SFs, two

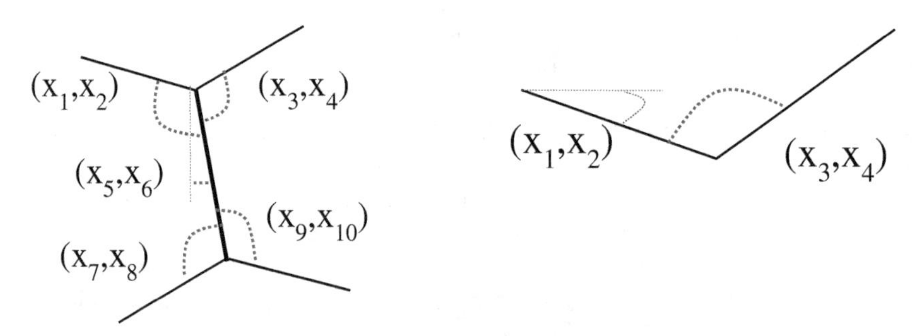

(a) Stick figure with 5 sticks that encodes 10 attributes.

(b) Stick figure with 2 sticks that encodes 4 attributes.

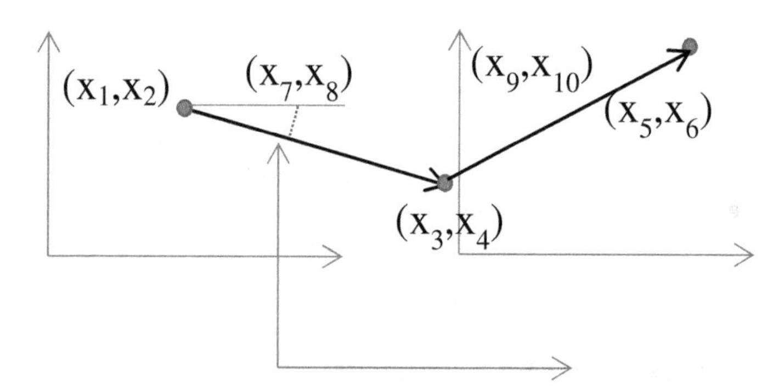

(c) Joint Shifted Paired Coordinates and a Stick figure with 2 sticks that encodes 10 attributes.

Fig. 13.2 Stick figures and Joint Shifted Paired Coordinates with a Stick figure

attributes are encoded by an *edge* ("stick") of the graph (length and angle of the edge).

In CFs two or more attributes are encoded by features of an open or closed line such as length, angle, curvature and others. SPCs allows representing a given n-D point as a *single* 2-D point losslessly by adapting shifts of pairs of coordinates as was shown above. CF and SF do not have such capability.

Next CF, SFs, and GLCs including Parallel Coordinates (PCs) are not invariant to the *order* of coordinates. Different orderings produce different figures in all of them. This is not necessary a deficiency because humans can discover patterns in some visualizations easier than in others. Once a pattern is discovered visually in one of orderings of coordinates, it can be converted to the *analytical* form that is "order free". See for instance Sect. 5.7. in Chap. 5.

Also in GLCs, coordinates can be labeled by actual names of attributes from the beginning. See Fig. 5.15 in Chap. 5 for health monitoring. It avoids memory overload to remember the meaning of indexed labels of coordinates X_i. Both CFs and SFs require remembering meaning of visual features in terms of attributes they encode, because commonly they are not labeled.

In CPCs, graphs are directed, but in SFs, the graphs are not directed. The directions of the graph edges can be beneficial, e.g., it shows a trend in World hunger data in Fig. 5.6 in Chap. 5. One of the benefits of SFs is familiarity of human body skeleton metaphor, which can be remembered faster. On the other side, this metaphor limits the number of features, which have a meaning in this metaphor, e.g., arms, legs and body.

Next, we propose a way to combine SFs and paired GLCs to *increase* the number of attributes to be encoded by the graph. It is based on the fact that SPCs and SFs use different parts of the graph to encode the attributes (SPCs use nodes and SFs edges). The idea is to use *both nodes and edges for encoding attributes*. SPCs do not use the length of the edges and angles between them to encode attributes, but use them for the simplification of graphs as it is done in Sect. 3. The length and the angles of the edges can be adjusted in SPCs to make their values to represent attributes.

To get a desired length of the edge a horizontal shifting a pairs of coordinates is sufficient. To get a desired angle of the edge shifting a pairs of coordinates along a given radial distance from the node where edge is originated is sufficient. In this way, a graph with two arrows will encode not 6 attributes as in the SPC, but 10 attributes as Fig. 13.2c shows. SF to represent 10 attributes requires 5 edges (sticks) (see Fig. 13.2a) and with two edges it encodes only 4 attributes (see Fig. 13.2b). While this method works for n-D points shown side-by-side as it is always done with SFs, it does not work for drawing graphs of multiple n-D points in the same SPC space. The reason is that adjusting the length and the angles for the second n-D point changes the length and the angle for the first n-D point already adjusted.

13.3 Comparison of Relational Information in GLCs and PC

The patterns representing the *relational information* in different GLCs such as CPCs, SPCs and Star CPCs in comparison with Parallel Coordinates (PCs) are shown in several chapters. Below we summarize these comparisons concentrated in Chaps. 3, 5, 6 and 7.

- PCs are a *special case* of GLCs when all coordinates are parallel. Thus, it is logical to use PCs as one of GLCs (not as opposing to GLCs) in the situations where PCs is *more intuitive and simpler* than other GLCs in discovering relations.
- Each edge of the graph in CPCs and SPCs directly visualizes a relation of *four dimensions*. In PCs it directly visualizes only a relation between *two adjacent dimensions*.
- PCs require *two times more nodes* to represent a relation between *n* dimensions as a graph than CPCs and SPCs require that leads to more *occlusion*.
- In PCs, for each value x_i, a different line must be drawn to show the linear relations $x_j = mx_i + b$ for the two adjacent dimensions. For all values of $x_{i,}$ this leads to an *infinite set of lines* for this linear relation. CPCs and SPCs allow a *single line* in a classical Cartesian form.
- In PCs, the *infinite set of lines* for linear relations $x_j = mx_i + b$ (regression) creates an extreme case of *full occlusion* (no line visible). Therefore, this drawing is *not scalable* for large datasets. A *single line* in CPCs and SPCs for the same dataset has *no occlusion* and is *scalable.*
- Classical Cartesian visualization of linear relations $x_j = mx_i+b$ used CPCs, and SPCs is *familiar* to everyone. It *does not require learning a new visualization* in contrast with PCs for this relation.
- Compact representations of the linear relation $y = kx+m$ (that do not directly map individual points x to y) have *the same expressiveness* in PCs, CPCs and SPCs requiring a single 2-D point.
- The SPC visualization of 4-D Health monitoring relations is much *simpler* and *more familiar* than in PCs in Fig. 5.15 in Sect. 5. It shows the relation between the initial health status, and its change over time to the goal state.
- Linear discrimination relations between Iris classes produced in SPC are *highly accurate,* while PCs and RadVis do not reveal such a linear discrimination relation, as Sect. 5.7 in Chap. 5 shows.
- CPS Stars allowed *more accurate* results (94%) than PCs (79%) in discovering noisy 160-D linear relations by humans, as Figs. 6.7 and 6.9 in Chap. 6 show.
- Visualization of a noisy 23-D linear relation in CPC is *simpler, more familiar and less occluded* in Fig. 5.5, than in PCs in Fig. 5.4 in Chap. 5.
- The GLC-L visulization method has the capabilities to represent *weighted discriminating linear relations* between *n* dimensions as shown in multiple case studies in Sect. 7.3 in Chap. 7. Such capabilities are not known for the PCs.
- Commonly the non-linear relations are modeled by interpolating them by a set of linear relations. The listed capabilites of the different GLCs, to represent the

linear relations with noise, show an opportunity to use them for *interpolating non-linear relations* in the future.

13.4 Fusion GLC with Other Methods

While reversible GLCs are the focus of this book, many other visualization methods exist and some of them are reversible too as was discussed in this book. The fusion of GLCs with these methods produces **hybrid methods**.

The hybrid approach was outlined in Chap. 1. It contains two aspects: (1) *combining point-to-point and point-to-graph visual representations* of the n-D data (i.e., non-reversible, lossy representations with reversible lossless representations) when separately these representations are not sufficient, and (2) combining *visual and analytical means* of knowledge discovery to get deeper knowledge.

The *combination* of lossless and lossy visual representations includes providing means for *evaluating* the weaknesses of each representation and *mitigating* them by sequential use of them for knowledge discovery. Combining *visual and analytical means* of knowledge discovery also guides:

- Discovering the information about the *structure* of data and patterns that separate the classes of data, and
- Finding the splits of data into the training–validation pairs that will allow the most complete evaluation of the discovered patterns. This includes guiding in finding the worst, best, and median splits.

The *hybrid methods* allow radically improve quality of knowledge discovery results by analyzing more information. In applying these methods we first reduce dimensionality with acceptable and controllable loss of information by using non-reversible methods. Then we apply reversible methods to represent remaining dimensions in 2-D losslessly. In Chap. 7 in Sect. 7.3.3 it was done with 484 original dimensions reduced to 38 dimensions with loss of some information and then these 38-D data are visualized losslessly in 2-D and classified with high accuracy.

13.5 Capabilities

In many engineering application 10% improvement in efficiency is considered as a valuable progress provided by a new technology. For GLC the benchmarks of current technology are Parallel and Radial (Star) Coordinates. Relative to these methods the progress in efficiency can be measured by decreasing the occlusion, which is indicated by decreasing the number of 2-D points and lines per n-D point.

As it is shown in this book such GLC as CPC, SPC, and Star CPC improve this measure two times (100%).

We have shown that *Lossless Visual Representation* (LVR) methods for n-D data are important complements to the non-reversible lossy visualizations methods. We expanded the methods of lossless visualization of n-D data and demonstrated their promising efficiency, using the modeled and real data. LVR allows the better interpretation of their features in terms of n-D data properties than some lossy visualizations such as the Multidimensional Scaling. LVR allows the efficient use of human shape perception capabilities in line with Gestalt laws and recent psychological experiments. LVR are naturally expandable to a collaborative framework.

The LVR is justified by deficiencies of *lossy visualizations* that map n-D data into 2-D data with significant loss of the information. Lossy visualizations not only drop information, but commonly do not control which n-D properties are dropped. The need in *multiple LVRs* is dictated by a very limited number of available LVRs of n-D data, and by the absence and likely the impossibility of a "silver bullet visualization", that can be ideal for all possible datasets.

The *General Line Coordinates,* as a class of LVP methods presented in this book, provide a *common visualization framework,* and *a large number of new visual representations* of multidimensional data, without dimension reduction. It is important that the GLC class is a very large and diverse class of coordinate systems. This increases the chances to capture diverse patterns/regularities in a variety of multidimensional data.

This book presented

- new methods for decreasing occlusion and simplifying visual patterns for classification tasks,
- demonstrated efficiency of new compact lossless representation by Parametric Shifted Paired Coordinates (PSPC) on real iris and health monitoring data,
- proposed a new two layer GLC concept and demonstrated its efficiency on real data,
- demonstrated advantages of closed contour lossless visual representations over Parallel Coordinates for high-dimensional data in the experiment with several about 70 participants for classification of modelled data (linear hyper-tubes),
- clarified limits of high-dimensionality of data for human visual classification of modelled n-D data (linear hyper-tubes) in Parallel Coordinates, star CPC and Radial Coordinates.

This creates an opportunity to design the advanced hybrid data mining/machine learning methods that integrate the advantages of analytical and visual methods to get higher accuracy, interpretability, and avoiding the overgeneralization and overfitting of discovered patterns. In the future such hybrid exploration may provide end users with "**n-D glasses**" to conduct deep n-D data exploration with less extensive involvement of data scientists.

13.6 Future Research

The challenge for the further studies is progressing to higher dimensions and larger datasets with GLCs. We envision three approaches.

The first approach is a *hybrid approach,* which combines the advantages of lossless and lossy methods. The attempt to visualize, say, 400-D in the first two principal components directly without GLCs will often lead to very significant loss of information.

The second approach is expanding the *GLC side-by-side approach* used in Chap. 5, where each n-D point is shown as a separate graph (figure) preferably as a *closed contour* to leverage the human perceptual abilities with closed contours. This visualization is free from occlusion, but suffers from switching gazing from graph to graph. Currently it can handle a quite limited number of graphs analyzed at each given time.

The third approach is splitting dimensions by clustering them and visualizing data in each subset of dimensions separately, with combining patterns found in such subsets of dimension into a joint pattern. For instance, data with 1000 dimensions can be split to 10 clusters of 100 dimensions. All three approaches are topics of future exploration. While we expect progress in all of them, we do not expect that GLC will provide a "silver bullet" for all possible tasks and data, as it is the case with all current methods.

For years Parallel Coordinates have been developed in multiple directions to enhance them (Heinrich and Weiskopf 2013). Most of these enhancements are applicable to the General Line Coordinates, and can be applied to develop their more advanced versions. These enhancements include supporting *unstructured datasets* with millions of points, multi-timepoint volumetric datasets with tens of millions of points per time step (Blass et al. 2008) and large document corpora (Candan et al. 2012). Next, to decrease the clutter from crossing lines and to deal with large datasets, multiple methods have been developed such as: parallel *hierarchical* coordinates (Candan et al. 2012; Fua et al. 1999), *smooth* parallel coordinates (Moustafa and Wegman 2002), *higher order* parallel coordinates (Theisel 2000), *continuous* parallel coordinates (Heinrich and Weiskopf 2013), reordering, spacing and filtering PC (Yang et al. 2003), and others. These developments deal with larger datasets and decreased clutter from crossing lines.

In (Chen et al. 2013; Viau et al. 2010; Yuan et al. 2009) parallel coordinates are combined with scatter-plot matrixes and histograms to produce the multiple coordinated views. The combination with the statistical analysis formed the *enhanced* parallel coordinates (Yuan et al. 2009). A significant effort has been devoted to controlling the ordering and the scaling of parallel coordinates (Andrienko and Andrienko 1999), including locating the variables of interest in adjacent axes because the ordering of the axes influences the shape of the lines and their interpretation. Significant effort also was devoted to exploring the mathematical properties of parallel coordinates (Inselberg 2009). The same is needed for GLC along with developing advanced GLC and applying them to challenging datasets.

As was pointed out above most of these approaches can be used to develop more advanced versions of GLC and hybrid methods for knowledge discovery in Big data. Several such options have been presented in this book. The explanatory power of visualization was recognized and demonstrated for a long time (Tufte and Robins 1997). The GLC contributes to it for multidimensional data.

A full classification of the General Line Coordinates for the cognitively efficient n-D data visualization and knowledge discovery is a task for future research as well as the deeper links with Machine Learning to be able to build visually the learning algorithms using visual means in GLC.

While many GLC challenges in knowledge discovery need to be resolved in the future research, this book shows that more complete preservation of multidimensional data in 2-D visualization and more efficient use of preserved information for visual and hybrid knowledge discovery is feasible.

References

Andrienko, G., Andrienko, N.: GIS visualization support to the C4.5 classification algorithm of KDD. In: Proceedings of the 19th International Cartographic Conference, pp. 747–755 (1999)

Blaas, J., Botha, C., Post, F.: Extensions of parallel coordinates for interactive exploration of large multi-timepoint data sets. IEEE Trans Visual Comput Graphics **14**(6), 1436–1443 (2008)

Candan, K.S., Di Caro, L., Sapino, M.L., PhC: Multiresolution visualization and exploration of text corpora with parallel hierarchical coordinates. ACM Trans. Intell. Syst. Technol. **3**(2), art. 22 (2012)

Chen, Y., Cai, J.-F., Shi, Y.-B., Chen, H.-Q.: Coordinated visual analytics method based on multiple views with parallel coordinates. Xitong Fangzhen Xuebao/J Syst Simul **25**(1), 81–86 (2013)

Chernoff, H.: The use of faces to represent points in k-dimensional space graphically. J. Am. Stat. Assoc. **68**, 361–68 (1973)

Chuah, M.C., Eick, S.G.: Information rich glyphs for software management data. IEEE Comput. Graphics Appl. **18**(4), 24–29 (1998)

De Soete, G.: A perceptual study of the Flury-Riedwyl faces for graphically displaying multivariate data, International J. Man-Machine Stud. **25**(5), 549–555 (1986). Montreal, Canada, ACM Press

Fua, Y., Ward, M.O., Rundensteiner, A.: Hierarchical parallel coordinates for exploration of large datasets. In: Proceedings of IEEE Visualization, pp. 43–50 (1999)

Grinstein, G., Pickett, R., Williams, M.G.: Exvis: an exploratory visualization environment. In: Graphics Interface 1989 Jun (Vol. 89, pp. 254–261)

Heinrich, J., Weiskopf, D.: State of the Art of Parallel Coordinates, EUROGRAPHICS 2013/ M. Sbert, L Szirmay-Kalos, 95–116 (2013)

Inselberg, A.: Parallel Coordinates: Visual Multidimensional Geometry and Its Applications. Springer (2009)

Kovalerchuk, B.: Visualization of multidimensional data with collocated paired coordinates and general line coordinates, In: SPIE Visualization and Data Analysis 2014, Proceedings of SPIE 9017, Paper 90170I, https://doi.org/10.1117/12.2042427. p. 15

Morris, C., Ebert, D., Rhengans, P.: An experimental analysis of the pre-attentiveness of features in Chernoff faces. In: Proceedings of Applied Imagery Pattern Recognition: 3D Visualization for Data Exploration and Decision Making (1999)

Moustafa, R., Wegman, E.: On some generalization to parallel coordinate plots. In: Seeing a Million—A Data Visualization Workshop, 41–48 (2002)

Pickett, R., Grinstein, G.: Iconographic displays for visualizing multidimensional data. In: Systems, Man, and Cybernetics, 1988. Proceedings of the 1988 IEEE International Conference on, pp. 514–519, (1988)

Schroeder, M.: Intelligent information integration: from infrastructure through consistency management to information visualization. In: Dykes J, MacEachren A.M, Kraak MJ. (eds.) Exploring Geovisualization 477–494 Elsevier (2005)

Spence, R.: Information Visualization, Harlow, London: Addison Wesley/ACM Press Books, p. 206 (2001)

Theisel, H.: Higher order parallel coordinates, In: Proceedings of the 5th Fall Workshop on Vision, Modeling, and Visualization, 415–420, Saarbrucken, Germany (2000)

Tominski, C., Abello, J., Schumann, H.: Axes-based visualizations with radial layouts, In: Proceedings of the 2004 ACM symposium on Applied computing, 1242–1247 (2004)

Tufte, E.R., Robins, D.: Visual explanations, Graphics Press (1997)

Viau, C., McGuffin, M.J., Chiricota, Y., Jurisica, I.: The FlowVizMenu and parallel scatterplot matrix: Hybrid multidimensional visualizations for network exploration. IEEE Trans Visual Comput Graphics **16**(6), 1100–1108 (2010)

Yang, J., Peng, W., Ward, M.O.: Rundensteiner EA. Interactive hierarchical dimension ordering, spacing and filtering for exploration of high dimensional datasets. In: Information Visualization, 2003. INFOVIS 2003. IEEE Symposium on 2003, pp. 105–112. IEEE

Yuan, X., Guo, P., Xiao, H., Zhou, H., Qu, H.: Scattering points in parallel coordinates. IEEE Trans Visual Comput Graphics **15**(6), 1001–1008 (2009)

CPSIA information can be obtained
at www.ICGtesting.com
Printed in the USA
LVHW082137060619
620471LV00004B/357/P

9 783319 892306